The Practical Approach in Chemistry Series

SERIES EDITORS

L. M. Harwood
Department of Chemistry
University of Reading

C. J. Moody
Department of Chemistry
University of Exeter

The Practical Approach in Chemistry Series

Organocopper reagents
Edited by Richard J. K. Taylor

Macrocycle synthesis
Edited by David Parker

High-pressure techniques in chemistry and physics
Edited by Wilfried B. Holzapfel and Neil S. Isaacs

Preparation of alkenes
Edited by Jonathan M. J. Williams

Transition metals in organic synthesis
Edited by Susan E. Gibson (née Thomas)

Matrix-isolation techniques
Ian R. Dunkin

Lewis acid reagents
Edited by Hisashi Yamamoto

Organozinc reagents
Edited by Paul Knochel and Philip Jones

Amino acid derivatives
Edited by Graham C. Barrett

Asymmetric oxidation reactions
Edited by Tsutomu Katsuki

Asymmetric Oxidation Reactions
A Practical Approach in Chemistry

Edited by
TSUTOMU KATSUKI

*Kyushu University
Department of Chemistry,
Faculty of Science,
Hakozaki,
Higashi-ku,
Fukuoka 812–81,
Japan*

This book has been printed digitally and produced in a standard specification in order to ensure its continuing availability

OXFORD
UNIVERSITY PRESS

Great Clarendon Street, Oxford OX2 6DP

Oxford University Press is a department of the University of Oxford.
It furthers the University's objective of excellence in research, scholarship,
and education by publishing worldwide in

Oxford New York

Auckland Bangkok Buenos Aires Cape Town Chennai
Dar es Salaam Delhi Hong Kong Istanbul Karachi Kolkata
Kuala Lumpur Madrid Melbourne Mexico City Mumbai Nairobi
São Paulo Shanghai Singapore Taipei Tokyo Toronto

with an associated company in Berlin

Oxford is a registered trade mark of Oxford University Press
in the UK and in certain other countries

Published in the United States
by Oxford University Press Inc., New York

©Oxford University Press, 2001

The moral rights of the author have been asserted
Database right Oxford University Press (maker)

Reprinted 2002

All rights reserved. No part of this publication may be reproduced,
stored in a retrieval system, or transmitted, in any form or by any means,
without the prior permission in writing of Oxford University Press,
or as expressly permitted by law, or under terms agreed with the appropriate
reprographics rights organization. Enquiries concerning reproduction
outside the scope of the above should be sent to the Rights Department,
Oxford University Press, at the address above

You must not circulate this book in any other binding or cover
and you must impose this same condition on any acquirer

ISBN 0 19 850201 X (Hbk)

Preface

This volume in the Practical Approach in Chemistry Series is devoted to asymmetric oxidation, a rapidly expanding area in chemistry. In the past two decades, many useful asymmetric oxidation reactions have been introduced and some have already become potent standard laboratory methodologies for the synthesis of a variety of enantiopure compounds, in particular highly functionalized ones. However, some asymmetric oxidations, such as asymmetric C—H oxidation that is very important from an academic and a practical point of view, remain underdeveloped. Thus, asymmetric oxidation reactions have attracted much attention not only as useful synthetic tools but also as challenging research targets in academic and industrial laboratories. Some laboratories, however, are not familiar with asymmetric oxidation, and there is a demand for standard protocols for asymmetric oxidation reactions which can be followed safely even by inexperienced chemists. In compliance with the demand, this volume was planned for publication. The main aim of this volume is to provide details of general procedures for asymmetric oxidation reactions, not only well-established ones but also promising ones, and to convince more researchers that asymmetric oxidation reactions, as potent tools for enantiopure syntheses of complex molecules, can be performed simply and safely if the correct protocols are followed. This volume describes important synthetic and biological oxidation reactions in five chapters and should be found to be a valuable guide to asymmetric oxidation chemistry.

I am grateful to all the authors, who are leading chemists in their respective fields, for their excellent contributions. I would like to thank Professor Moors and Professor Harwood, series editors of The Practical Approach in Chemistry Series, who gave useful comments and suggestions for this book. I also thank to Dr Azerad for his helpful information on the researchers in the field of biological oxidation, and Yukiko Shiraishi for her typographical work.

Finally, I sincerely hope that this book guides many new chemists to step into oxidation chemistry and to stimulate the future development of asymmetric oxidation.

A journey of a thousand miles starts with but a single step.

Fukuoka, Japan　　　　　　　　　　　　　　　　　　　　　　　　Tsutomu Katsuki

Contents

List of contributors	xiii
Abbreviations	xvii

Introduction   1
Tsutomu Katsuki

1. Oxidation of the C—H bond   5

1.1 Asymmetric oxidation of the C—H bond   5
Beena Bhatia, T. Punniyamurthy, and Javed Iqbal

    1. Introduction   5
    2. Asymmetric benzylic hydroxylation   6
    3. Asymmetric allylic oxidation   10
    References   16

2. Oxidation of the C=C bond   19

2.1 Metal-catalysed asymmetric epoxidation of simple olefins   19
Yoshio N. Ito and Tsutomu Katsuki

    1. Introduction   19
    2. Metalloporphyrin-catalysed asymmetric epoxidation   20
    3. Metallosalen-catalysed asymmetric epoxidation   25
    4. Aerobic asymmetric epoxidation   33
    References   36

2.2 Asymmetric epoxidation using peroxides and related reagents   37
Bang-Chi Chen, Ping Zhou, and Franklin A. Davis

    1. Introduction   37
    2. Diastereoselective epoxidations   37
    3. Enantioselective epoxidation   45

Contents

4. Double asymmetric differentiation	48
References	49

2.3 Asymmetric epoxidation of olefins bearing precoordinating functional groups — 50
Victor S. Martin

1. Introduction	50
2. Asymmetric epoxidation of allylic alcohols	51
3. Catalytic asymmetric epoxidation	56
4. Kinetic resolution of racemic secondary allylic alcohols	61
References	69

2.4 Enantioselective epoxidations of electron-deficient olefins — 70
Ivan Lantos

1. Phase-transfer catalysed epoxidation	70
2. Epoxidations promoted by organometallic agents	77
References	79

2.5 Asymmetric dihydroxylation — 81
Heinrich Becker and K. Barry Sharpless

1. Introduction	81
2. Scope and limitations	82
3. Choice of ligand and reaction conditions	84
4. Protocols for the asymmetric dihydroxylation reaction	87
5. Determination of the enantiomeric excess	101
References	104

2.6 Asymmetric aminohydroxylation — 104
Gunther Schlingloff and K. Barry Sharpless

1. Introduction	104
2. Sulfonamide-based nitrogen sources	105
3. Carbamate-based nitrogen sources	107
4. Acetamide-based nitrogen sources	111
References	114

2.7 Asymmetric aziridination   115

Margaret M. Faul and David A Evans

    1. Introduction   115
    2. Asymmetric aziridination of olefins using chiral bis(oxazoline) ligands   117
    3. Asymmetric aziridination of olefins using chiral bis(imine) ligands   122
    4. Conclusions   126
    References   126

2.8 Asymmetric hydroxylations of enolates and enol derivatives   128

Ping Zhou, Bang-Chi Chen, and Franklin A. Davis

    1. Introduction   128
    2. Asymmetric α-hydroxylation of enolates   128
    3. Asymmetric α-hydroxylation of enol derivatives   141
    References   145

3. Oxidation of carbonyl compounds   147

3.1 Asymmetric Baeyer–Villiger oxidation   147

Carsten Bolm, T. Kim Khanh Luong, and Oliver Beckmann

    1. Introduction   147
    2. Asymmetric metal-catalysed Baeyer–Villiger oxidations   148
    References   151

4. Oxidation of hetero atoms   153

4.1 Asymmetric oxidation of sulfides and selenides   153

Henri B. Kagan

    1. Introduction   153
    2. Reactions   154
       2.1 Asymmetric sulfoxidation (stoichiometric)   154
          2.1.1 Chiral oxaziridines   154
          2.1.2 Hydroperoxides in the presence of chiral titanium complexes   157

Contents

2.2	Asymmetric sulfoxidation (catalytic)	162
	2.2.1 Hydroperoxides in the presence of chiral titanium complexes	162
	2.2.2 Hydroperoxides in the presence of chiral Schiff base complexes	164
2.3	Asymmetric selenoxidation	166
References		169

4.2 Asymmetric oxidation of other heteroatoms — 171

Sotaro Miyano and Tetsutaro Hattori

1. Introduction — 171
2. Asymmetric oxidation of imines — 172
3. Kinetic resolution of β-hydroxy amines by *N*-oxide formation — 175

References — 179

5. Oxidation using a biocatalyst — 181

5.1 Hydroxylation at a saturated carbon atom — 181

Robert Azerad

1. Introduction — 181
2. Fungal hydroxylations — 181
3. General methods for biohydroxylations — 183
4. Examples of biohydroxylations at saturated carbon atoms — 187
   - 4.1 Hydroxylation of terpenes — 187
   - 4.2 Hydroxylation of synthetic compounds — 189
   - 4.3 Hydroxylation of drugs — 195

References — 198

5.2 Oxidation of alcohols — 200

Giancarlo Fantin and Paola Pedrini

1. Introduction — 200
2. Enzymatic oxidations — 200
3. Microbial oxidations — 210

References — 212

Contents

5.3 Asymmetric Baeyer–Villiger oxidation using biocatalysis 214
Véronique Alphand and Roland Furstoss

    1. Introduction 214
    2. Mechanism and classification 215
    3. Purified enzymes versus whole cells 216
    4. Some examples of biocatalysed BV oxidations 217
        4.1 BV oxidation of enantiopure ketones 217
        4.2 BV oxidation of racemic ketones 221
        4.3 BV oxidation of prochiral ketones 225
    5. Conclusion 225
    References 226

5.4 Oxidation of sulfides 227
Stefano Colonna, Nicoletta Gaggero, Giacomo Carrea, and Piero Pasta

    1. Introduction 227
    2. Chloroperoxidase catalysed oxidation of methyl p-tolyl sulfide to (R)-methyl p-tolyl sulfoxide 228
    3. Cyclohexanone monooxygenase catalysed oxidation of 1,3-dithiane to (R)-1,3-dithiane monosulfoxide 230
    References 234

A1 List of suppliers 237

Reaction Index 239

Reagent Index 240

Subject Index 242

Contributors

VÉRONIQUE ALPHAND
Groupe 'Biocatalyse et Chimie fine', UMR 6111 associé au CNRS, Faculté des Sciences de Luminy, Case 901, 163 avenue de Luminy, F-13288 Marseille Cedex 9, France

ROBERT AZERAD
UMR 8601 (ex URA 400 CNRS), Laboratoire de Chimie et Biochimie Parmacologiques et Toxicologiques, Université R. Descartes-Paris V, 45 rue des Saints-Pères, 75270-Paris Cedex 06, France

HEINRICH BECKER
Covion Organic Semiconductors GmbH, Industrial Park Höchst, D-65926 Frankfurt am Main, Germany

OLIVER BECKMANN
Institut für Organische Chemie der RWTH Aachen, Professor-Pirlet-Straße 1, D-52056 Aachen, Germany

BEENA BHATIA
Department of Chemistry, Princeton University, Princeton, New Jersey 08544, USA

CARSTEN BOLM
Institut für Organische Chemie der RWTH Aachen, Professor-Pirlet-Straße 1, D-52056 Aachen, Germany

GIACOMO CARREA
Istituto di Biocatalisi e Riconoscimento Molecolare CNR, via Mario Bianco 9, 20131 Milano, Italy

BANG-CHI CHEN
Discovery Chemistry, Bristol-Myers Squibb Pharmaceutical Research Institute, Princeton, New Jersey 08543, USA

STEFANO COLONNA
Centro CNR and Istituto di Chimica Organica, Facoltà di Farmacia, via Venezian 21, 20133 Milano, Italy

FRANKLIN A. DAVIS
Department of Chemistry, College of Science and Technology, Temple University, Philadelphia, Pennsylvania 19122-2585, USA

DAVID A. EVANS
Department of Chemistry, Harvard University, 12 Oxford Street, Cambridge, Massachusetts 02138, USA

Contributors

GIANCARLO FANTIN
Dipartimento di Chimica, Univesità di Ferrara, Via Borsari 46, I-44100 Ferrara, Italy

MARGARET M. FAUL
Chemical Process Research & Development, Eli Lilly & Company, Indianapolis, Indiana 46285-4813, USA

ROLAND FURSTOSS
Groupe 'Biocatalyse et Chimie fine', UMR 6111 associé au CNRS, Faculté des Sciences de Luminy, Case 901, 163 avenue de Luminy, F-13288 Marseille Cedex 9, France

NICOLETTA GAGGERO
Centro CNR and Istituto di Chimica Organica, Facoltà di Farmacia, via Venezian 21, 20133 Milano, Italy

TETSUTARO HATTORI
Department of Biomolecular Engineering, Graduate School of Engineering, Tohoku University, Aoba-ku, Sendai 980-8579, Japan

JAVED IQBAL
Department of Chemistry, Indian Institute of Technology, Kanpur, Kanpur 208 016, India

YOSHIO N. ITO
Department of Chemistry, Graduate School of Science, Kyushu University Higashi-ku, Fukuoka 812-8581, Japan

HENRI B. KAGAN
Laboratoire de Synthése Asymmétrique, Institut de Chimie Moléculaire d'Orsay, Université de Paris-Sud, F-91405 Orsay, France

TSUTOMU KATSUKI
Department of Chemistry, Graduate School of Science, Kyushu University Higashi-ku, Fukuoka 812-8581, Japan

IVAN LANTOS
47 Militia Hill, Dr. Wayne, Pennsylvania 19406-0939, USA

T. KIM KHANH LUONG
Institut für Organische Chemie der RWTH Aachen, Professor-Pirlet-Straße 1, D-52056 Aachen, Germany

VICTOR S. MARTIN
Instituto Universitario de Bio-Organica 'Antonio Gonzalez', Universidad de La Laguna, Carretera de La Esperanza, 2, 38206 La Laguna, Tenerife, Spain

Contributors

SOTARO MIYANO
Department of Biomolecular Engineering, Graduate School of Engineering, Tohoku University, Aoba-ku, Sendai 980-8579, Japan

PIERO PASTA
Istituto di Biocatalisi e Riconoscimento Molecolare CNR, via Mario Bianco 9, 20131 Milano, Italy

PAOLA PEDRINI
Dipartimento di Chimica, Univesità di Ferrara, Via Borsari 46, I-44100 Ferrara, Italy

T. PUNNIYAMURTHY
UMR 5637 Laboratoire De Chemie Moleculaire Ét Organisation de Solide, Universite Montpellier 2, F-34095 Montpellier Cedex, France

GUNTHER SCHLINGLOFF
Ciba Specialty Chemicals Inc., Consumer Care/Oxidation Catalysis, PO Box 1266, D-79630 Grenzach-Wyhlen, Germany

K. BARRY SHARPLESS
Department of Chemistry, The Scripps Research Institute, 10550 North Torrey Pines Road, La Jolla, California 92037, USA

PING ZHOU
Chemical Sciences, CNS Disorders, Wyeth-Ayerst Research, CN 8000, Princeton, New Jersey 08543-8000, USA

Abbreviations

AA	asymmetric aminohydroxylation
AD	asymmetric dihydroxylation
ADH	alcohol dehydrogenase
BINAP	2,2'-bis(diphenylphosphino)-1,1'-binaphthyl
BV oxidation	Baeyer–Villiger oxidation
CHMO	cyclohexanone monooxygenase
CHP	cumene hydroperoxide
m-CPBA	m-chloroperoxybenzoic acid
CPO	chloroperoxidase
DBU	1,8-diazabicyclo[5.4.0]undec-7-ene
DEAD	diethyl azodicarboxylate
DET	diethyl tartrate
DIPT	diisopropyl tartrate
DHQ	dihydroquinine
DHQD	dihydroquinidine
(DHQ)$_2$PHAL	1,4-bis(dihydroquinine)phthalazine
(DHQD)$_2$PHAL	1,4-bis(dihydroquinidine)phthalazine
DMAP	N,N-dimethylaminopyridine
DMDO	dimethyldioxirane
DMF	N,N-dimethylformamide
DMSO	dimethyl sulfoxide
EDTA	ethylenediaminetetraacetic acid
FAD	flavine-adenine dinucleotide
FMN	flavin mononucleotide; riboflavin monophosphate
GLC	gas–liquid chromatography
HLADH	horse liver alcohol dehydrogenase
HMDS	hexamethyldisilazane
HPLC	high-performance liquid chromatography
KHMDS	potassium bis(trimethylsilyl)amide
LDA	lithium diisopropylamide
MoOPH	oxodiperoxymolybdenum(pyridine) hexamethylphosphoric triamide
MS	molecular sieves
NAD	adenosine 5'-(trihydrogen diphsphate) P'→5'-ester with 3-(aminocarbonyl)-1-β-D-ribofuranosylpyridinium inner salt; nicotonamide–adenine dinucleotide
NADH	the reduced form of NAD
NADP	adenosine 5'-(trihydrogen diphosphate 2'-dihydrogenphosphate) P'→5'-ester with 3-

Abbreviations

	(aminocarbonyl)-1-β-D-ribofuranosylpyridinium inner salt; nicotonamide–adenine dinucleotide phosphate
NADPH	the reduced form of NADP
NaHMDS	sodium bis(trimethylsilyl)amide
NCA	*N*-carboxy anhydride
NMO	*N*-methylmorpholine *N*-oxide
TBHP	*t*-butyl hydroperoxide
TEA buffer	triethanolamine hydrochloride buffer
THF	tetrahydrofuran
TLC	thin-layer chromatography
Tris HCl buffer	tris(hydroxymethyl)aminomethane HCl buffer

Introduction

T. KATSUKI

Oxidation is an electron transfer reaction from a substrate to a reagent which is called as oxidant. Even if limited to organic reactions, a wide variety of reactions, such as hydroxylation, epoxidation, dihydroxylation, aziridination, hydroxyamination, halogenation, etc., are classified into oxidation reactions. As the names of each reaction indicate, these reactions are related to the functionalization of the substrates. For example, hydroxylation of alkanes produces alcohols and epoxidation and dihydroxylation of alkenes affords epoxides and diols, respectively. In addition to this, most oxidation reactions endow oxidized carbon(s) with chirality (Eqs 1–3). Newly introduced oxygen functionalities can be transformed stereospecifically into various other useful functional groups. Thus, stereocontrolled oxidation reactions are valuable and potent tools for the synthesis of optically active and functionalized compounds. For these reasons, much effort has long been devoted to the study of asymmetric oxidation. This book starts with a brief review of the development of stereocontrol in oxidation reactions and the following chapters are devoted to the practical approaches to various asymmetric oxidation reactions.

$$R^1\text{-CH}_2\text{-}R^2 \xrightarrow{\text{hydroxylation}} R^1\text{-CH(OH)-}R^2 \qquad (1)$$

$$R^1R^2C=CR^3R^4 \xrightarrow{\text{epoxidation}} R^1R^2C\text{-}O\text{-}CR^3R^4 \qquad (2)$$

$$R^1R^2C=CR^3R^4 \xrightarrow{\text{dihydroxylation}} R^1R^2C(OH)\text{-}C(OH)R^3R^4 \qquad (3)$$

∗ = asymmetric carbon

As in many other asymmetric reactions,[1] the first stereocontrolled oxidation was carried out in a diastereoselective manner. In 1907, Mckenzie and Wren reported the asymmetric synthesis of tartaric acid by dihydroxylation of chiral fumaric acid esters with potassium permanganate as an oxidant (Eq. 4).[2]

$$RO_2C\text{-CH=CH-}CO_2R \xrightarrow[\text{2) H}^+]{\text{1) KMnO}_4} HO_2C\text{-CH(OH)-CH(OH)-}CO_2H \qquad (4)$$

R = (−)-menthyl   <20% de

Several decades later, this was followed by enantioselective oxidation, that is oxidation of an achiral substrate with a chiral oxidant. In 1960, asymmetric oxidation of achiral sulfides with chiral (+)-monoperoxycamphoric acid was reported independently by Balenovic et al.[3] and by Montanari et al.[4] (Eq. 5). The same chiral peroxy acid was also used as an oxidant for the asymmetric epoxidation of styrene by Henbest et al. (Eq. 6).[5] Furthermore, copper salts bearing optically active carboxylate ligands were found to serve as catalysts for asymmetric allylic C—H oxidation.[6] Although enantioselectivities observed in these early experiments were low, these pioneering studies demonstrated that oxidation could be carried out in an enantioselective manner. Since then, many chiral oxidants or chiral catalysts were developed for oxidation, and enantioselectivity in oxidation was gradually improved.

In 1977, Yamada et al.[7] and Sharpless et al.[8] independently reported metal-catalysed asymmetric epoxidation of allylic alcohols. Three years later, this procedure was improved to a useful practical level by Katsuki and Sharpless:[9] the introduction of a system of titanium tetraisopropoxide, dialkyl tartrate, and tert-butyl hydroperoxide realized a high enantioselectivity of >90% ee (Eq. 7). This high-selective reaction has been widely used as a key step(s) in the synthesis of numerous optically active compounds, demonstrating the important use of asymmetric oxidation in organic synthesis.[10] This brought about a change in organic synthesis; before 1980, many total syntheses of natural products had ended up with the synthesis of their racemates but, after 1980, total synthesis actually means the synthesis of a target molecule in an optically pure form.

Following this discovery, various types of well-designed optically active oxidants and oxidation catalysts have been introduced and have expanded the scope of asymmetric oxidation. Today, high enantioselectivity exceeding 90% ee has been achieved in many oxidation reactions including epoxidation, dihydroxylation, hydroxyamination, aziridination, oxidation of sulfides, and

Introduction

hydroxylation of activated C—H bonds.[11] Proper use of these asymmetric reactions enables efficient short-step synthesis of a large variety of optically active and useful compounds. However, some reactions remain to be explored, as exemplified by the asymmetric hydroxylation of non-activated C—H bonds.

On the other hand, it is well known that oxidation reactions participate widely in metabolism, that is in biological transformations. These oxidation reactions are catalysed by oxidizing enzymes or microorganisms and generally proceed with excellent and frequently complete stereo-, chemo-, and regioselectivity. The spectrum of biological oxidations is very broad. Not only biotic substances but also exobiotic substances can be oxidized by biological methods. Even non-activated C—H bonds can be selectively oxidized under very mild conditions by using a suitable biocatalyst.[12] For example, hydroxylation of progesterone by *Rhizopus nigricans* or more effectively by a mutant strain of *Aspergillus ochraceus* takes place stereo-, regio-, and chemoselectively to give 11α-hydroxyprogesterone as a single product (Eq. 8). Although substrate specificity of biocatalysts explained by rock-and-key theory limits the scope of biological oxidation, it should be emphasized that various bio catalysts which show different stereo- and regioselectivity are available today and can be used as a chemical reagents in organic synthesis. Therefore, this book also deals with biological oxidations.

$$\text{progesterone} \xrightarrow[\text{or}\atop\text{A mutant strain of } Asperugillus\ ocharaceus]{Rhizopus\ nigricans} 11\alpha\text{-hydroxyprogesterone} \quad (8)$$

As described above, many excellent chemical and biological methods are available for asymmetric oxidation today, but these methods have their own characteristics, merits, and disadvantages. Furthermore, the optimized reaction conditions in terms of stereo- and regioselectivities, chemical yield of the desired product, and turnover number often vary with substrates. Therefore, the use of the reaction of choice under the optimized reaction conditions is a critical factor for achieving an efficient synthesis. The main object of this book is to provide readers with the features of individual asymmetric (mostly enantioselective but some diastereselective) oxidations and their optimized reaction conditions, which include the experimental precautions, detailed laboratory protocols, manipulation with example results, lists of equipments, and expert advice necessary for performing asymmetric oxidation reactions. This knowledge should be of great help for readers to choose an oxidation reaction suitable for a synthetic purpose, and to perform the reaction safely.

Before discussing individual asymmetric oxidation reactions, general precautions for performing oxidations safely are described here.

Safety glasses and proper gloves must always be worn in the laboratory.

Oxidation reactions can be carried out safely under proper reaction conditions but experiments must be undertaken in a well-ventilated hood with the sash closed. Some oxidants are explosive, flammable, corrosive, and self-reactive. Therefore, great care must be taken when oxidants are handled and used in the laboratory. Careless experiments may cause hazards such as explosion, fire, and uncontrollable exothermic reactions. These undesired reactions are very often caused by the contact of oxidants with acids or metal ions, and handling of oxidants and operation of oxidation reactions must be conducted with clean glassware. Do not insert a sampling device such as a syringe or a metal spoon into the container of an oxidant and do not return unused oxidant to the container, so avoiding contamination. The container of the oxidant should be kept in a dark and chilled place, possibly in a refrigerator.

Use of a pure or unnecessarily concentrated oxidant except for some stable oxidants must be avoided. Most hazards are caused by radical-chain decomposition of the oxidant. Use of a diluted oxidant will alleviate these problems. In connection with this, the oxidant remaining in the reaction mixture is recommended to be reduced before the work-up procedure, to avoid its concentration together with the desired oxidation product.

Oxidation reactions are generally exothermic. Whenever possible, mix oxidant and substrate slowly. Control the reaction temperature as indicated in the protocol. A higher temperature may make the reaction difficult to control. A lower temperature slows the reaction and it may cause accumulation of the oxidant.

Asymmetric oxidation can be carried out safely, if chemists follows the correct protocol. The following chapters are invitations to useful and practical asymmetric oxidation reactions.

References

1. Morrison, J. D.; Mosher, H., ed., *Asymmetric Organic Reactions*, Prentice-Hall Inc. Englewood Cliffs (**1971**).
2. Mckenzie, A.; Wren, H. *J. Chem. Soc.* **1907**, *9*, 1907.
3. Balenovic, K.; Bregant, N.; Francetic, D. *Tetrahedron Lett.* **1960**, *6*, 20.
4. Mayr, A.; Montanari, F.; Tramontini, M. *Gazz. Chim. Ital.,* **1960**, *90*, 739.
5. Ewins, R. C.; Henbest, H. B.; McKervey, M. A. *J. Chem. Soc., Chem. Commun.,* **1967**, 1085.
6. Denney, D. B.; Napier, R.; Cammarata, A. *J. Org. Chem.* **1965**, *30*, 3151.
7. Yamada, S.; Mashiko, M.; Terashima, S. *J. Am. Chem. Soc.* **1977**, *99*, 1988.
8. Michaelson, R. C; Palermo, R. E.; Sharpless, K. B. *J. Am. Chem. Soc.* **1977**, *99*, 1990.
9. (a) Katsuki, T.; Sharpless, K. B. *J. Am. Chem. Soc.* **1980**, *102*, 5974. (b) Johnson, R.; Sharpless, K. B. In *Comprehensive Organic Synthesis*, Trost, B. M., ed.; Pergamon: Oxford, **1991**, vol. 7, Ch. 3.2, p. 389.
10. Katsuki, T.; Martin, V. S. *Organic Reactions* **1996**, *48*, 1–299.
11. Beller, M.; Bolm, C., ed., *Transition Metal for Organic Synthesis*, vol. II; Wiley-VCH, Weinheim, **1998**.
12. Brown, S. M. In *Comprehensive Organic Synthesis*, Trost, B. M., ed.; Pergamon: Oxford, **1991**, vol. 7, Ch. 1.4, p. 53.

1

Oxidation of the C—H bond

1.1 Asymmetric oxidation of the C—H bond

B. BHATIA, T. PUNNIYAMURTHY, and J. IQBAL

1. Introduction

The development of methodology for asymmetric functionalization of the C—H bond offers a challenge in the strive for acquiring optically active building-blocks from simple starting materials. Asymmetric synthesis has emerged as an exciting area of chemistry by judiciously employing the principles of organic synthesis, molecular recognition, metal coordination chemistry, and catalysis. Among the various strategies available for exploiting the pools of chiral compounds, catalytic asymmetric induction offers a distinct advantage in achieving a high level of enantioselectivity during bond formation mainly due to the influence of asymmetry possessed by the ligands in the catalyst. The titanium-catalysed epoxidation[1-3] of allylic alcohols, the osmium-catalysed dihydroxylation[4-7] of alkenes, the copper-catalysed cyclopropanation[8-10] of alkenes, and the ruthenium-catalysed asymmetric hydrogenation[11,12] of alkenes are outstanding achievements in the catalytic asymmetric induction approach. Another approach is inspired by the analogy derived from nature, where the monooxygenase enzymes are, for example, known to effect stereospecific oxidation of organic compounds.[13] This biomimetic approach has led to widespread research activity in this area and as a result an impressive start has been made in achieving good levels of selectivity in the oxidation of organic substrates by employing well-crafted small molecular catalysts. The possibility of mimicking the enantioselective reaction of monooxygenases is a highly viable proposition from the point of view of synthetic organic chemists. Indeed, the design and development of synthetic monooxygenase mimics that catalyse enantioselective epoxidation of unfunctionalized alkenes have already made an impressive impact on stereoselective synthesis mainly due to the pioneering efforts of the research groups of Groves,[14] Jacobsen,[15-17] and Katsuki,[18-22] respectively. On the other hand, the enantioselective oxidation of the saturated C—H bond has proven more difficult to accomplish with synthetic catalysts. Consequently, there has been continual effort to achieve a high level of enantioselectivity in the oxidation of the saturated C—H bond over the past few years. The following sections deal with the results achieved so far on the enantioselective oxidation of the C—H bond by employing chiral catalysts derived from ligands easily prepared from simple and easily accessible starting materials. The emphasis is on the experimental procedures of those reactions which are operationally simple and easy to perform on a synthetically useful scale.

2. Asymmetric benzylic hydroxylation

In 1989, Groves and Viski,[14b] using the concept of synthetic monooxygenase mimics, reported the asymmetric hydroxylation of the benzylic C—H bond in reasonably good enantioselectivity by employing a chiral iron–porphyrin complex (Fe-**1**) as a catalyst (Scheme 1.1.1). This complex was prepared from optically active vaulted binaphthyl porphyrin ligand **1** and iron pentacarbonyl. By this method, 1,2,3,4-tetrahydronaphthalene **2** can be oxidized to (*R*)-1,2,3,4-tetrahydro-1-naphthol **3** with 72% ee in the presence of iodosylbenzene and the catalyst (Fe-**1**) in dichloromethane. Derivatization of (*R*)-1,2,3,4-tetrahydro-1-naphthol with (*R*)-2-phenylpropionyl chloride and pyridine provided the diastereomeric naphthyl ester **4** which can be separated completely by GLC analysis. The derivatization did not affect the enantiomeric composition of the original alcohol **3**. This is the first reported efficient catalytic asymmetric hydroxylation of hydrocarbons. This approach is exquisite and the preparation of this fascinating porphyrin derivative **1** poses no practical limitations in its use in large-scale applications.

1

1.1: Asymmetric oxidation of the C—H bond

Protocol 1.
Synthesis of (*R*)-1,2,3,4-tetrahydro-1-naphthol (Structure 3, Scheme 1.1.1)

Caution! Carry out all procedures in a well-ventilated hood, and wear protective chemical-resistant gloves and safety goggles.

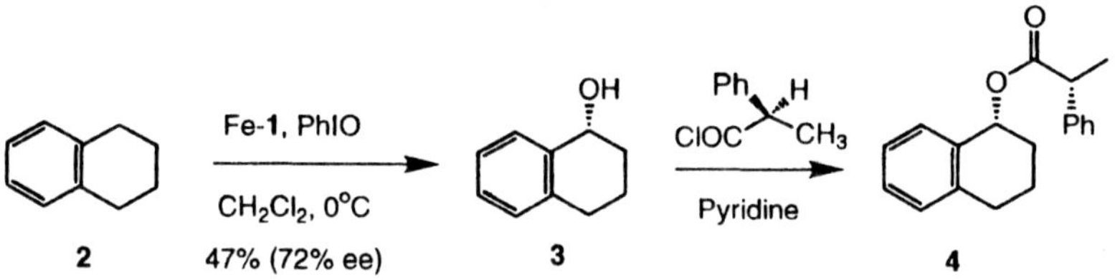

Scheme 1.1.1

This reaction[a] is a representative example of enantioselective benzylic oxidation using iron–porphyrin complex (Fe-1) as a catalyst.[14]

Equipment

- Two round-bottomed flasks (5 mL and 100 mL)
- Dry glass syringes with stainless-steel needles
- Two Teflon-coated magnetic stirrer bars
- Syringe (250 μL)
- Magnetic stirrer
- Oil bath
- Inert gas supply

Materials

- Chiral ligand **1**,[b] 0.1 g, 0.07 mmol
- Iodine, 0.5 g, 2 mmol — corrosive, highly toxic
- Iron pentacarbonyl, 0.138 g, 0.7 mmol — flammable, highly toxic
- Toluene — flammable, toxic
- Benzene — flammable, carcinogenic
- Ethyl acetate — flammable, irritant
- Hydrochloric acid, 10% — corrosive, toxic
- Calcium chloride — hygroscopic, irritant
- Dichloromethane — toxic, irritant
- 1,2,3,4-Tetrahydronaphthalene, 132 mg, 136 μL, 1 mmol — toxic, irritant
- Iodosylbenzene, 22 mg, 0.1 mmol — oxidizer, self-reacting substance
- Pyridine, 1 mL — toxic, flammable liquid
- (*R*)-2-Phenylpropionyl chloride, 0.119 g, 0.71 mmol — corrosive, lachrymator
- Diethyl ether — flammable liquid, toxic

Preparation of the catalyst (Fe-1)

1. To a solution of **1** (0.1 g, 0.07 mmol) and iodine (0.5 g, 2 mmol) in toluene (40 mL) under nitrogen, add iron pentacarbonyl (0.138 g, 0.7 mmol).
2. Stir the reaction mixture for 4 h at 100 °C and expose it to air at room temperature for 3 h.
3. Filter the solid on a sintered-glass filter funnel and wash with benzene.

Protocol 1. Continued

4. Dissolve the solid in ethyl acetate and shake the solution with 10% HCl until the visible spectrum is stabilized.
5. Dry the ethyl acetate solution over calcium chloride and concentrate under reduced pressure on a rotary evaporator.
6. Chromatograph the residue on silica gel using ethyl acetate as eluent to give the iron–porphyrin complex (Fe-1, 97 mg, 91%).

Hydroxylation of 1,2,3,4-tetrahydronaphthalene

1. Place Fe-1 (1.5 mg, 1 μmol) in a round-bottomed flask containing degassed dichloromethane (2 mL) under nitrogen.
2. Add 1,2,3,4-tetrahydronaphthalene (132 mg, 136 μL, 1 mmol) via syringe to the above reaction mixture.
3. Cool the reaction mixture to 0 °C, add iodosylbenzene (22 mg, 0.1 mmol) and stir for 1 h.
4. Filter off the insoluble materials and remove the solvent under reduced pressure on a rotary evaporator.
5. Chromatograph the resultant residue on silica gel using dichloromethane as eluent. Concentrate the alcohol-containing fractions (47%,[c] 72% ee).
6. Dissolve the alcohol **3** in pyridine (1 mL). Add (*R*)-2-phenylpropionyl chloride (0.119 g, 0.71 mmol) and stir the reaction mixture at 70 °C for 1 h.
7. Pour the reaction mixture into ice–water and extract the diastereomeric ester products with diethyl ether.
8. Dry the ether solution over sodium sulfate and concentrate under reduced pressure on a rotary evaporator.
9. Determine the diastereomeric excess of **4** by GLC analysis employing SPB-35, or a chiral fused silica capillary column.[d]

[a] The relative ratio of catalyst, PhIO, and substrate is 1:100:1000.
[b] Prepared according to Ref. 14.
[c] The yield was calculated on the basis of the amount of iodosylbenzene used.
[d] Prepared according to Ref. 41.

This is the first reported efficient method for the enantioselective hydroxylation of the benzylic C—H bond. The reaction has been proposed to proceed via a radical intermediate[14b] and enantioselectivity has been explained to depend both on the selectivity of the hydrogen abstraction step and the relative rate of radical cage collapse as the enantiomeric radical intermediates escape from the vaulted ligand cage at different rates.

Katsuki and co-workers have developed[23] a (salen)manganese(III) complex for enantioselective C—H oxidation which seems more viable and practical to

1.1: Asymmetric oxidation of the C—H bond

adopt for a large-scale preparation. They used (salen)manganese(III) complex **5** as the catalyst in a solvent of high viscosity to suppress the formation of free radical intermediates. According to their protocol, the oxidation of 1,1-dimethylindan **6** with iodosylbenzene in the presence of the catalyst **5** in chlorobenzene afforded the corresponding hydroxy compound **7** in 64% ee (Scheme 1.1.2). The preparation of the catalyst **5** is simple and it can be obtained in large quantities in optically pure form. Recently they reported that the remarkably improved enantioselectivity in benzylic hydroxylation (up to 84% ee) was realized by using a new (salen)manganese(III) complex of concave type.[23b] The enantiomeric excesses of the resulting benzylic alcohols were further improved (up to 90% ee) as the reaction time was increased, since the minor enantiomers of the alcohols were oxidized preferentially to the corresponding ketones under these conditions.

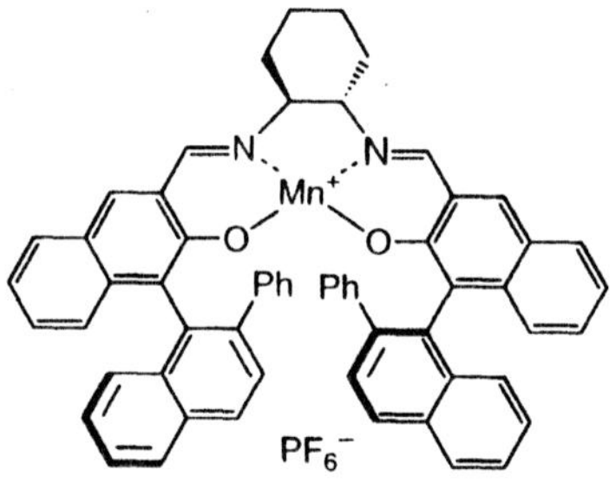

5

Protocol 2.
Synthesis of (R)-3,3-dimethylindan-1-ol (Structure 7, Scheme 1.1.2)

Caution! Carry out all procedures in a well-ventilated hood, and wear protective chemical-resistant gloves and safety goggles.

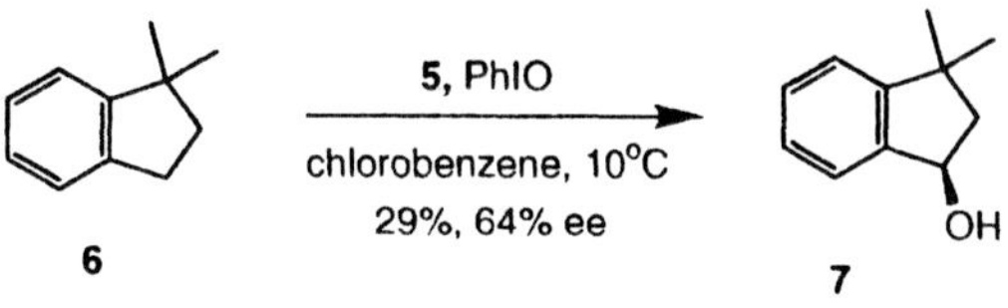

6 → **7**
5, PhIO
chlorobenzene, 10°C
29%, 64% ee

Scheme 1.1.2

This is also a representative example of enantioselective benzylic oxidation using the (salen)manganese(III) complex **5**.[23a]

Equipment

- Two round-bottomed flasks (5 mL)
- Two Teflon-coated magnetic stirrer bars
- Magnetic stirrer
- Two three-way stopcocks
- Inert gas supply
- Canula
- Dry glass syringes with stainless-steel needles

Protocol 2. Continued

Materials

- 1,1-Dimethylindan,[a] 14.6 mg, 0.1 mmol   flammable
- (Salen)manganese(III) complex 5, 2.0 mg, 2.0 μmol
- Dry chlorobenzene,[b] 1.0 mL   flammable liquid, irritant
- Iodosylbenzene, 22.0 mg, 0.1 mmol   oxidizer, self-reacting substance
- Dimethyl sulfide   flammable liquid, stench
- Hexane   flammable liquid, irritant
- Ethyl acetate   flammable liquid, irritant

Method

1. Place 1,1-dimethylindan **6** (14.6 mg, 0.1 mmol), (salen)manganese(III) complex **5** (2.0 mg, 2.0 μmol), and chlorobenzene (1.0 mL) in a 5-mL round-bottomed flask under an atmosphere of nitrogen.

2. Cool the mixture with a liquid-nitrogen bath and deoxygenate the mixture by alternately evacuating and pressurizing with argon (three times).

3. Warm the mixture to 10°C.

4. Transfer the mixture through a canula to another flask containing iodosylbenzene (22.0 mg, 0.1 mmol) under an atmosphere of nitrogen.

5. Stir the mixture for 1.5 h at 10°C.

6. Quench the reaction mixture by adding several drops of dimethyl sulfide and concentrate under reduced pressure on a rotary evaporator.

7. Dissolve the residue in the minimum volume of 9:1 hexane–ethyl acetate and load onto the silica gel column. Elute the column with 9:1 hexane–ethyl acetate to give (R)-3,3-dimethylindan-1-ol **7** (4.7 mg, 29 %).

8. The enantiomeric excess of this product was 64% as determined by GLC using an optically active column (SUPELCO β-DEX 120 fused silica capillary column, 30 m × 0.25 mm i.d., 0.25 μm film).

[a] Prepared according to Ref. 42.
[b] Freshly distilled from benzophenone ketyl under an inert atmosphere.

---

3. Asymmetric allylic oxidation

The Kharasch reaction,[24–26] in which the allylic carbon of an alkene is acyloxylated in the presence of copper and peracid derivatives, provides an exciting possibility for the synthesis of enantiopure allylic alcohols (Scheme 1.1.3). This protocol has been the focus of several studies in recent years and a majority of them have exploited copper complexes bearing α-amino acid-based chiral ligands. Mechanistically, this reaction is believed to proceed via the formation of a C—Cu bond, and this has led to the assumption that the reaction might be rendered enantioselective by using the copper catalyst with a suitable chiral

1.1: Asymmetric oxidation of the C—H bond

ligand. Indeed, this assumption has been vindicated as several groups have reported the formation of highly optically active products using chiral non-racemic copper complexes as catalysts or stoichiometric reagents.[27-29]

Pfaltz and co-workers have utilized[10,30] Cu(I) complexes, prepared *in situ* from chiral bisoxazoline **8**[31] and copper(I) trifluoromethanesulfonate, for the allylic oxidation of cycloalkenes. They have used 5 mol% of the catalyst, and *t*-butyl peroxybenzoate as oxidant to convert cycloalkenes to optically active 2-cycloalkenyl benzoate in moderate to good yields. According to this protocol, cyclopentene **9** is converted to (*S*)-2-cyclopentenyl benzoate **10** in 84% ee at $-20\,°C$ with good chemical yield (61%) (Scheme 1.1.3). The most promising results were obtained with Cu(I) complexes prepared *in situ* from copper(I) trifluoromethanesulfonate or $[Cu(CH_3CN)_4](PF_6)$ and bisoxazoline. These complexes were found to undergo 10–15 turnovers without any loss of enantioselectivity. Either acetonitrile or acetone proved to be the solvent of choice depending on the substrate and the ligand used.

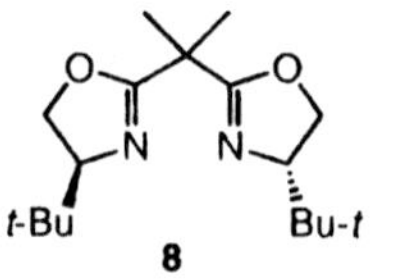

Protocol 3.
Synthesis of (*S*)-2-cyclopentenyl benzoate (Structure 10, Scheme 1.1.3)

Caution! Carry out all procedures in a well-ventilated hood, and wear protective chemical-resistant gloves and safety goggles.

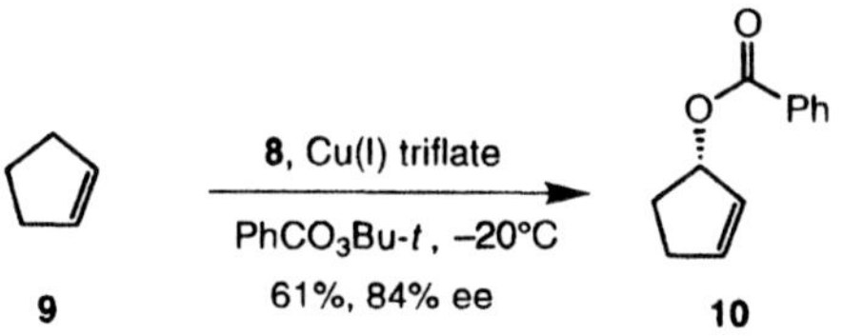

Scheme 1.1.3

This is an example of enantioselective allylic oxidation of alkenes using a chiral bisoxazoline **8**–copper complex.[30]

Equipment

- A round-bottomed flask (5 mL)
- A Schlenk flask
- Two Teflon-coated magnetic stirrer bars
- Two three-way stopcocks
- Inert gas supply
- Canula
- Dry glass syringes with stainless-steel needles
- Magnetic stirrer

Protocol 3. Continued

Materials

- Bisoxazoline **8**, 220 mg, 0.75 mmol
- Copper(I) trifluoromethanesulfonate benzene complex,[a]   flammable solid
  140 mg, 0.5 mmol, (ca. 90%)
- Cyclopentene **9**, 2.7 g, 40 mmol   flammable liquid, irritant
- t-Butyl peroxybenzoate,[a] 2.03 g, 9.4 mmol, (ca. 90%)   oxidizer, self-reacting substance
- Chloroform, 2 mL   highly toxic
- Diethyl ether   flammable liquid
- Hexane   flammable liquid, irritant
- Ethyl acetate   flammable liquid, irritant

Method

1. Place copper(I) trifluoromethanesulfonate–benzene complex (140 mg, 0.5 mmol, ca. 90%) and a stirrer bar in a 5-mL round-bottomed flask under nitrogen.
2. Add a solution of ligand **8** (220 mg, 0.75 mmol) in chloroform (2 mL) via syringe.
3. Stir the solution at room temperature for 1 h.
4. Transfer the catalyst solution through a chromafil disposable filter to a Schlenk tube containing cyclopentene (2.7 g, 40 mmol) and acetonitrile (6 mL) under nitrogen.
5. Add t-butyl peroxybenzoate (2.03 g, 9.4 mmol, ca. 90%) to the reaction mixture over a period of 10 min.
6. Keep the reaction mixture at −20°C for 22 days.
7. Add water by syringe to the mixture and extract twice with diethyl ether.
8. Wash the combined organic layers with 2 N HCl and water.
9. Remove the solvent under reduced pressure on a rotary evaporator.
10. Purify the resultant residue by flash chromatography on silica gel (3.5 × 30 cm column) using 99:1 hexane–ethyl acetate as eluent to give (S)-2-cyclopentenyl benzoate (61%[b], 84% ee).[c]
11. Determine the enantioselectivity by HPLC (t_R = 32.3 min (S), 39.8 min (R); Chiralcel OD, 0.46 × 30 cm, hexane–i-PrOH 1000:1; 0.5 mL/min).

[a] [Cu(I)OTf·0.5(C$_6$H$_6$)] and t-butyl peroxybenzoate were purchased from Fluka.
[b] Yield is based on the consumed peroxyester.
[c] The spectroscopic data were in accordance with data in Ref. 43.

A similar protocol reported by Andrus and co-workers also afforded a good selectivity for cyclopentene oxidation (81% ee) using copper-**11** catalyst. Unlike the ligand **8**, the ligand **11** from the catalyst can be recovered from the reaction mixture and recycled.[32]

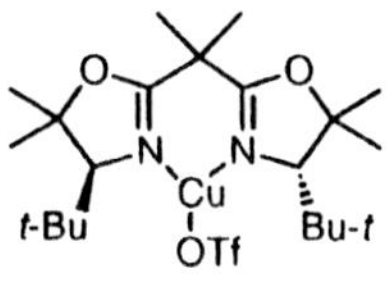

1.1: Asymmetric oxidation of the C—H bond

11

Muzart and co-workers have reported[33,34] enantioselectivity up to 54% during the oxidation of cyclopentene with *t*-butyl peroxybenzoate or a mixture of *t*-butyl hydroperoxide and benzoic acid in the presence of a copper salt of (*S*)- or (*R*)-proline. Later, Zondervan and Feringa[35] used a combination of anthraquinone, *t*-butyl hydroperoxide, *in situ* prepared bis-(*S*)-prolinato-copper(II) complex and propionic acid to oxidize cyclohexene, and obtained enantioselectivity up to 63%. Interestingly, the addition of anthraquinone brought about not only the improvement in enantioselectivity but also the reversal of the non-linear relationship between the ee of the product and the ee of the (*S*)-proline ligand.

Södergren and Andersson[36] have used synthetic bicyclic amino acids as chiral ligands. They prepared a catalytic system *in situ* from copper(II) acetate monohydrate, copper bronze, and the ligand. With the copper complex of bicyclic α-amino acid **12** as a catalyst, cyclohexene was oxidized to (*R*)-2-cyclohexenyl benzoate in 63% yield with enantioselectivity of 65% (Scheme 1.1.4).

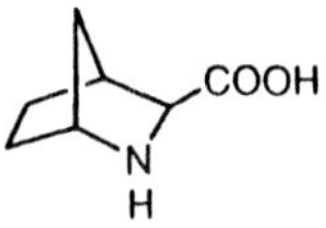

12

Protocol 4.
Synthesis of (*R*)-2-cyclohexenyl benzoate (Structure 14, Scheme 1.1.4)

Caution! Carry out all procedures in a well-ventilated hood, and wear protective chemical-resistant gloves and safety goggles.

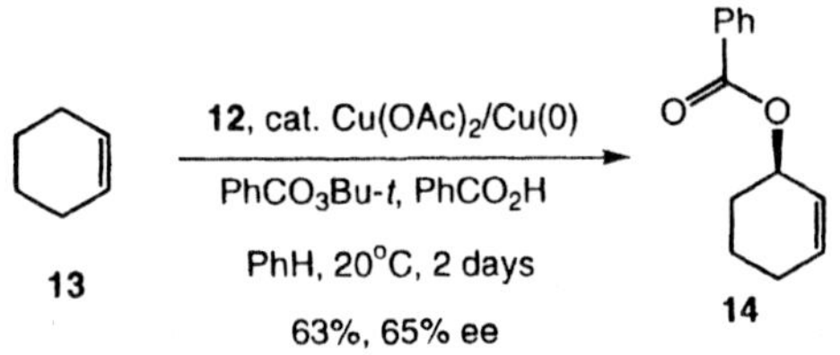

13 → **12**, cat. Cu(OAc)$_2$/Cu(0), PhCO$_3$Bu-*t*, PhCO$_2$H, PhH, 20°C, 2 days → **14**

63%, 65% ee

Scheme 1.1.4

This is a representative example of allylic oxidation using a chiral copper catalyst obtained from the bicyclic amino acid **12**.

Protocol 4. Continued

Equipment
- A round-bottomed flask (5 mL)
- A Teflon-coated magnetic stirrer bar
- Magnetic stirrer
- Inert gas supply
- Dry glass syringes with stainless-steel needles
- Rubber septum
- Separating funnel

Materials[a]
- Cyclohexene 13, 127 mg, 1.55 mmol — flammable liquid, irritant
- Copper(II) acetate monohydrate, 3.1 mg, 16 μmol — irritant
- Copper bronze, 10 mg, 0.16 mmol — flammable solid, irritant
- Ligand 12, 10.9 mg, 77 μmol
- Benzoic acid, 35.4 mg, 0.29 mmol — irritant
- Benzene, 0.8 mL — flammable liquid, carcinogenic
- t-Butyl peroxybenzoate, 56 mg, 55 μl, 0.29 mmol — oxidizer, self-reacting substance
- Sodium hydrogencarbonate
- Diethyl ether — flammable, toxic
- Pentane — flammable, irritant

Method

1. Place copper(II) acetate monohydrate (3.1 mg, 16 μmol), copper bronze (10 mg, 0.16 mmol), ligand 12 (10.9 mg, 77 μmol), benzoic acid (35.4, 0.29 mmol), and a stirrer bar in a round-bottomed flask (5 mL).
2. Attach a rubber septum and purge the flask three times with argon using a gas adapter.
3. Add benzene (0.8 mL) and cyclohexene (127 mg, 1.55 mmol) to the reaction mixture via a syringe.
4. Cool the resulting suspension in a water–ice bath.
5. Add t-butyl peroxybenzoate (56 ml, 55 μl, 0.29 mmol) dropwise via syringe over a period of 5 min.
6. Stir the reaction mixture at room temperature for 2 days.
7. Quench the reaction mixture with saturated sodium hydrogencarbonate solution (2 mL).
8. Extract the aqueous layer with diethyl ether (2 × 2 mL).
9. Combine the organic phases and dry over anhydrous magnesium sulfate.
10. Concentrate and purify the residue by silica gel column chromatography[b] using 9:1 pentane–diethyl 1 ether as eluent (R_f = 0.40) to afford (R)-2-cyclohexeny 1 benzoate 14 (63%, 65% ee).
11. Determine the enantiomeric excess by HPLC analysis employing a Chiralcel-AD column and hexane–i-PrOH (99:1, 0.4 mL/min) as the mobile phase. The enantiomers eluted at 16.0 and 17.4 min were detected by a UV-detector (254 nm).

[a] Benzene, cyclohexene, and t-butyl peroxybenzoate were purchased from Lancaster and used as-received, except for the benzene which was stored over 4-Å activated molecular sieves.
[b] Analytical TLC was carried out on Macherey–Nagel silica gel 60 precoated plates and column chromatography was performed using Merck Kiesel gel 60 (230–400 mesh).

1.1: Asymmetric oxidation of the C—H bond

Duttagupta and Singh[37] have employed copper(I) complexes of chiral bis(oxazolinyl)pyridine ligands[38] in the enantioselective oxidation of cycloalkenes using *t*-butyl peroxybenzoate in the presence of molecular sieves. The highest enantioselectivity (81%) was achieved for the oxidation of cyclohexene using the *in situ* prepared (Cu(I)-**15** complex in a reasonably good chemical yield (58%). A mechanistic model has been proposed which invokes a radical intermediate in these transformations.

15

Optically active copper(II)-tris(oxazoline) complexes have been used by Katsuki and co-workers[39,40] as model compounds of the active site of non-heme oxygenase to oxidize cycloalkenes to the corresponding enantiomerically enriched allylic benzoates. Although both the copper complexes obtained from copper(I) or copper(II) trifluoromethanesulfonates and ligand **16** were found to be efficient, the copper(II) complex showed higher asymmetric induction than the copper(I) complex. For example, the reaction using copper(II) trifluoromethanesulfonate and *t*-butyl peroxybenzoate gave (*S*)-2-cyclopentenyl benzoate **10** (68%, 74% ee) at room temperature, while the reaction using copper(I) trifluoromethanesulfate and *t*-butyl peroxybenzoate provided the same product with 66% ee in 67% yield. They have also demonstrated that the reaction rates were enhanced by addition of molecular sieves (81%, 83% ee at 0 °C). These workers also examined the effect of the substituent on the ligand and peroxyesters, but no profound influence of the manipulations was observed on the enantioselectivity.

16

In summary, the iron-porphyrin complex, (salen)manganese(III) complexes and copper complexes derived from chiral bis- or tris-oxazoline ligands and bicyclic α-amino acids are the best catalysts known so far for asymmetric oxidation of the C–H bond of benzylic and allylic substrates. Although the enantioselectivity and catalytic efficiency still need to be improved, a first step has been taken towards a practical catalyst system for these transformations.

References

1. (a) Sharpless, K. B.; Woodard, S. S.; Finn, M. G. *Pure Appl. Chem.* **1983**, *55*, 1823. (b) Katsuki, T.; Martín, V. S. In *Organic Reactions*; Paquette, L. A., ed.; Wiley: New York, 1996; vol. 48, ch. 1, pp. 1–299.
2. Finn, M. G.; Sharpless, K. B. *J. Am. Chem. Soc.* **1991**, *113*, 113.
3. Carlier, P. R.; Sharpless, K. B. *J. Org. Chem.* **1989**, *54*, 4016.
4. Hartmuth, C. K.; VanNeieuwenhze, M. S.; Sharpless, K. B. *Chem. Rev.* **1994**, *94*, 2483.
5. Sharpless, K. B.; Teranishi, A. Y.; Bäckvall, J.-E. *J. Am. Chem. Soc.* **1977**, *99*, 3120.
6. Jorgensen, K. A.; Schiot, B. *Chem. Rev.* **1990**, *90*, 1483.
7. Göbel, T.; Sharpless, K. B. *Angew. Chem., Int. Ed. Engl.* **1993**, *32*, 1329.
8. Evans, D. A.; Woerpel, K. A.; Hinman, M. M.; Faul, M. M. *J. Am. Chem. Soc.* **1991**, *113*, 726.
9. Lowenthal, R. E.; Masamune, S. *Tetrahedron Lett.* **1991**, *32*, 7273.
10. Pfaltz, A. *Acc. Chem. Res.* **1993**, *26*, 339.
11. Noyori, R. *Chem. Soc. Rev.* **1989**, *18*, 187.
12. Noyori, R. *Science* **1990**, *248*, 1194.
13. Collman, J. P.; Zhang, X.; Lee, V. J.; Uffelman, E. S.; Brauman, J. I. *Science* **1993**, *261*, 1404.
14. (a) Groves, J. T.; Viski, P. *J. Org. Chem.* **1990**, *55*, 3628. (b) Groves, J. T.; Viski, P. *J. Am. Chem. Soc.* **1989**, *111*, 8537.
15. Zhang, W.; Loebach, J. L.; Wilson, S. R.; Jacobsen, E. N. *J. Am. Chem. Soc*, **1990**, *112*, 2801.
16. Lee, N. H.; Muci, A. R.; Jacobsen, E. N. *Tetrahedron Lett*, **1991**, *32*, 5055.
17. Jacobsen, E. N. In *Catalytic Asymmetric Synthesis*; Ojima, I., ed.; VCH Inc.: New York, **1993**; p. 159.
18. Irie, R.; Noda, K.; Ito, Y.; Matsumoto, N.; Katsuki, T. *Tetrahedron Lett*, **1990**, *31*, 7345.
19. Hosoya, N.; Irie, R.; Ito, Y,; Katsuki, T. *Synlett* **1991**, 691.
20. Hosoya, N.; Hatayama, A.; Irie, R.; Sasaki, H.; Katsuki, T. *Tetrahedron* **1994**, *50*, 4311.
21. Katsuki, T. *Coord. Chem. Rev.* **1995**, *140*, 189.
22. Katsuki, T. *J. Synth. Org. Chem. Jpn*, **1995**, *53*, 940.
23. (a) Hamachi, K.; Irie, R.; Katsuki, T. *Tetrahedron Lett.* **1996**, *37*, 4979. (b) Hamada, T.; Irie, R.; Mihara, J.; Hamachi, K.; Katsuki, T. *Tetrahedron* **1998**, *54*, 10017.
24. Kharasch, M. S.; Sosnovsky, G. *J. Am. Chem. Soc.* **1958**, *80*, 756.
25. Kharasch, M. S.; Fono, A. *J. J. Org. Chem.* **1958**, *23*, 324.
26. Kharasch, M. S.; Sosnovsky, G.; Vang, N. C. *J. Am. Chem. Soc.* **1959**, *81*, 5819.
27. Denney, D. B.; Napier, R.; Cammarata, A. *J. Org. Chem.* **1965**, *30*, 3151.
28. Araki, M.; Nagase, T. *Chem. Abstr.* **1977**, *86*, 120776r.
29. Rawlinson, D. J.; Sosnovsky, G. *Synthesis* **1972**, 1.
30. Gokhale, A. S.; Minidis, A. B. E.; Pfaltz, A. *Tetrahedron Lett.* **1995**, *36*, 1831.
31. Gant, T. G.; Meyers, A. I. *Tetrahedron: Asymmetry* **1994**, *8*, 2297.
32. Andrus, M. B.; Argade, A. B.; Chen, X.; Pamment, M. G. *Tetrahedron Lett.* **1995**, *36*, 2945.
33. Levina, A.; Muzart, J. *Tetrahedron: Asymmetry* **1995**, *6*, 147.

1.1: Asymmetric oxidation of the C—H bond

34. Levina, A.; Muzart, J. *Synth. Commun.* **1995**, *25*, 1789.
35. Zondervan, C.; Feringa, B. L. *Tetrahedron: Asymmetry* **1996**, *7*, 1895.
36. Södergren, M. J.; Andersson, P. G. *Tetrahedron Lett*, **1996**, *37*, 7577.
37. Dattagupta, A.; Singh, V. K. *Tetrahedron Lett.* **1996**, *37*, 2633.
38. Nishiyama, H.; Yamaguchi, S.; Kondo, M.; Itoh, K. *J. Org. Chem.* **1992**, *57*, 4306.
39. Kawasaki, K.; Tsumura, S.; Katsuki, T. *Synlett* **1995**, 1245.
40. Kawasaki, K.; Katsuki, T. *Tetrahedron* **1997**, *53*, 6337.
41. Schurig, V.; Weber, R. *J. Chromatogr.* **1981**, 217, 51.
42. Reetz, M. T.; Westermann, J.; Kyung, S.-H. *Chem. Ber.* **1985**, *118*, 1050.
43. Gupta, A. K.; Kazlauskas, R. J. *Tetrahedron: Asymmetry* **1993**, *4*, 879.

2

Oxidation of the C=C bond

2.1 Metal-catalysed epoxidation of simple olefins

Y. N. ITO and T. KATSUKI

1. Introduction

Metal complexes promote catalytic epoxidation in two different ways: (i) by activation of an oxidant and subsequent migration of an oxygen from the oxidant to olefins; (ii) by oxidation of a metal ion by an oxidant to the high-valent metal oxide and subsequent oxygen transfer to olefins with the reduction of metal ion which can be reoxidized to the metal oxide *in situ*. The former types of reactions are catalyzed mostly by early transition metal complexes and the latter type of reactions mainly by group VII–X transition metal complexes.[1] For example, t-butyl hydroperoxide is activated by $O=V(OR)_3$ or $O_2Mo(OR)_2$ and undergoes epoxidation.[2] During these reactions, the oxidation states of the vanadium and molybdenum ions do not change (Scheme 2.1.1). In general, olefins bearing a precoordinating functional group are good substrates for this type of reaction (see Section 2.3), while epoxidation of

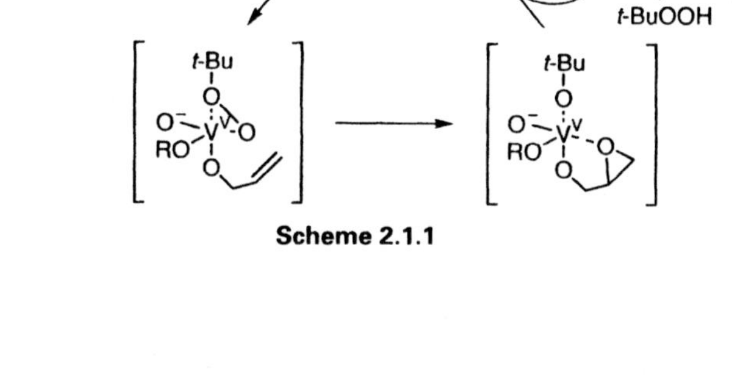

Scheme 2.1.1

isolated olefins is rather slow. On the other hand, stereospecific epoxidation of olefins and C—H hydroxylation of alkanes are realized in living cells with the aid of oxidizing enzymes. Based on the clarification of the catalysis of an iron porphyrin complex at the active site of the representative oxidizing enzyme cytochrome P450, investigations into the enantioselective epoxidation of simple olefins with metalloporphyrin complexes or its equivalents as catalysts were commenced.[3] The P-450-catalysed reactions transfer oxygen atoms to substrates via a high-valent oxo iron porphyrin complex that is produced by the complex redox procedure containing the enzymatic reaction with molecular oxygen. However, the formation of the same oxo species can be chemically effected by treatment of an iron porphyrin complex with stoichiometric oxidants such as iodosylbenzene (Scheme 2.1.2).[4] Besides iron porphyrins, many other metal complexes have been found to catalyse oxygen-transfer reactions via the corresponding high-valent metal oxides.[1]

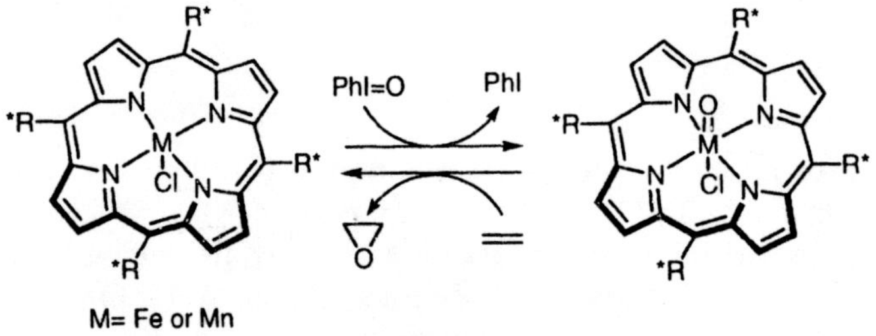

Scheme 2.1.2

To achieve a high level of asymmetric epoxidation chemically, the recognition of olefins to force one of their enantiofaces to an active metal centre, which is effected by the protein-cavity in biological reactions, should be realized through non-covalent interactions between olefins and the chiral ligand of a metal complex. Various types of metal complexes were designed, synthesized, and examined as chiral catalysts for asymmetric epoxidation and excellent results have been reported. In this section, metalloporphyrin-catalysed asymmetric epoxidation is introduced first.[5] Then, metallosalen complexes will be shown as the most efficient catalysts to date.[6] Study of aerobic enantioselective epoxidations that employ molecular oxygen as the terminal oxidant will also be presented.

2. Metalloporphyrin-catalysed asymmetric epoxidation

After the pioneering works of Groves *et al.* on epoxidation and hydroxylation catalysed by iron porphyrins in 1979[4] and the catalytic asymmetric epoxidation of simple olefins catalysed by a chirally modified porphyrin in 1983,[3] various types of chiral metalloporphyrins were synthesized and examined as catalysts.[7] Iodosylbenzene, hydrogen peroxide, or hypochlorite derivatives were employed as stoichiometric oxidants for the oxidation of metalloporphyrins to the

2: Oxidation of the C=C bond

corresponding oxo species which constructs the shunt route in the enzymatic catalytic cycle. However, the substrates that exhibit high enantioselectivity are limited to some styrene derivatives.

Later, Inoue and co-workers reported that the optically active manganese (III)–porphyrin complex, which is endowed with facial chirality by a strap introduced onto the porphyrin ring, shows moderate enantioselectivity in the presence of imidazole, despite its simple structure.[8] The role of imidazole is considered to block the unhindered face of the porphyrin ring by coordinating to the manganese ion.

Collman et al. have recently designed a new threitol-strapped manganese (III)–porphyrin complex **1** taking advantage of the features of the above two types of chiral metalloporphyrin complexes and achieved high enantioselectivity (70–90% ee) in the epoxidation of conjugated mono- and cis-disubstituted olefins, when the reaction is carried out in the presence of 1,5-dicyclohexylimidazole (Scheme 2.1.3).[9] Although various oxidants described above promote this epoxidation, use of iodosylbenzene leads to the best enantioselectivity. This difference of enantioselectivity with the oxidants has been attributed to the scrambling of the axial ligand on the unhindered face of the complex **1**.

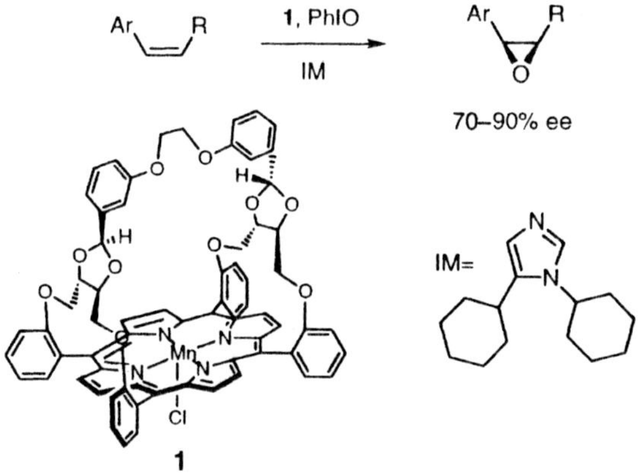

Scheme 2.1.3

Metalloporphyrin-catalysed epoxidation can be carried out in dichloromethane, acetonitrile, toluene, and so on, but dichloromethane is the most generally used solvent.

Protocol 1.
Synthesis of optically active manganese–porphyrin complex 1 (Structure 1, Scheme 2.1.3)

Caution! Carry out all procedures in a well-ventilated hood, and wear safety glasses and disposable vinyl or latex gloves.

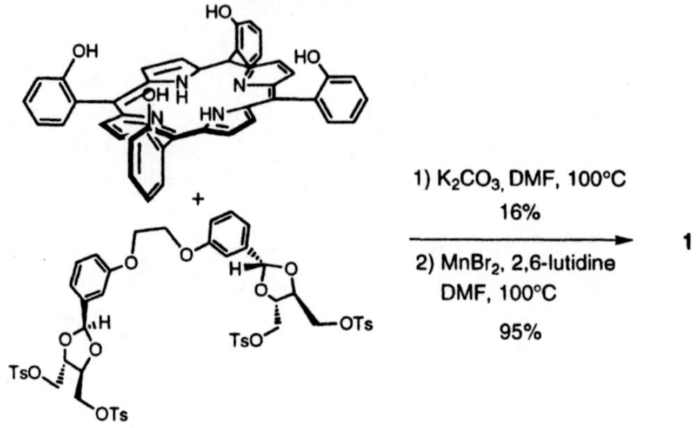

Scheme 2.1.4

Equipment

- Magnetic hotplate stirrer
- Three-necked round-bottomed flasks (50 mL and 500 mL)
- Teflon-coated magnetic stirring bar
- Nitrogen inlet
- Water bath (1 L)
- Separatory funnels (2 L and 200 mL)

Materials

5,10,15,20-Tetrakis(2-hydroxyphenyl)porphyrin[a] (FW 680.75), 611 mg, 0.90 mmol	unknown toxicity, treat as toxic
Tetratosylated bis(threitol) derivative 2[b] (FW 1095.24), 1.18 g, 1.1 mmol	unknown toxicity, treat as toxic
Dry N,N-dimethylformamide (DMF) (230 mL)	cancer-suspect, irritant
Dichloromethane	toxic, irritant, harmful by inhalation
Ethyl acetate	irritant, flammable
K_2CO_3 (FW 138.21), 1.49 g, 11 mmol	irritant, hygroscopic
Na_2SO_4	irritant, hygroscopic
Silica gel	harmful by inhalation
Methanol	toxic, flammable
2,6-Lutidine	toxic, flammable
Anhydrous manganese(II) bromide (FW 214.76) (Alfa) 120 mg, 0.56 mmol	hygroscopic
NaCl	irritant, hygroscopic
Diethyl ether	toxic, flammable
Heptane	irritant, flammable

2: Oxidation of the C=C bond

Method

1. Equip a 500-ml three-necked round-bottomed flask with a nitrogen inlet, a nitrogen bubbler, and a magnetic stirring bar. Maintain the apparatus under a positive pressure of nitrogen.
2. Place tetratosylated bis(threitol) derivative 2 (1.18 g, 1.1 mmol), 5,10,15,20-tetrakis(2-hydroxyphenyl)porphyrin (611 mg, 0.90 mmol), and dry DMF (200 mL) in the flask.
3. Heat the mixture to 100 °C with stirring.
4. Add K_2CO_3 (1.49 g, 11 mmol) and continue stirring at 100 °C for 16 h.
5. Cool the mixture to room temperature and dilute with dichloromethane (300 mL).
6. Transfer the reaction mixture into a separatory funnel and wash with water (350 mL).
7. Separate the phases and dry the organic phase over Na_2SO_4.
8. Filter and concentrate on a rotary evaporator to give a purple solid.
9. Purify the product by flash chromatography (silica gel) eluting with 2% ethyl acetate–dichloromethane. The first purple band includes bridged porphyrin. Collect the fractions including the porphyrin and concentrate on a rotary evaporator to give a solid. UV–vis (CH_2Cl_2): λ 410 (sh), 430 (Soret), 492 (sh).
10. Recrystallize the solid from dichloromethane and methanol to give the pure ligand (151 mg, 16%).
11. Place the porphyrin ligand (30 mg, 28 μmol), anhydrous manganese(II) bromide (120 mg, 0.56 mmol), 2,6-lutidine (3 drops), and dry DMF (30 mL) in a 50-mL three-necked round-bottomed flask equipped with a nitrogen inlet, a nitrogen bubbler, and a magnetic stirring bar.
12. Heat the mixture at 100 °C (24–96 h) in air until the Soret band of the free ligand is replaced by that of the manganese porphyrin at 488 nm.
13. Cool the mixture to room temperature, transfer to a separatory funnel, and wash with a brine solution (2 × 50 mL).
14. Separate the phases and dry the organic phase over NaCl.
15. Filter and concentrate the filtrate on a rotary evaporator to a solid.
16. Dissolve the solid in dichloromethane (5 mL) and filter through a pad of silica gel using dichloromethane–diethyl ether (1:1) as the eluent. Concentrate the filtrate on a rotary evaporator and crystallize the residue from dichloromethane–heptane to yield a manganese(III)–porphyrin complex 1 (30 mg, 95%).

[a] Prepared according to Ref. 19
[b] Prepared according to Ref. 9b.

Protocol 2.
Synthesis of (1*S*,2*R*)-1,2-epoxy-1-phenylpropane (Structure 3, Scheme 2.1.5)

Caution! Carry out all procedures in a well-ventilated hood, and wear safety glasses and disposable vinyl or latex gloves.

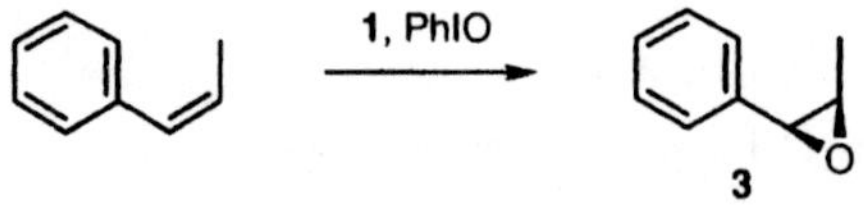

1, PhIO →

3
76%, 80% ee

Scheme 2.1.5

Equipment

- Magnetic hotplate stirrer
- Two-necked round-bottomed flask (20 mL)
- Teflon-coated magnetic stirring bar
- Nitrogen inlet
- Syringes (2 mL, 250 µL, 2 × 50 µL, and 10 µL)

Materials

- Complex **1** (FW 1174.28), 1.2 mg, 1.0 µmol — unknown toxicity, treat as toxic
- 1,5-Dicyclohexylimidazole (FW 232.37), 58 mg, 0.25 mmol — irritant
- (*Z*)-1-Phenyl-1-propene (FW 118.18), 118 mg, 130 µL, 1.0 mmol — cancer suspect, flammable
- Iodosylbenzene (FW 220.01), 22 mg, 0.10 mmol — oxidizer, self-reacting substance
- Nonane — flammable, irritant
- Dichloromethane — toxic, irritant, harmful by inhalation
- Triphenylphosphine (FW 262.29) — irritant
- Silica gel — harmful by inhalation
- Pentane — flammable, irritant
- Diethyl ether — toxic, flammable

Method

1. Equip a 20-mL two-necked round-bottomed flask with a nitrogen inlet, a nitrogen bubbler, and a magnetic stirring bar and purge it with nitrogen.

2. Place complex **1** (1.2 mg, 1.0 µmol), 1,5-dicyclohexylimidazole (58 mg, 0.25 mmol), (*Z*)-1-phenyl-1-propene (118 mg, 130 µL, 1.0 mmol), nonane (an internal standard, 20 µL), and dichloromethane (2 mL) into the flask.

3. Add iodosylbenzene (22 mg, 100 µmol) and stir the mixture rapidly for 1 h at 0 °C. Remove aliquots (5 µL) from the reaction mixture at appropriate intervals and quench them with a 2% triphenylphosphine solution in dichloromethane (50 µL). Monitor the formation of the product by GLC analysis of the aliquots.

4. Filter the mixture through a plug of glass wool and quickly condense the filtrate on a rotary evaporator at 0 °C. The produced epoxide is volatile and does not raise the bath temperature of the evaporator beyond 0 °C.

2: Oxidation of the C=C bond

5. Chromatograph the residue on silica gel with pentane to remove the remaining olefin and with 20% diethyl ether–pentane to elute the epoxide. Concentrate the fractions on a rotary evaporator at 0 °C to yield the epoxide **3** (10.2 mg, 80% ee, 76% yield)[a].

6. The enantiomeric excess of **3** can be determined by GLC analysis using SUPELCO β-DEX™ 120 fused silica capillary column (30 m × 0.25 mm i.d., 0.25 mm film, column temp. 95 °C).

[a] The yield was based upon consumption of iodosylbenzene.

3. Metallosalen-catalysed asymmetric epoxidation

Metallosalen complex [salen = N,N'-ethylenebis(salicylideneimine)] has a structure similar to metalloporphyrin. For example, the central metal ions in both complexes take a square planar geometry. As expected from these common structures, chiral metallosalen complexes are also oxidized to the corresponding oxo species in the presence of stoichiometric oxidants and undergo various oxygen atom transfer reactions such as epoxidation.[10] However, there are some structural differences between porphyrin and salen complexes: the porphyrin ligand encircles a central metal ion and its peripheral carbons are all sp^2. Therefore, the chiral auxiliaries in an optically active metalloporphyrin are placed outside the porphyrin ring (vide supra). This requires a sophisticated device in the construction of an efficient porphyrin catalyst. On the other hand, the salen ligand surrounds a central metal partially and it includes two sp^3 carbons in its ethylenediamine moiety (Fig. 2.1.1). By taking advantage of these structural features, chiral metallosalen complexes that bear chiral

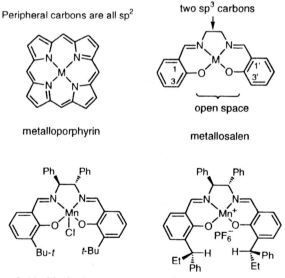

Fig. 2.1.1 Design of chiral (salen)manganese complexes

carbons at the ethylenediamine parts and bulky and/or chiral substituents at C3- and C3'-carbons were synthesized independently by Jacobsen and co-workers and the present authors and co-workers in 1990.[11]

Another advantage of the salen ligand is its pliability due to the presence of the two sp^3 carbons. Thus, oxo (salen)manganese(V) complexes take a non-planar stepped conformation and olefins preferentially approach the metal–oxo bond from the downward side of the ligand (Fig. 2.1.2).[12] The orientation of the incoming olefin is mainly dictated by the repulsion between the C3-substituent (R^1) in the salen ligand and the olefinic substituent (L) but, when the substrate is a conjugated olefin, π,π-antibonding repulsion between the salen ligand and the olefinic substituent (L) also contributes to the regulation of the orientation.[13] The latter factor explains why conjugated olefins, especially cis- and trisubstituted conjugated olefins, are good substrates for the epoxidation using the chiral (salen)manganese complex as the catalyst. Among various C3-substituents, the 2-phenylnaphthyl group serves as the best controller of the orientation probably because it disposes the 2-phenyl group close to the incoming olefin and intensifies the steric and π,π-antibonding repulsions (Fig. 2.1.2). In fact, the (salen)manganese(III) complex 4 bearing the 2-phenylnaphthyl group as the C3- and C3'-substituent showed high enantioselectivity in the epoxidation of conjugated cis-di- and trisubstituted olefins (Table 2.1.1). Iodosylbenzene or sodium hypochlorite is generally used as the terminal oxidant. The choice of the oxidant depends on the substrate used. Although various organic solvents such as dichloromethane, acetonitrile, ether, fluorobenzene, and ethyl acetate can be used for the present epoxidation, dichloromethane and acetonitrile are generally used in a combination with aqueous sodium hypochlorite and iodosylbenzene, respectively, in the epoxidation using 4 or 5 as the catalyst. The active oxo species carries an axial donor ligand (X) (Fig. 2.1.2) and its nature also affects the enantioselectivity of the reaction. In general, salen-catalysed epoxidation is performed in the presence of excess amount (5–10 equivalents to the catalyst) of a donor ligand such as 4-phenylpyridine N-oxide (PPNO).

Complexes 4 and 5 can be readily prepared from aldehyde 6 and 1,2-bis(3,5-dimethylphenyl)ethylenediamine 7 or 1,2-diphenylethylenediamine, respectively, according to the reported procedure (Scheme 2.1.6).[14] Complexes 4 and 5 can be used equally under the same reaction conditions, although 5 shows slightly reduced enantioselectivity (1–4% ee).

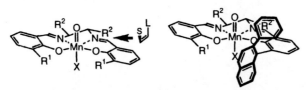

Fig. 2.1.2 Conformation of oxo (salen)manganese(V) complexes and orientation of incoming olefins

2: Oxidation of the C=C bond

Table 2.1.1 Epoxidation of conjugated cis-di- and trisubstituted olefins using Mn-salen complex **4** as a catalyst

Entry	Olefin	Oxidant	Solvent	Temp. (°C)	Yield	ee (%)
1	(O$_2$N, AcNH-substituted chromene)	PhIO	CH$_2$Cl$_2$	0	80	>99
2	(chromene)	PhIO	CH$_3$CN	20	60	>99
3	(indene)	NaOCl	CH$_2$Cl$_2$	0	55	98
4	(dihydronaphthalene)	NaOCl	CH$_2$Cl$_2$	0	78	98
5	(cis-β-methylstyrene, Ph)	NaOCl	CH$_2$Cl$_2$	0	80[a]	96 (trans) 92 (cis) (94)[b]
6	(cyclopentadiene)	NaOCl	CH$_2$Cl$_2$	−18	40	93
7	(1-methyl-dihydronaphthalene)	PhI)	CH$_3$CN	−20	41	96
8	(2-methyl-dihydronaphthalene)	PhIO	CH$_3$CN	−20	48	92
9	(methyl indene)	NaClO	CH$_2$Cl$_2$	0	91	88
10	(methyl chromene)	PhIO	CH$_3$CN	−20	81	>99

[a] A mixture of cis- and trans-epoxides in a ratio of 2:1.
[b] The number in parentheses stands for the face selectivity. Face selectivity = ee$_{trans}$ × %trans + ee$_{cis}$ × %cis (see Ref. 20).

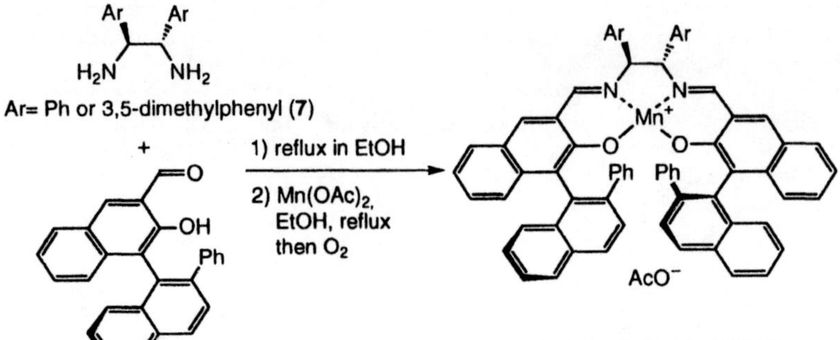

6

4 Ar= 3,5-dimethylphenyl
5 Ar= phenyl

Scheme 2.1.6

Protocol 3.
Synthesis of (3*S*,4*S*)-3,4-epoxy-2,2-dimethylchromane: oxidation with iodosylbenzene (Structure 8, Scheme 2.1.7)

Caution! Carry out all procedures in a well-ventilated hood, and wear safety glasses and disposable vinyl or latex gloves.

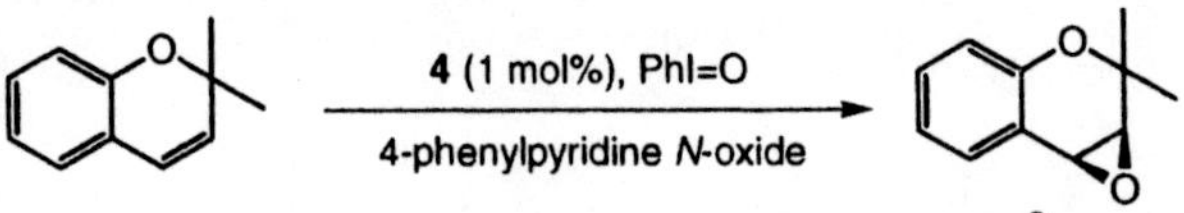

8

>99% ee, 60%

Scheme 2.1.7

This is representative of the epoxidation using iodosylbenzene as a terminal oxidant.

Equipment

- Magnetic hotplate stirrer
- Two-necked round-bottomed flask (10 mL)
- Adapter with three-way stopcock
- Teflon-coated magnetic stirring bar
- Nitrogen inlet
- Syringes (5 mL and 50 μL)
- Cooling machine (EYELA COOL LECS 30)

Materials

• (Salen)manganese(III) complex **4**[a] (FW 1037.09), 5.4 mg, 5.2 μmol	**unknown toxicity, treat as toxic**
• 2,2-Dimethylchromene (FW 160.21), 32.0 mg, 32.0 μL, 0.20 mmol	**unknown toxicity, treat as toxic**
• 4-Phenylpyridine *N*-oxide (FW 171.20), 8.6 mg, 50 μmol	**irritant**
• Iodosylbenzene (FW 220.01), 88.0 mg, 0.40 mmol	**oxidizer, self-reacting substance**
• Acetonitrile	**toxic, flammable**
• Diethyl ether	**toxic, flammable**
• Pentane	**flammable**
• Celite	
• Silica gel	**harmful by inhalation**

2: Oxidation of the C=C bond

Method
1. Assemble a 10-mL two-necked round-bottomed flask, a stopper, a magnetic stirring bar, and an adapter with a three-way stopcock, one way of which is connected to nitrogen inlet and the other to a vacuum line.
2. Evacuate the flask, dry with a heat gun, and back-fill with nitrogen. Remove the vacuum line and cap with septum under a positive pressure of nitrogen.
3. Introduce 2,2-dimethylchromene (32.0 mg, 32.0 μL, 0.20 mmol) and acetonitrile (2.5 mL) via syringes through the septum cap.
4. Add 4-phenylpyridine N-oxide (8.6 mg, 50 μmol) and complex 4 (5.4 mg, 5.2 μmol) rapidly against a positive nitrogen counterflow.
5. Cool the mixture to −20 °C.
6. Add iodosylbenzene (88.0 mg, 0.40 mmol) rapidly against a positive nitrogen counterflow.
7. Stir for 24 h at −20 °C and warm to room temperature.
8. Filter the reaction mixture through a pad of Celite, wash the pad with diethyl ether (1 mL) and concentrate the filtrate on a rotary evaporator at 0 °C.
9. Chromatograph the residue on silica gel with pentane–diethyl ether (1:0–19:1) to yield (3S,4S)-3,4-epoxy-2,2-dimethylchromane (21.2 mg, >99% ee, 60%). The enantiomeric excess of this sample is determined by HPLC (Daicel Chiralcel OJ, hexane–2-propanol 15:1).

[a] Prepared according to Ref. 14.

Protocol 4.
Synthesis of (1S,2R)-1,2-epoxy-1,2,3,4-tetrahydronaphthalene: oxidation with sodium hypochlorite (Structure 9, Scheme 2.1.8)

Caution! Carry out all procedures in a well-ventilated hood, and wear safety glasses and disposable vinyl or latex gloves.

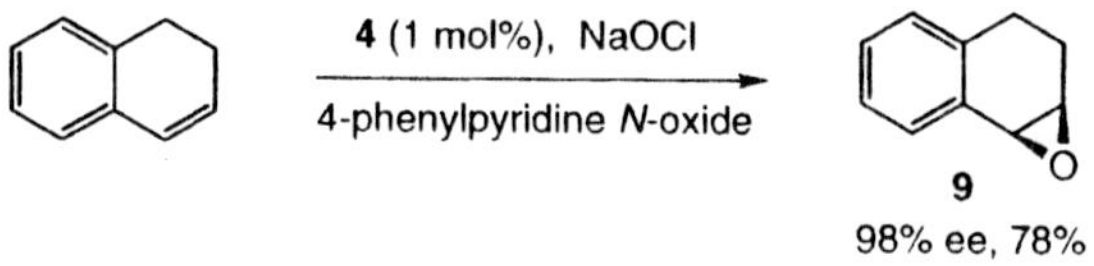

Scheme 2.1.8

4 (1 mol%), NaOCl
4-phenylpyridine N-oxide

9
98% ee, 78%

This is representative of the epoxidation using aqueous sodium hypochlorite as a terminal oxidant.

Equipment
- Magnetic hotplate stirrer
- Two-necked round-bottomed flask (10 mL)
- Separatory funnel (10 mL)
- Nitrogen inlet

Protocol 4. Continued

- Adapter with three-way stopcock
- Teflon-coated magnetic stirring bar
- Syringes (2 × 5 mL, 50 μL)
- Cooling machine (EYELA COOL ECS 30)

Materials

- (Salen)manganese(III) complex **4**[a] (FW 1037.09), 6.6 mg, 6.4 μmol — unknown toxicity, treat as toxic
- 1,2-Dihydronaphthalene (FW 130.19), 31.3 mg, 0.24 mmol — unknown toxicity, treat as toxic
- 4-Phenylpyridine *N*-oxide (FW 171.20), 10.3 mg, 60 μmol — irritant
- NaOCl in phosphate buffer (0.588 M, pH 11.37), 2.1 mL — oxidizer, corrosive
- Dichloromethane — toxic, irritant, harmful by inhalation
- Na_2SO_4 — irritant, hygroscopic
- Diethyl ether — toxic, flammable
- Pentane — irritant, flammable
- Celite
- Silica gel — harmful by inhalation

Method

1. Assemble a 10-mL two-necked round-bottomed flask, a stopper, a magnetic stirring bar, and an adapter with three-way stopcock, one way of which is connected to nitrogen inlet and the other to a vacuum line.
2. Evacuate the flask, dry with a heat gun, and back-fill with nitrogen. Remove the vacuum line and cap with septum under a positive pressure of nitrogen.
3. Introduce 1,2-dihydronaphthalene (31.3 μL, 0.24 mmol) and dichloromethane (1.5 mL) via syringes through the septum cap.
4. Add 4-phenylpyridine *N*-oxide (10.3 mg, 60 μmol) and complex **4** (6.6 mg, 6.4 μmol) rapidly against a positive nitrogen counterflow. Cool the mixture to 0°C and stir.
5. Add NaOCl in phosphate buffer (0.59 M, pH 11.37) (2.1 mL) via a syringe and continue vigorous stirring for 4 h at 0°C.
6. Warm the mixture to room temperature and transfer to a 10-mL separatory funnel.
7. Separate the phases and extract the aqueous phase with dichloromethane (2 mL).
8. Dry the combined organic phases over Na_2SO_4 and concentrate on a rotary evaporator at 0°C.
9. Chromatograph the residue on silica gel with pentane–diethyl ether (1:0–49:1) to yield (1*S*,2*R*)-1,2-epoxy-1,2,3,4-tetrahydronaphthalene (27.3 mg, 98% ee, 78%). The enantiomeric excess of the product is determined by GLC (SUPELCO β-DEX 120 fused silica capillary column, 30 m × 0.25 mm i.d., 0.25 μm film, column temp. 120°C) or by HPLC (Daicel Chiralcel OB-H, hexane–2-propanol 9:1). Use of HPLC is more reliable because the rearrangement of a small amount of the epoxide to ketone is observed in the GLC conditions.

[a] Prepared according to Ref. 14.

2: Oxidation of the C=C bond

Complexes **10a,b** can also be applied to the epoxidation of conjugated cis- and tri-substituted olefins under the conditions described in Protocols 3 and 4, although the enantioselectivity suffers to some extent (80–98% ee).[15] Complex **10a** is commercially available from Aldrich and Fluka.

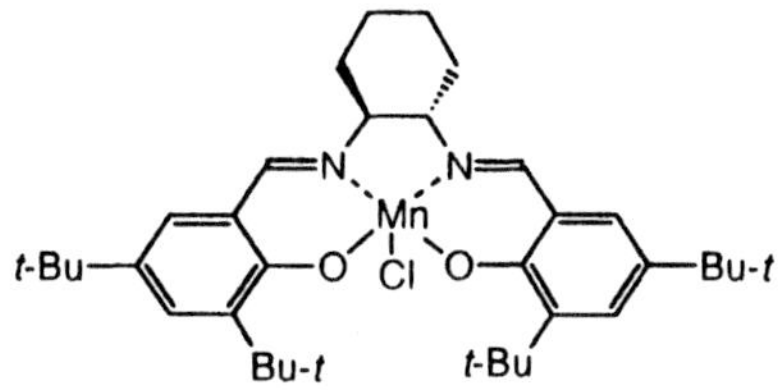

10a

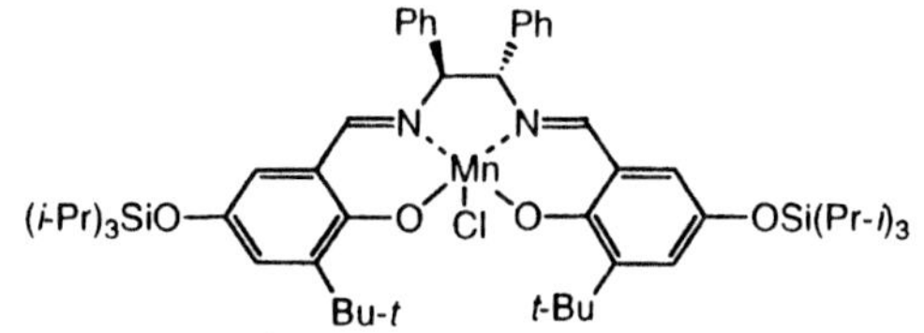

10b

Mn-salen-catalysed epoxidation of conjugated olefins has been considered to proceed through a radical intermediate in which the carbon–carbon bond between the radical and its neighbouring oxygenated carbons can rotate leading to the formation of a diastereomeric epoxide. Thus, epoxidation of cis-β-methylstyrene (R = Me) gives a mixture of cis- and trans-epoxides (Scheme 2.1.9).[11,13] However, when the substrate is styrene (R = H), the rotation leads to the formation of the enantiomeric epoxide and diminishes the enantioselectivity of the reaction. Jacobsen and co-workers[16] have reported that this undesired rotation is suppressed by performing epoxidation at low temperature, remarkably improving enantioselectivity. In fact, epoxidation of styrene using m-chloroperoxybenzoic acid in the presence of N-methylmorpholine N-oxide at −78°C shows enantioselectivity as high as 86% ee (Scheme 2.1.9).[16]

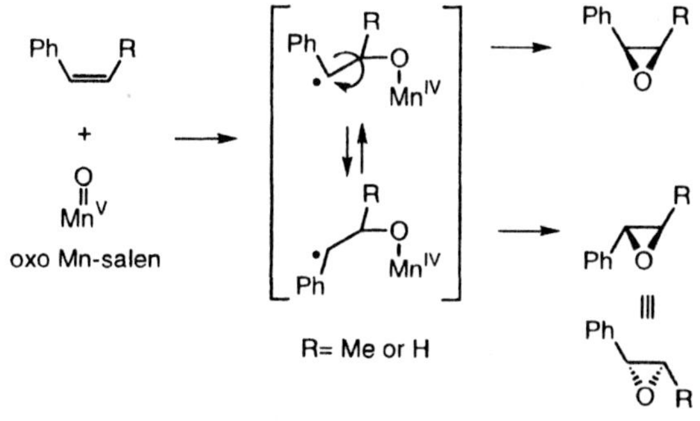

Scheme 2.1.9

Protocol 5.
Synthesis of (S)-styrene oxide (Structure 11, Scheme 2.1.10)

Caution! Carry out all procedures in a well-ventilated hood, and wear safety glasses and disposable vinyl or latex gloves.

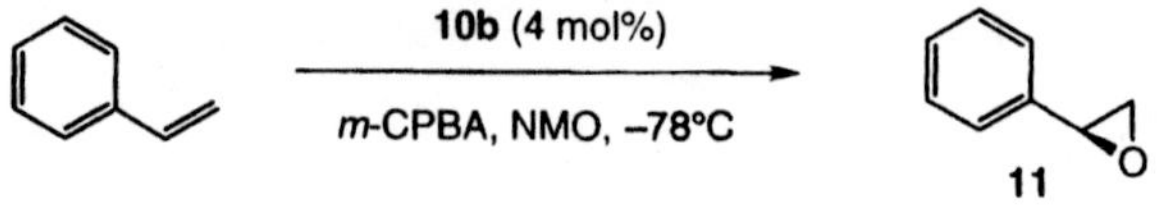

Scheme 2.1.10

Styrene → (via 10b (4 mol%), m-CPBA, NMO, −78°C) → **11** (86% ee, 88%)

This is representative of the epoxidation using m-chloroperoxybenzoic acid as a terminal oxidant at low temperature.

Equipment

- Magnetic hotplate stirrer
- Two-necked round-bottomed flask (25 mL)
- Teflon-coated magnetic stirring bar
- Adapter with three-way stopcock
- Separatory funnel (50 mL)
- Nitrogen inlet
- Syringes (10 mL and 250 μL)

Materials

(Salen)manganese(III) complex **10b** (FW 964.46), 37 mg, 38 μmol	unknown toxicity, treat as toxic
Styrene (FW 104.15), 110 μL, 100 mg, 0.96 mmol	cancer-suspect agent, flammable
N-Methylmorpholine N-oxide (NMO) (FW 117.15), 562 mg, 4.8 mmol	irritant
m-Chloroperoxybenzoic acid (m-CPBA) (FW 172.57), 343 mg, 1.9 mmol	oxidizer, irritant
1 M NaOH (FW 40.00)	corrosive, toxic
Dichloromethane	toxic, irritant, harmful by inhalation
Na_2SO_4	irritant, hygroscopic
Diethyl ether	toxic, flammable
Pentane	flammable
Silica gel	harmful by inhalation

Method

1. Assemble a 25-mL two-necked round-bottomed flask, a stopper, a magnetic stirring bar, and an adapter with three-way stopcock, one way of which is connected to nitrogen inlet and the other to a vacuum line.
2. Add NMO (562 mg, 4.8 mmol) and (salen)manganese(III) complex **10b** (37 mg, 38 μmol) rapidly against a positive nitrogen counterflow.
3. Add dichloromethane (8 mL) and styrene (110 μL, 0.96 mmol) subsequently via syringes, cool the mixture to −78°C (dry ice–acetone bath) and stir.
4. Add m-CPBA (343 mg) in four roughly equal portions over a 2-min period against a positive nitrogen counterflow. Continue stirring for 45 min at −78°C.

2: Oxidation of the C=C bond

5. Add 1 M NaOH (10 mL), remove the cooling bath and transfer the mixture to a 50-mL separatory funnel.
6. Separate the phases and extract the aqueous phase with dichloromethane (2 × 10 mL).
7. Dry the combined organic phases over Na_2SO_4, filter and concentrate to approximately to 2 mL by distillation.
8. Filter the residual oil through a short pad of silica gel to remove the catalyst, wash the pad with pentane and concentrate the filtrate completely on a rotary evaporator at 0 °C to yield (S)-styrene oxide (102 mg, 86% ee, 88%). The enantiomeric excess of the product is determined by HPLC using Daicel Chiralpak AD (Hexane–i-PrOH 1000:1, elution rate 0.5 mL/min).

4. Aerobic asymmetric epoxidation

Molecular oxygen is an economical and environmentally acceptable oxidant. Mukaiyama and co-workers have reported that molecular oxygen can be used as a terminal oxidant for transition metal-catalysed epoxidation in the presence of an aldehyde (Scheme 2.1.11).[17] In this process, the aldehyde serves as a reductant and is oxidized to generate an acylperoxo-metal species (**12**) which epoxidizes olefins.

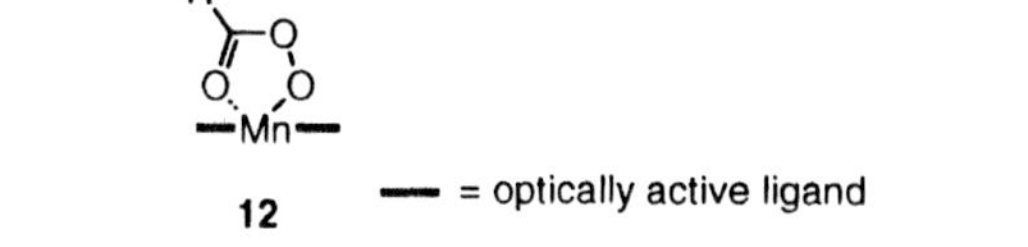

Scheme 2.1.11

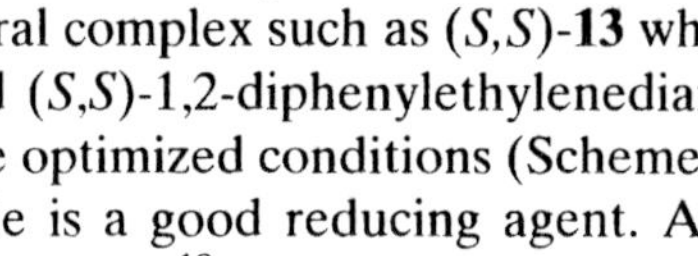

12

━ = optically active ligand

This aerobic epoxidation can be performed in an enantioselective manner with an appropriate chiral complex such as (S,S)-**13** which is readily prepared from compound **14** and (S,S)-1,2-diphenylethylenediamine (Scheme 2.1.12), as the catalyst under the optimized conditions (Scheme 2.1.13).

An aliphatic aldehyde is a good reducing agent. Aromatic hydrocarbons are preferable as the solvents.[18] Mn-salen complexes are also employable as catalysts although the asymmetric induction is lowered. It is noteworthy that the sense of enantioface selection in this aerobic epoxidation is opposite to that observed in the epoxidation using iodosylbenzene or NaClO as the oxidant.

Protocol 6.
Synthesis of complex (*S,S*)-13 (Structure 13, Scheme 2.1.12)

Caution! Carry out all procedures in a well-ventilated hood, and wear safety glasses and disposable vinyl or latex gloves.

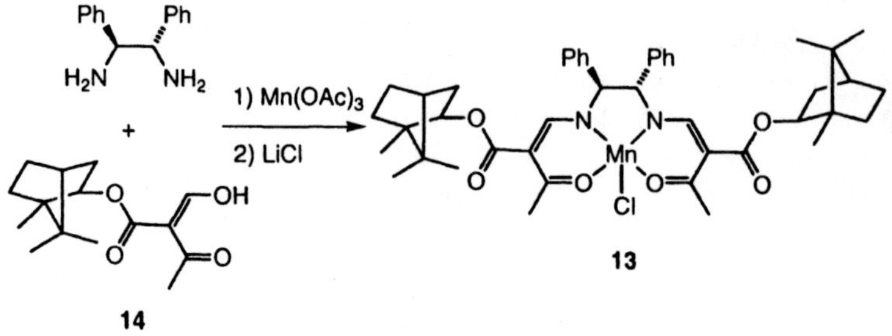

Scheme 2.1.12

Equipment

- Magnetic hotplate stirrer
- Three-necked round-bottomed flask (100 mL)
- Teflon-coated magnetic stirring bar
- Nitrogen inlet
- Syringe (20 mL)
- Oil bath (500 mL)

Materials

- (±)-Isobornyl 2-formyl-3-oxobutyrate **14** (FW 266.33), 1.76 g, 6.6 mmol — unknown toxicity, treat as toxic
- (*S,S*)-1,2-Diphenylethylenediamine (FW 212.29), 0.64 g, 3.0 mmol — irritant
- 1,2-Dichloroethane — cancer suspect, flammable
- Ethanol — irritant, flammable
- Mn(III)(OAc)$_3$•2H$_2$O (FW 268.10), 1.63 g, 6.0 mmol — irritant, hygroscopic
- Lithium chloride (FW 42.39) 0.32 g, 7.6 mmol — irritant, hygroscopic
- Dichloromethane — toxic, irritant, harmful by inhalation
- Acetone — irritant, flammable
- Diethyl ether — toxic, flammable
- Na$_2$SO$_4$ — irritant, hygroscopic
- Silica gel — harmful by inhalation

Method

1. Equip a three-necked round-bottomed flask with a reflux condenser, a nitrogen inlet, a nitrogen bubbler, and a magnetic stirring bar.
2. Add Mn(III)(OAc)$_3$•2H$_2$O (1.63 g, 6.0 mmol), ethanol (5mL), and 1,2-dichloroethane (15 mL).
3. Add a solution of (±)-isobornyl 2-formyl-3-oxobutyrate (1.76 g, 6.6 mmol) and (*S,S*)-1,2-diphenylethylenediamine (0.64 g, 3.0 mmol) in 1,2-dichloroethane (10 mL).
4. Heat the mixture at reflux for 3 h with stirring.

2: Oxidation of the C=C bond

5. Add lithium chloride (0.32 g, 7.6 mmol) at once and stir for an additional 1 h.
6. Cool the flask to room temperature, evacuate the solvent.
7. Extract the residue with CH_2Cl_2.
8. Dry the combined extract over Na_2SO_4 and concentrate the solution. Purify the residue by silica gel column chromatography eluting with CH_2Cl_2–acetone and reprecipitation from CH_2Cl_2–ether to give the complex (S,S)-**13** (1.10 g, 45%) as a dark brown powder.

Protocol 7.
Synthesis of (1R,2S)-1,2-epoxy-1,2,3,4-tetrahydronaphthalene by aerobic epoxidation (Structure 9, Scheme 2.1.13)

Caution! Carry out all the procedures in a well-ventilated hood, and wear safety glasses and disposable vinyl or latex gloves.

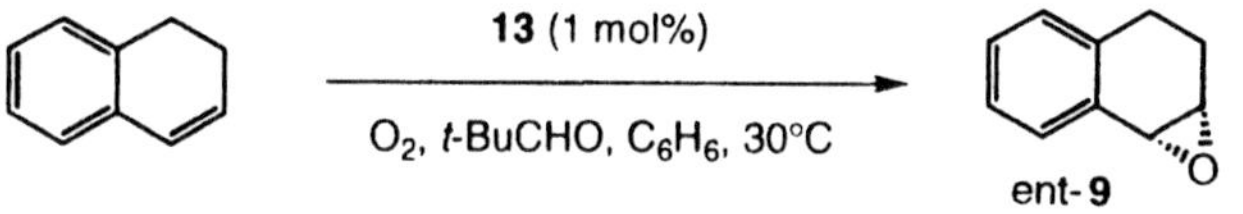

13 (1 mol%)

O_2, *t*-BuCHO, C_6H_6, 30°C

ent-**9**

64% ee, 70%

Scheme 2.1.13

Equipment
- Magnetic hotplate stirrer
- Three-necked round-bottomed flask (50 mL)
- Teflon-coated magnetic stirring bar
- Water bath (100 mL)
- Nitrogen inlet
- Oxygen inlet
- Separatory funnel (100 mL)
- Syringes (2 × 2 mL)

Materials

(S,S)-**13** (FW 811.33), 83 mg, 0.104 mmol	unknown toxicity, treat as toxic
Dihydronaphthalene (FW 130.19), 104 mg, 104 μL, 0.80 mmol	irritant, flammable
2,2-Dimethylpropanal (FW 86.13), 303 μL, 241 mg, 2.8 mmol	flammable
Benzene	cancer suspect, flammable
Oxygen	supports combustion
$NaHCO_3$	irritant, hygroscopic
Diethyl ether	toxic, flammable
Na_2SO_4	irritant, hygroscopic
Silica gel	harmful by inhalation
Hexane	irritant, flammable
Ethyl acetate	irritant, flammable

Method
1. Equip a three-necked round-bottomed flask with a nitrogen inlet, a nitrogen bubbler, and a magnetic stirring bar.
2. Add (S,S)-**13** (83 mg, 0.10 mmol) in benzene (1.0 mL) via syringe.

Protocol 7. Continued

3. Add a solution of dihydronaphthalene (104 mg, 104 μL, 0.80 mmol) and 2,2-dimethylpropanal (241 mg, 303 μL, 2.8 mmol) in benzene (1.0 mL) via syringe.
4. Replace nitrogen inlet with oxygen inlet.
5. Stir the mixture at 30°C for 1 h under an atmospheric pressure of oxygen.
6. Dilute the mixture with aqueous $NaHCO_3$ and extract with diethyl ether.
7. Dry the combined extract over anhydrous Na_2SO_4 and concentrate it in vacuo.
8. Purify the residue with silica gel column chromatography eluting with hexane–ethyl acetate to give the epoxide as a colourless oil (81 mg, 70%). The enantiomeric excess of this sample is 64%. $[\alpha]_D^{30} = +81.8°$ ($c = 0.45$, $CHCl_3$).

References

1. Trost, B. M., ed.; *Comprehensive Organic Synthesis*; Pergamon Press: Oxford, **1991**; vol. 7.
2. Sharpless, K. B.; Verhoeven, T. R. *Aldrichimica Acta* **1979**, *12*, 63–74.
3. Groves, J. T.; Myers, R. S. *J. Am. Chem. Soc.* **1983**, *105*, 5791–5796.
4. Groves, J. T.; Nemo, T. E.; Myers, R. S. *J. Am. Chem. Soc.* **1979**, *101*, 1032–1033.
5. Collman, J. P.; Zhanng, X.; Lee, V J.; Uffelman, E. S.; Brauman, J. I. *Science* **1993**, *261*, 1404–1411.
6. Katsuki, T. *Coord. Chem. Rev.* **1995**, *140*, 189–214.
7. (a) Mansuy, D.; Battioni, P.; Renaud, J.-P.; Guerin, P. *J. Chem. Soc., Chem. Commun.* **1985**, 155–156. (b) O'Malley, S.; Kodadek, T. *J. Am. Chem. Soc.* **1989**, *111*, 9116–9117. (c) Naruta, Y.; Tani, F.; Ishihara, N.; Maruyama, K.; *J. Am. Chem. Soc.* **1991**, *113*, 6865–6872. (d) Halterman, R. L.; Jan, S.-T. *J. Org. Chem.* **1991**, *56*, 5253–5254. (e) Groves, J. T.; Viski, P. *J. Org. Chem.* **1990**, *55*, 3628–3634. (f) Collman, J. P.; Lee, V. J.; Zhang, X.; Ibers, J. A.; Brauman, J. I. *J. Am. Chem. Soc.* **1993**, *115*, 3834–3835. (g) Collman, J. P.; Lee, V. J.; Kellen-Yuen, C. J.; Zhang, X.; Ibers, J, A.; Brauman, J. I. *J. Am. Chem. Soc.* **1995**, *117*, 692–703. (h) Collman, J. P.; Wang, Z.; Straumanis, A.; Quelquejeu, M. *J. Am. Chem. Soc.* **1999**, *121*, 460–461.
8. Konishi, K.; Oda, K.; Nishida, K.; Aida, T.; Inoue, S. *J. Am. Chem. Soc.* **1992**, *114*, 1313–1317.
9. (a) Collman, J. P.; Lee, V. J.; Zhang, X.; Ibers, J. A.; Brauman, J. I. *J. Am. Chem. Soc.* **1993**, *115*, 3834–3835. (b) Collman, J. P.; Lee, V. J.; Kellen-Yuen, C. J.; Zhang, X.; Ibers, J. A.; Brauman, J. I.; *J. Am. Chem. Soc.* **1995**, *117*, 692–703. (c) Collman, J. P.; Wang, Z.; Straumanis, A.; Quelquejeu, M. *J. Am. Chem. Soc.* **1999**, *121*, 460–461.
10. Srinivasan, K.; Michaud, P.; Kochi, J. K. *J. Am. Chem. Soc.* **1986**, *108*, 2309–2320.
11. (a) Zhang, W.; Loebach, J. L.; Wilson, S. R.; Jacobsen, E. N. *J. Am. Chem. Soc.* **1990**, *112*, 2801–2803. (b) Irie, R.; Noda, K.; Ito, Y.; Matsumoto, N.; Katsuki, T. *Tetrahedron Lett.* **1990**, *31*, 7345–7348.
12. (a) Hamada, T.; Fukuda, T.; Imanishi, H.; Katsuki, T. *Tetrahedron* **1996**, *52*, 515–530. (b) Hashihayata, T.; Ito, Y.; Katsuki, T. *Tetrahedron* **1997**, *53*, 9541–9552.

13. Hamada, T.; Irie, R.; Katsuki, T. *Synlett* **1994**, 479–481.
14. Sasaki, H.; Irie, R.; Hamada, T.; Suzuki, K.; Katsuki, T. *Tetrahedron* **1994**, *50*, 11827–11838.
15. (a) Jacobsen, E. N. Asymmetric catalytic epoxidation of unfuctionalized olefins. In *Catalytic Asymmetric Synthesis*; Ojima, I., ed.; VCH: New York, **1993**; pp. 159–202. (b) Katsuki, T. *J. Mol. Cat. A, Chem.* **1996**, *113*, 87–107.
16. Palucki, M.; Pospisil, P. J.; Zhang, W.; Jacobsen, E. N. *J. Am. Chem. Soc.* **1994**, *116*, 9333–9334.
17. Yamada, T.; Takai, T.; Rhoode, O.; Mukaiyama, T. *Bull. Chem. Soc. Jpn.* **1991**, *64*, 2109–2117.
18. Nagata, T.; Imagawa, K.; Yamada, T.; Mukaiyama, T. *Bull. Chem. Soc. Jpn.* **1995**, *68*, 1455–1465.
19. Momenteau, M.; Mispelter, J.; Loock, B.; Bisagni, E. *J. Chem. Soc., Perkin Trans. 1*, **1983**, 189.
20. Zhang, W.; Lee, N. H.; Jacobsen, E. N. *J. Am. Chem. Soc.* **1994**, *116*, 425–426.

2.2 Asymmetric epoxidation using peroxides and related reagents

B.-C. CHEN, P. ZHOU, and F. A. DAVIS

1. Introduction

The asymmetric epoxidation of alkenes with peroxides is a facile method for bis-functionalization of a carbon–carbon double bond in a single operation. For the asymmetric epoxidation of alkenes, the peroxide-based reagents including peracids, *N*-sulfonyloxaziridines and dioxiranes are the most commonly utilized.

2. Diastereoselective epoxidations

For simple olefins without precoordinating functional groups in the molecule, the diastereoselectivity of epoxidation is controlled by steric and stereoelectronic effects between reagent and substrate in the transition state. Stereochemically, direct epoxidation usually occurs from the less-hindered face of the carbon–carbon double bond. For example, epoxidation of (*S*)-3-methyl-1-pentene (**1**) with *m*-chloroperoxybenzoic acid (*m*-CPBA) affords (2*R*,3*S*)-1,2-epoxy-3-methylpentane (**2**) as the major isomer.[1] As might be expected, increasing the size difference between the adjacent allylic substituents will result in improved diastereoselectivity. Epoxidation of ecdysone precursor **3** with *m*-CPBA apparently afforded a single epoxide **4** in 79% yield.[2]

High diastereoselectivity has been reported for certain vinylcyclohexanes where a strong bias exists for a single conformation. For example, epoxidation of **5** with *m*-CPBA gives **6** in 80% yield and 80% de.[3] The large bromine atom in **5** hinders epoxidation from the back side of the carbon–carbon double bond.

Allylsilanes strongly favour an antiperiplanar attack of the peracid with respect to the C—Si bond due to σ–π conjugation.[4] Thus reaction of **7** with *m*-CPBA in dichloromethane at −20°C gives epoxide **8** in 85% yield and >90% de.[4]

Group size difference also plays an important role in the facial differentiation in the epoxidation of exocyclic alkenes; e.g. the epoxidation of **9** affords **10** in 75% yield as a single diastereomer (Scheme 2.2.1).[5]

Protocol 1.
Synthesis of [1R-(1α,2α,3β,5α)]-3,6,6-trimethylspirobicyclo[3.1.1] heptane-2,2'-oxirane (Structure 10, Scheme 2.2.1)

Caution! Peroxy acids are explosive in nature. Handle these chemicals with extreme caution. Carry out all procedures in a well-ventilated hood, and wear disposable vinyl or latex gloves and chemical-resistant safety goggles.

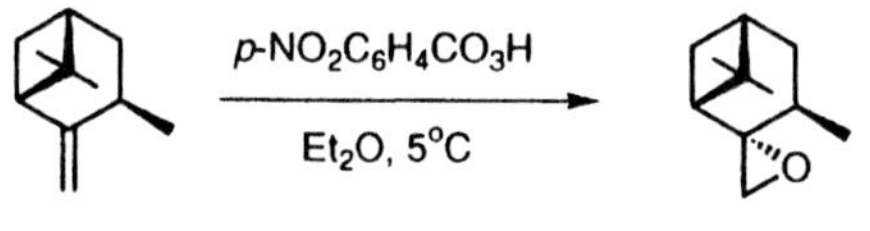

Scheme 2.2.1

This reaction is representative of the diastereoselective epoxidation of chiral alkenes with peracids such as *p*-nitroperoxybenzoic acid.

Equipment

- Three-necked, round-bottomed flask (250 mL)
- Cooling bath
- Magnetic stirrer
- Source of dry argon
- Teflon-coated magnetic stirrer bar
- Sintered glass filter funnel
- Pressure-equalizing addition funnel (50 mL)
- Rubber septum

Materials

Dry diethyl ether, 110 mL	flammable, irritant
p-Nitroperoxybenzoic acid, 7.3 g, 0.04 mol	explosive
3,6,6-Trimethyl-2-methylenebicyclo[3.1.1] heptane, 4.5 g, 0.03 mmol	unknown toxicity, treat as toxic
Water, as needed	
Ice, as needed	

Method

1. Equip a 250-mL three-necked round-bottomed flask with an argon bubbler, a 100-mL dropping funnel, a rubber septum, and a magnetic stirring bar. Maintain the apparatus under a positive pressure of argon.

2. Add *p*-nitroperoxybenzoic acid (7.3 g, 0.04 mol) and anhydrous diethyl ether (80 mL).

3. Stir the mixture and cool to 0–5 °C with an ice–water bath.

4. Add 3,6,6-trimethyl-2-methylenebicyclo[3.1.1]heptane (4.5 g, 0.03 mol) in the pressure-equalizing addition funnel. Add anhydrous diethyl ether (30 mL). Mix well to give a solution.

5. Add dropwise the solution obtained in step 4 to the reaction mixture in step

Protocol 1. Continued

3 while keeping the reaction temperature below 5°C. Stir the reaction mixture until the reaction is complete.

6. Filter to remove p-nitrobenzoic acid.

7. Remove solvent on a rotary evaporator and distil the residue under vacuum to give 3.74g of [1R-(1α,2α,3β,5α)]-3,6,6-trimethylspirobicyclo[3.1.1]heptane-2,2'-oxirane (75%), b.p. 40°C/0.1 mmHg, $[\alpha]_D = -34.5°$ ($c = 5$, cyclohexane).

The stereochemical outcome of the asymmetric epoxidation of endocyclic alkenes is also largely governed by the steric effect between the epoxidizing reagent and the alkene. For example, epoxidation of 4-methylcyclopentene (**11**) with peroxylauric acid gives *trans*- and *cis*-**12** in a ratio of 76:24 (Scheme 2.2.2).[6] As might be expected, increasing the bulkiness of the 4-substituent on the cyclopentene ring as well as in the epoxidizing reagent improves the diastereoselectivity. An example is the epoxidation of 4-*t*-butylcyclopentene (**13**) with peroxyphthalic acid where the ratio of *trans*- to *cis*-epoxide **14** was increased to 92:8.[7]

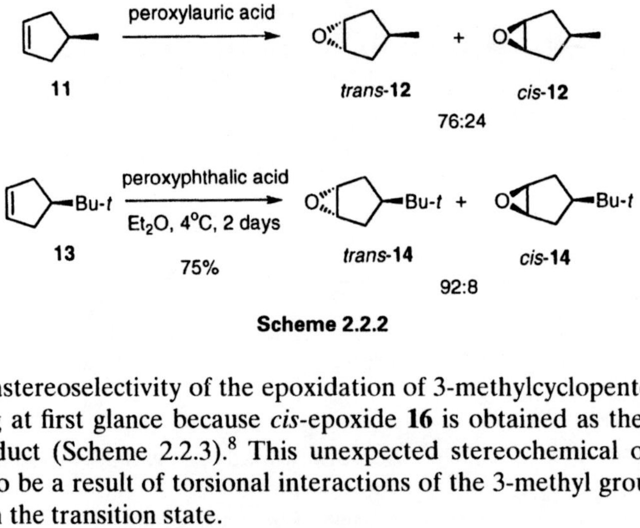

Scheme 2.2.2

The diastereoselectivity of the epoxidation of 3-methylcyclopentene (**15**) is surprising at first glance because *cis*-epoxide **16** is obtained as the predominant product (Scheme 2.2.3).[8] This unexpected stereochemical outcome is thought to be a result of torsional interactions of the 3-methyl group and the peracid in the transition state.

Protocol 2.
Synthesis of *cis*-3-methylcyclopenetene-1,2-oxide (Structure 16, Scheme 2.2.3)

Caution! Peroxy acids are explosive in nature. Handle these chemicals with extreme caution. Carry out all procedures in a well-ventilated hood, and wear disposable vinyl or latex gloves and chemical-resistant safety goggles.

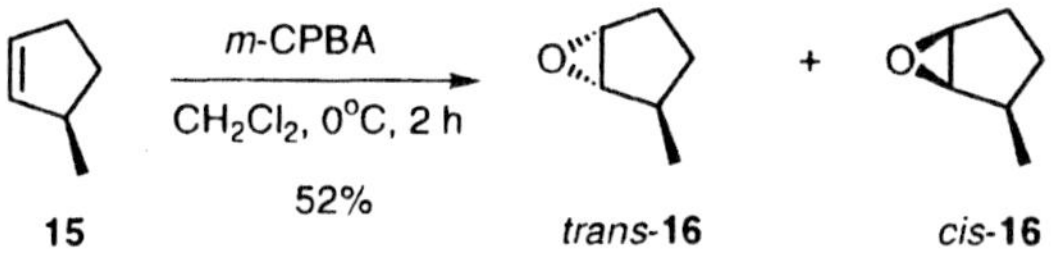

Scheme 2.2.3

This reaction is representative of the diastereoselective epoxidation of alkenes with *m*-chloroperoxybenzoic acid.

Equipment

- Three-necked, round-bottomed flask (250 mL)
- Syringe (15 mL)
- Magnetic stirrer
- Cooling bath
- Teflon-coated magnetic stirrer bar
- Source of dry argon
- Pressure-equalizing addition funnel (100 mL)
- Sintered glass filter funnel
- Rubber septum

Materials

- Dichloromethane, 150 mL — toxic, irritant
- *m*-Chloroperoxybenzoic acid, 85%, 25.0 g, 123 mmol — explosive
- 3-Methylcyclopent-1-ene, 10 g, 122 mmol — unknown toxicity, treat as toxic
- Water, as needed
- Ice, as needed

Method

1. Equip a 250-mL three-necked round-bottomed flask with an argon bubbler, a rubber septum, and a magnetic stirring bar. Maintain the apparatus under a positive pressure of argon.
2. Add *m*-chloroperoxybenzoic acid (85%, 25.0 g, 123 mmol) and dichloromethane (150 mL).
3. Stir the mixture and cool with an ice–water bath.
4. Add 3-methylcyclopent-1-ene (10 g, 124 mmol) dropwise via syringe while keeping the reaction temperature below 5 °C. The addition is complete in 30 min.
5. Stir the reaction mixture at 4 °C for 1.5 h until the reaction is complete. As the reaction proceeds, *m*-chlorobenzoic acid precipitates out.

Protocol 2. Continued

6. Filter to remove the benzoic acid.
7. Remove the solvent on a rotary evaporator to give a residue and distil the product through a 60 × 0.7 cm tantalum spiral column to give 6.21 g (52%) of 3-methylcyclopenetene-1,2-oxide; b.p. 115–116°C (atmospheric pressure). The *trans/cis* ratio is 33:67 as determined by GLC (6' × 0.25" 10% Carbowax 20M on Chromsorb W, 60cc/min, 70°C, retention times *cis*-epoxide, 3.7 min, *trans*-epoxide, 4.2 min). The following spectra are diagnostic: IR (film) 1470, 1390, 1370, 1310, 1290, 1230, 1200, 1010, 980, 960, 920, 905, 845, 797, 730, and 655 cm^{-1}; ^{1}H NMR δ 1.05 (d, J = 7 Hz, 3H), 1.2–2.2 (m, 5H), 3.08 (d, J = 3 Hz, 1H), 3.25 (d, J = 3 Hz, 1H).

The steric effects of both 3- and 4-substituents in cyclopentenes may act synergistically. This is true in the epoxidation of *trans*-3,4-dimethylcyclopentene (**17**) where the *endo*- and *exo*-epoxides **18** are obtained in a ratio of 85:15.[9] In the more rigid *trans*-fused cyclopentenes **19**, the synergetic effect is even more pronounced where the *endo*-epoxide **20** is obtained as single diastereomer.[9]

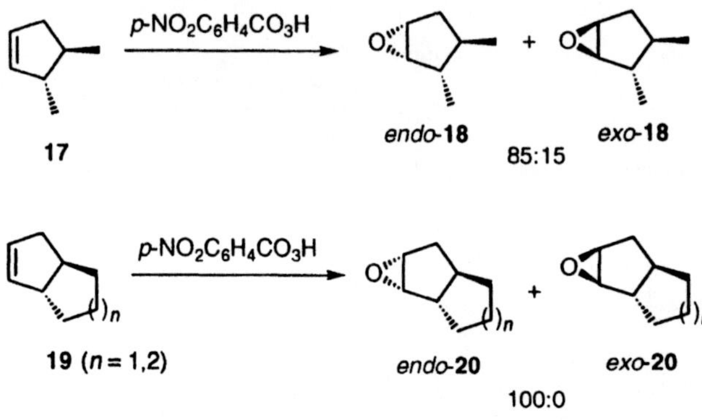

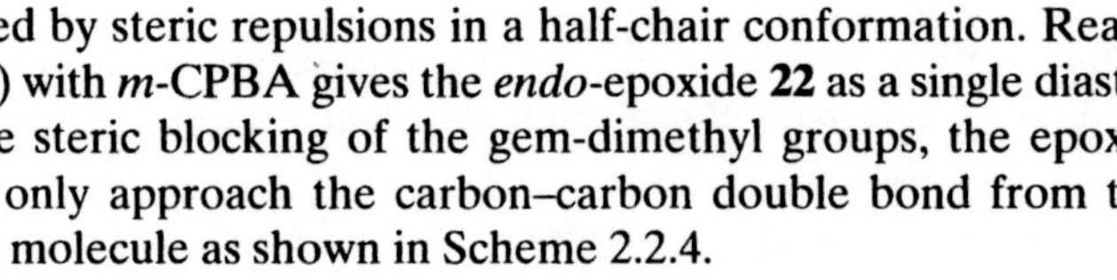

The diastereoselectivity observed in the epoxidation of cyclohexene systems is controlled by steric repulsions in a half-chair conformation. Reaction of α-pinene (**21**) with *m*-CPBA gives the *endo*-epoxide **22** as a single diastereomer.[10] Due to the steric blocking of the gem-dimethyl groups, the epoxidizing reagent can only approach the carbon–carbon double bond from the bottom face of the molecule as shown in Scheme 2.2.4.

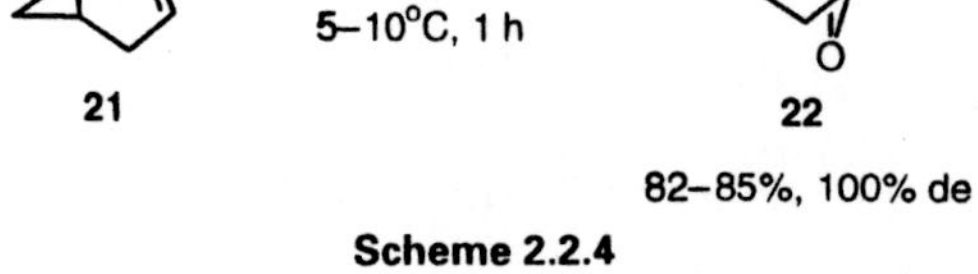

Scheme 2.2.4

2: Oxidation of the C=C bond

Epoxidation of **23** with *m*-CPBA gives *endo*-epoxide **25** in 92% yield and 98% de (Scheme 2.2.5).[11] The high stereoselectivity cannot be easily explained by steric control but is consistent with electrophilic attack to minimize torsional strain. Similar results are obtained with the aza-analogue **24**.[11,12]

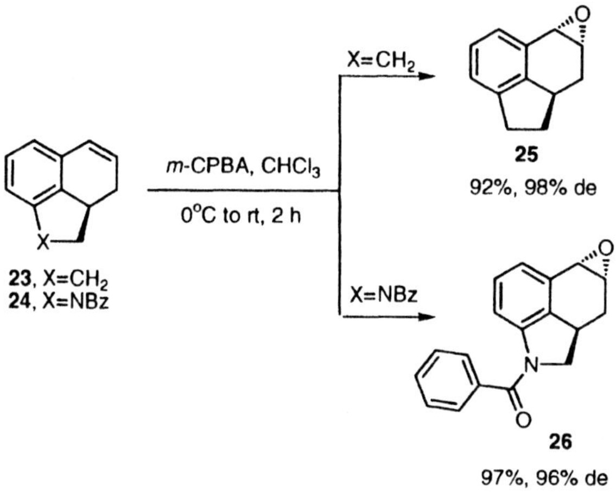

Scheme 2.2.5

Epoxidation of compound **27** with *m*-CPBA gives epoxide **28** in 98% yield as a single diastereomer.[13] The three methyl substituents in the 10-membered ring system favour the ring conformation depicted by all taking equatorial positions. The epoxidation thus occurs from the front side of the carbon–carbon double bond to give product **28** as the back side is shielded by the lactone functionality.

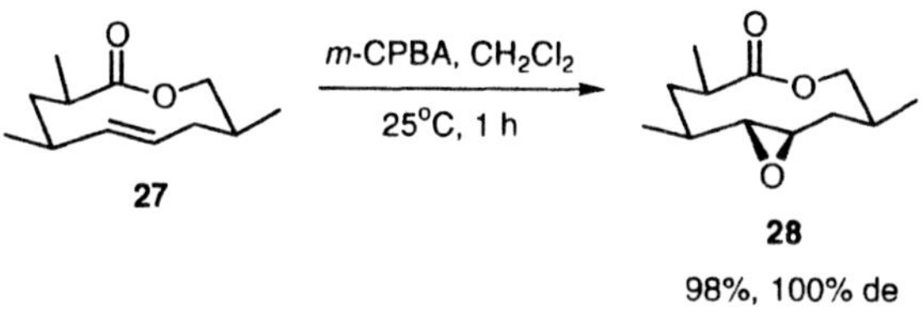

Dioxiranes are another class of important epoxidizing reagents.[14] These unstable reagents are usually prepared from ketones and an oxidizing reagent such as Oxone® and used *in situ*.[15–17] Akin to peroxy acids, dioxiranes perform *syn* stereospecific *O*-transfer to the alkenes, i.e. Z-alkenes give Z-epoxides and E-alkenes give E-epoxides via a concerted transition state similar to that proposed for peroxy acid epoxidation.[14] In the epoxidation of 3β-acetyl vitamin D₃ (**29**) with dimethyldioxirane (**30**), epoxide **31** is obtained in 60% yield (Scheme 2.2.6).[18] Significantly, use of the more reactive methyl(trifluoromethyl)dioxirane (**32**) gives a single diastereomeric triepoxide **33** with an average stereoselectivity for the three epoxidations >96%.[18]

Protocol 3.
Synthesis of (5Z)-(7R,8R)-3β-acetoxy-7,8-epoxy-9,10-secocholesta-5,10(19)-diene (Structure 31, Scheme 2.2.6)

Caution! Dimethyldioxirane is a volatile peroxide and should be treated as such. Carry out all procedures in a well-ventilated hood, and wear disposable vinyl or latex gloves and chemical-resistant safety goggles.

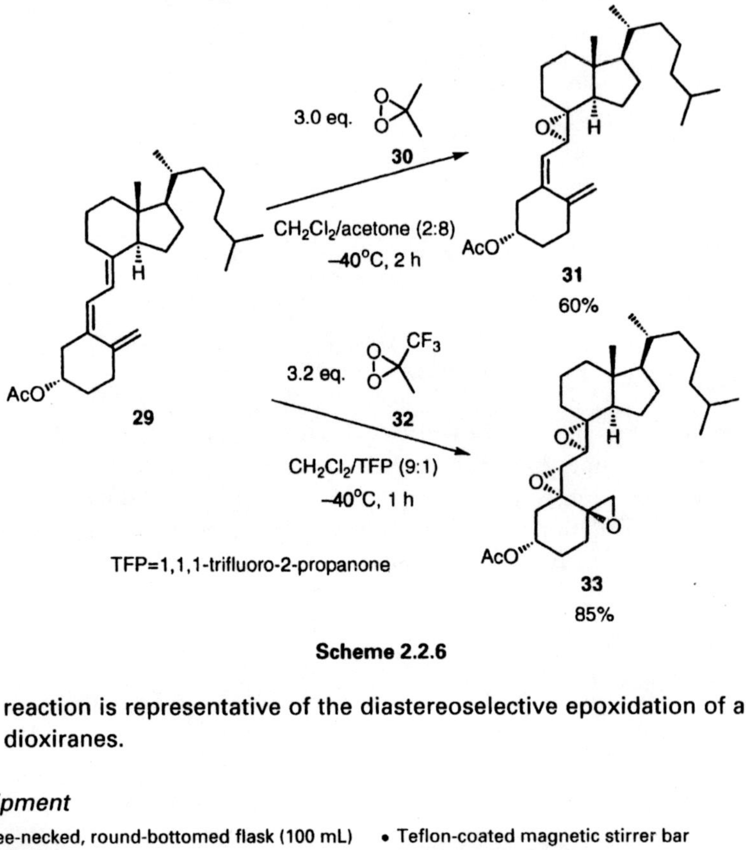

Scheme 2.2.6

This reaction is representative of the diastereoselective epoxidation of alkenes with dioxiranes.

Equipment
- Three-necked, round-bottomed flask (100 mL)
- Rubber septum
- Magnetic stirrer
- Cooling bath
- Teflon-coated magnetic stirrer bar
- Source of dry nitrogen
- Syringes (25 mL)

Materials

Dichloromethane, 12 mL	flammable, irritant
Dimethyldioxirane, 0.08 M in acetone, 48.3 mL, 3.86 mmol	flammable, explosive
(5Z,7E)-3β-Acetoxy-9,10-secocholesta-5,7,10(19)-triene, 0.55 g, 1.29 mmol	unknown toxicity, treat as toxic
Acetonitrile, as needed	flammable, toxic
Dry ice, as needed	

2: Oxidation of the C=C bond

Method
1. Equip a 100-mL 3-necked round-bottomed flask with a nitrogen bubbler, a rubber septum and a magnetic stirring bar. Maintain the apparatus under a positive pressure of nitrogen.
2. Add (5Z,7E)-3β-acetoxy-9,10-secocholesta-5,7,10(19)-triene (0.55 g, 1.29 mmol) and dichloromethane (12 mL).
3. Stir the mixture and cool to −40°C with a dry ice–acetonitrile bath.
4. Add a solution of dimethyldioxirane (0.08 M in acetone, 16.1 mL, 1.29 mmol)[15] via syringe. Stir the reaction mixture for 10 min.
5. Repeat step 4 twice for a total of three equivalents of dimethyldioxirane added to the reaction mixture. Stir the reaction mixture at −40°C until it is complete by TLC.
6. Remove solvent on a rotary evaporator to give a residue and purify the product by flash chromatography to give 0.34 g (60%) of (5Z)-(7R,8R)-3β-acetoxy-7,8-epoxy-9,10-secocholesta-5,10(19)-diene as a viscous oil, $[\alpha]_D$ = +24.3° (c = 0.50, acetone). The following spectra are diagnostic: IR (neat) 2957, 2942, 1739, 1640, 1466, 1378, 1245, 1165, 1082, 1049, 1035, 959, 916, 877 cm^{-1}; ^{1}H NMR (500 MHz, CDCl$_3$) δ 5.17 (d, J = 9.3 Hz, 1H), 5.01 (m, 1H), 4.94 (m, 1H), 4.81 (d, J = 2.3 Hz, 1H), 3.85 (d, J = 9.3 Hz, 1H), 2.01 (s, 3H), 0.89 (d, J = 6.6 Hz, 3H), 0.84 (d, J = 6.6 Hz, 3H), 0.83 (d, J = 6.6 Hz, 3H), 0.66 (s, 3H); ^{13}C NMR (50 MHz, CDCl$_3$) δ 170.64, 144.46, 144.31, 121.50, 112.62, 71.18, 65.54, 56.57, 56.31, 54.06, 45.87, 41.98, 39.41, 36.04, 35.57, 31.87, 31.72, 30.70, 27.97, 27.37, 23.79, 22.79, 22.53, 22.25, 21.34, 19.96, 18.74, 12.61.

3. Enantioselective epoxidation

The reports on enantioselective epoxidation of prochiral simple alkenes with peroxide-based reagents have been limited until recently due to the scarcity of chiral non-racemic epoxidizing reagents. The chiral peroxy acid available for the enantioselective epoxidation of alkenes is (+)-monoperoxycamphoric acid (**35**). However, the asymmetric induction using this reagent is generally low.[19]

A class of more efficient epoxidizing reagents developed for the epoxidation of simple olefins are chiral non-racemic N-sulfamyloxaziridines and N-sulfonyloxaziridines prepared by the peroxy acid oxidation of the corresponding imines (see Chapter 4, Section 4.2).[20] For example, reaction of **37** with N-sulfamyloxaziridine **38** affords epoxide **39** in 65% ee.[21]

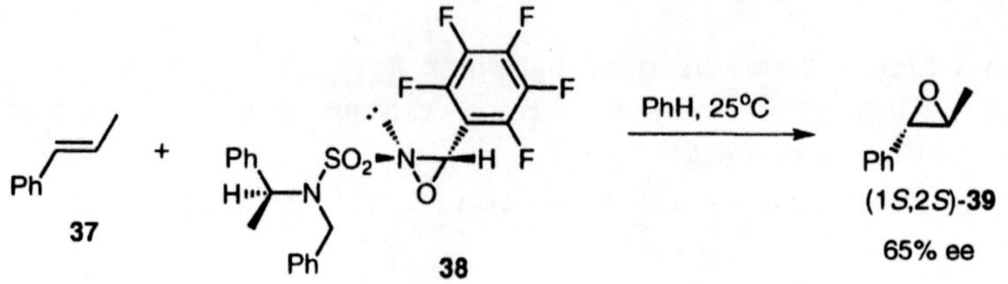

One of the most efficient methods for the enantioselective epoxidation of simple alkenes without a precoordinating group entails the use of chiral nonracemic dioxiranes generated catalytically *in situ* from chiral ketones using Oxone as the stoichiometric oxidant.[22–24] One example employs the dioxirane generated from fructose-based ketone **40**.[23] With *trans*-alkenes, very high enantiomeric excesses are achieved; e.g. the epoxidation of *trans*-β-methylstyrene (**37**) gave (1R,2R)-**39** in 92% ee and 93% yield (Scheme 2.2.7).[24]

Protocol 4.
Synthesis of (1R,2R)-β-methylstyrene oxide (Structure 39, Scheme 2.2.7)

Caution! Carry out all procedures in a well-ventilated hood, and wear disposable vinyl or latex gloves and chemical-resistant safety goggles.

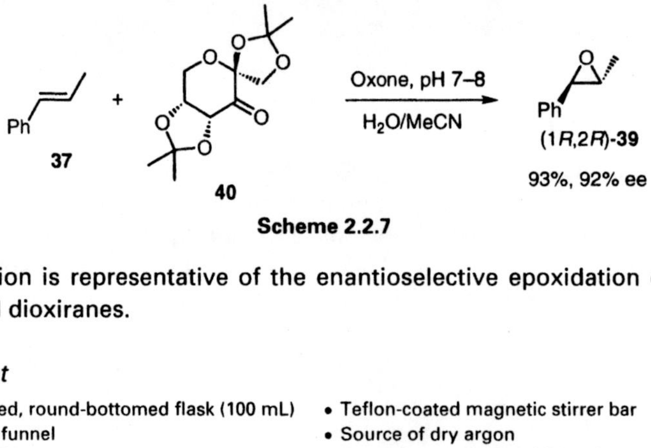

Scheme 2.2.7

This reaction is representative of the enantioselective epoxidation of alkenes with chiral dioxiranes.

Equipment

- Three-necked, round-bottomed flask (100 mL)
- Separating funnel
- Magnetic stirrer
- Cooling bath
- Teflon-coated magnetic stirrer bar
- Source of dry argon
- Pressure-equalizing addition funnels (25 mL)
- Sintered glass filter funnel

Materials

- Acetonitrile, 15 mL — **flammable, irritant**
- Buffer (0.05 M sodium tetraborate decahydrate in 4 × 10⁻⁴ M aq. ethylenediaminetetraacetic acid, disodium salt), 10 mL — **irritant**
- Tetrabutylammonium hydrogen sulfate, 0.015 g, 0.04 mmol — **irritant**
- *trans*-β-Methylstyrene, 0.118 g, 1 mmol — **irritant**

2: Oxidation of the C=C bond

- 1,2:4,5-Bis-O-isopropylidene-β-D-erythro-2,3-hexodiulo-2,6-pyranose, 0.0774 g, 0.3 mmol — **unknown toxicity, treat as toxic**
- Oxone®, 0.85 g, 1.38 mmol — **oxidizer, irritant**
- Aqueous ethylenediaminetetraacetic acid, disodium salt, 4×10^{-4} M, 6.5 mL — **irritant**
- Potassium carbonate, 0.8 g, 5.8 mmol — **irritant**
- Pentane, as needed — **flammable, irritant**
- Water, as needed
- Ice, as needed
- Brine, 30 mL
- Sodium sulfate, as needed — **irritant**
- Silica gel, 10 g
- Triethylamine, as needed — **flammable, corrosive**
- Diethyl ether, as needed — **flammable, toxic**

Method

1. Equip a 100-mL three-necked round-bottomed flask with two 25-mL addition funnels, an argon bubbler, and a magnetic stirring bar. Maintain the apparatus under a positive pressure of argon.
2. Add buffer (0.05 M $Na_2B_4O_7 \cdot 10H_2O$ in 4×10^{-4} M aq. Na_2EDTA, 10 mL), acetonitrile (15 mL), *trans*-β-methylstyrene (0.118 g, 1 mmol), tetrabutylammonium hydrogen sulfate (0.015 g, 0.04 mmol), and 1,2,4,5-bis-O-isopropylidene-β-D-*erythro*-2,3-hexodiulo-2,6-pyranose (0.0774 g, 0.3 mmol).
3. Stir the mixture and cool with an ice-water bath.
4. Add in one addition funnel Oxone® (0.85 g, 1.38 mmol) and an aqueous solution of Na_2EDTA (4×10^{-4} M, 6.5 mL). Mix well to give a solution.
5. Add in the other addition funnel potassium carbonate (0.8 g, 5.8 mmol) and water (6.5 mL). Mix well to give a solution.
6. Add the solution obtained in step 4 and the solution in step 5 at the same rate to the reaction mixture in step 3 over 1.5 h. After addition of these solutions is complete, add pentane (30 mL) and water (30 mL).
7. Transfer the reaction mixture into a separatory funnel and separate the phases.
8. Extract the aqueous phase obtained in step 7 with pentane (3×30 mL).
9. Combine the organic phase obtained in step 7 with the pentane extracts from step 8 and wash with brine (30 mL).
10. Dry the organic phases over anhydrous sodium sulfate.
11. Filter to remove sodium sulfate.
12. Remove the solvent on a rotary evaporator to give a residue and purify the product by flash chromatography (the silica gel was buffered with 1% triethylamine in pentane) eluting with mixtures of pentane–ether (1:0 to 50:1, v/v) to afford 0.124 g (93%) of *trans*-β-methylstyrene oxide as a colourless liquid. The enantiomeric excess is determined by chiral shift reagent to be 92% ee (8 mg of epoxide dissolved in 0.5 mL $CDCl_3$, epoxide:Eu(hfc)$_3$ 10:1.4 w/w).

4. Double asymmetric differentiation

Double asymmetric differentiation can often be used to improve the asymmetric induction in the reaction of a chiral substrate with a chiral reagent. This strategy has been employed in the asymmetric epoxidation of simple alkenes. For example, reaction of (R)-limonene (**41**) with N-sulfonyloxaziridine (+)-**42** results in a 90% yield of 55:45 *cis/trans* mixture of limonene oxides (Scheme 2.2.8). The *cis/trans* selectivity improves to 93:7 when (S)-limonene is epoxidized by (+)-**42**. Double asymmetric differentiation is apparently operating in these epoxidations because (S)-**41** and (+)-**42**, the matched pair, give a higher *cis/trans* ratio than (R)-**41** and (+)-**42**, the mismatched pair.[25]

Protocol 5.
Synthesis of (1S,2R,4S)-limonene oxide (Structure (S)-cis-43, Scheme 2.2.8)

Caution! Chloroform is a cancer-suspect agent. Carry out all procedures in a well-ventilated hood, and wear disposable vinyl or latex gloves and chemical-resistant safety goggles.

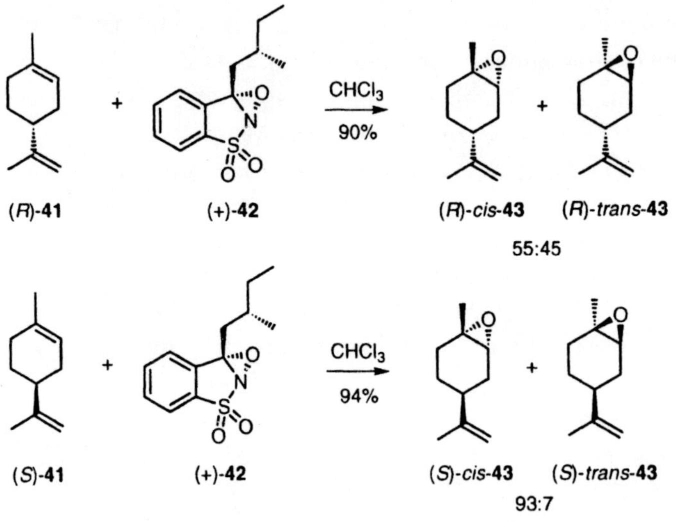

Scheme 2.2.8

This is an example of asymmetric epoxidation of chiral alkene with chiral nonracemic N-sulfonyloxaziridine oxidizing reagent via double asymmetric differentiation.

Equipment

- Round-bottomed flask (10 mL)
- Oil bath
- Magnetic stirrer
- Reflux condenser
- Teflon-coated magnetic stirrer bar
- Source of dry argon

Protocol 5. Continued

Materials

- (S)-Limonene, 0.019 g, 0.146 mmol — irritant
- (+)-(2R,3S)-3-[(S)-2-methylbutyl)]-2,3-epoxy-1,2-benzisothiazole-1,1-dioxide, 0.037 g, 0.146 mmol — unknown toxicity, treat as toxic
- Chloroform, 3 mL — highly toxic, cancer-suspect agent
- n-Pentane, 3 mL — flammable, irritant

Method

1. Equip a 10-mL round-bottomed flask with a condenser, an oil bath, an argon bubbler, and a magnetic stirring bar. Maintain the apparatus under a positive pressure of argon.
2. Add (S)-limonene (0.019 g, 0.146 mmol), (+)-(2R,3S)-3-[(S)-2-methylbutyl)]-2,3-epoxy-1,2-benzisothiazole-1,1-dioxide (0.037 g, 0.146 mmol), and chloroform (3 mL). Stir to give a solution.
3. Place the reaction flask in a preheated oil bath and stir the reaction mixture at 60°C for 48 h.
4. Remove the solvent on a rotary evaporator to dryness. Add n-pentane (3 mL) to the residue and cool the mixture to −20°C.
5. Decant carefully the solvent from the precipitated (+)-(2R,3S)-3-[(S)-2-methylbutyl)]-1,2-benzisothiazole-1,1-dioxide.
6. Evaporate the solvent to dryness to give 0.020 g (94%) of (1S,2R,4S)-limonene oxide with a diastereomeric ratio of (1S,2R)/(1R,2S) 93:7.

References

1. Schurig, V.; Leyrer, U.; Wistuba, D. *J. Org. Chem.* **1986**, *51*, 242.
2. Trost, B. M.; Matsumura, Y. *J. Org. Chem.* **1977**, *42*, 2036.
3. Corey, E. J.; Snider, B. B. *J. Am. Chem. Soc.* **1972**, *94*, 2549.
4. Murphy, P. J.; Russell, A. T.; Procter, G. *Tetrahedron Lett.* **1990**, *31*, 1055.
5. Bessiere-Chretien, Y.; Moncef El Gaied, M.; Meklati, B. *Bull. Soc. Chim. Fr.* **1970**, 1000.
6. Henbest, H. B. *Proc. Chem. Soc. London* **1963**, 159.
7. Bernath, G.; Svoboda, M. *Tetrahedron* **1972**, *28*, 3475.
8. Finnegan, R. A.; Wepplo, P. J. *Tetrahedron* **1972**, *28*, 4267.
9. Girard, J. P.; Vidal, J. P.; Granger, R.; Rossi, J. C.; Chapat, J. P. *Tetrahedron Lett.* **1974**, 943.
10. Crandall, J. K.; Crawley, L. C. *Org. Synth.* **1973**, *53*, 17.
11. Martinelli, M. J.; Peterson, B. C.; Khau, V. V.; Hutchinson, D. R.; Leanna, M. R.; Audia, J. E.; Droste, J. J.; Wu, Y.-D.; Houk, K. N. *J. Org. Chem.* **1994**, *59*, 2204.
12. Leanna, M. R.; Martinelli, M. J.; Varie, D. L.; Kress, T. J. *Tetrahedron Lett.* **1989**, *30*, 3935.
13. Schreiber, S. L.; Sammakia, T.; Hulin, B.; Schulte, G. *J. Am. Chem. Soc.* **1986**, *108*, 2106.

14. Curi, R.; Dinoi, A.; Rubino, M. F. *Pure Appl. Chem.* **1995**, *67*, 811.
15. Dembech, P.; Ricci, A.; Seconi, G.; Taddei, M. *Org. Synth.* **1997**, *74*, 84.
16. Adam, W.; Bialas, J.; Hadjiarapoglou, L. *Chem. Ber.* **1991**, *124*, 2377.
17. Gilbert, M.; Ferrer, M.; Sanchez-Baeza, F.; Messeguer, A. *Tetrahedron* **1997**, *53*, 8643.
18. Curi, R.; Detomaso, A.; Prencipe, T.; Carpenter, G. B. *J. Am. Chem. Soc.* **1994**, *116*, 8112.
19. Pirkle, W. H.; Rinaldi, P. L. *J. Org. Chem.* **1977**, *42*, 2080.
20. Davis, F. A.; Sheppard, A. C. *Tetrahedron* **1989**, *45*, 5703.
21. Davis, F. A.; Chattopadhyay, S. *Tetrahedron Lett.* **1986**, *27*, 5079.
22. Yang, D.; Wang, X.-C.; Wong, M.-K.; Yip, Y.-C.; Tang, M.-W. *J. Am. Chem. Soc.* **1996**, *118*, 11311.
23. Tu, Y.; Wang, Z.-X.; Shi, Y. *J. Am. Chem. Soc.* **1996**, *118*, 9806.
24. Wang, Z.-X.; Tu, Y.; Frohn, M.; Shi, Y. *J. Org. Chem.* **1997**, *62*, 2328.
25. Davis, F. A.; ThimmaReddy, R.; McCauley, J. P.; Przeslawski, R. M.; Harakal, M. E.; Carroll, P. J. *J. Org. Chem.* **1991**, *56*, 809.

2.3 Asymmetric epoxidation of olefins bearing precoordinating functional groups

V. S. MARTIN

1. Introduction

In 1970, it was found that the epoxidation of allylic alcohols with an alkyl hydroperoxide under catalysis by VO(acac)$_2$ proceeds smoothly in comparison with isolated olefins.[1] These metal-catalysed epoxidations of allylic alcohols are chemo-, regio-, and stereoselective, the reaction occurs in the coordination sphere of the metal ion, and the bystander ligened (L) in complex **1** is located adjacent to the reaction site (see also Section 2.1).[2,3] Therefore, the use of chiral bystander ligands places the reaction site in an asymmetric environment.[2]

2: Oxidation of the C=C bond

2. Asymmetric epoxidation of allylic alcohols

In 1980, Sharpless and Katsuki discovered a system for the asymmetric epoxidation of primary allylic alcohols that utilizes Ti(OPr-i)$_4$, a dialkyl tartrate, as a chiral ligand, and t-butyl hydroperoxide (TBHP) as the oxidant.[4] The reaction proceeds under mild conditions with good chemical yield and with high regio- and chemoselectivity. The stereochemistry of this reaction can be predicted by the empirical rule shown in Scheme 2.3.1. When an allylic alcohol is drawn in a plane with the hydroxymethyl group positioned at the lower right, the deliver of oxygen occurs from the bottom side of the olefin to give the (2S)-epoxide if an (R,R)-dialkyl tartrate is used as the chiral auxiliary. Obviously, when an (S,S)-auxiliary is employed, oxygen is delivered from the top side. There is no exception to the empirical rule in the asymmetric epoxidation of prochiral allylic alcohols, but the rule does not always predict correctly the stereochemistry of the epoxidation of chiral allylic alcohols, especially when the substrates bear bulky substituents near the site of epoxidation.

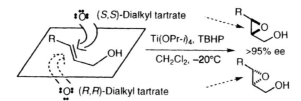

Scheme 2.3.1

An example of the stoichiometric asymmetric epoxidation reaction is the synthesis of (2S,3S)-2,3-epoxy-1-hexanol **2** (Scheme 2.3.2). The epoxidation of (E)-2-hexen-1-ol in dichloromethane in the presence of titanium tetraisopropoxide and (R,R)-diethyl tartrate with a dry isooctane solution of t-butyl hydroperoxide afforded **2** in 80% yield with an optical purity greater than 95% ee. Although t-butyl hydroperoxide is a fairly safe oxidant, it is important to emphasize the potential danger in the manipulation of peroxides as a possible subject for violent decomposition by adventitious catalysts. In this sense, never add strong acid (not even a drop) or transition-metal salts to high-strength TBHP solutions and never work with pure TBHP. t-Butyl hydroperoxide[5] is commercially available as an anhydrous 5.0–6.0 M solution in nonane that can be used as received. In order to use the commercial 70% or 90% aqueous solution it is necessary to remove the water and to prepare an anhydrous solution in an organic solvent, such as isooctane.[6]

Protocol 1.
Synthesis of (2S,3S)-2,3-epoxy-1-hexanol (Structure 2, Scheme 2.3.2)

Caution! The oxidant t-butyl hydroperoxide is susceptible to violent decomposition under the action of strong acids or certain metals. Titanium tetraisopropoxide decomposes very quickly when exposed to atmospheric moisture causing white fumes. Carry out all procedures in a well-ventilated hood, and wear disposable gloves and safety goggles.

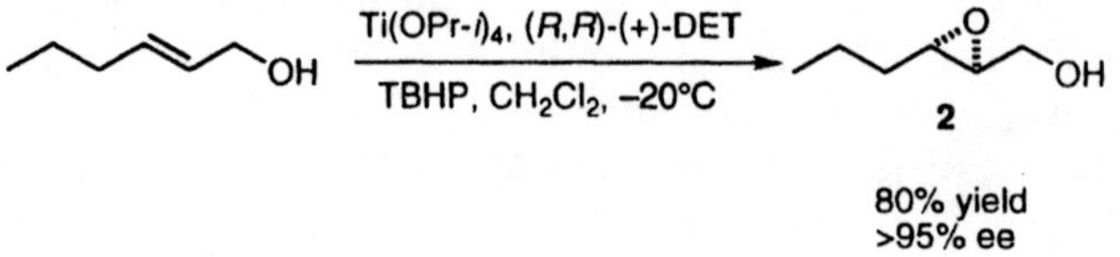

$$\text{CH}_3\text{CH}_2\text{CH}_2\text{CH=CHCH}_2\text{OH} \xrightarrow[\text{TBHP, CH}_2\text{Cl}_2,\ -20°\text{C}]{\text{Ti(OPr-}i)_4,\ (R,R)\text{-(+)-DET}} \underset{\mathbf{2}}{\text{epoxide-OH}}$$

80% yield
>95% ee

Scheme 2.3.2

This is a representative example of the synthesis of an enantiomerically enriched 2,3-epoxy alcohol by stoichiometric asymmetric epoxidation of an allylic alcohol.

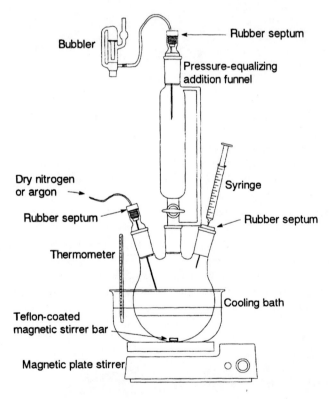

Fig. 2.3.1

2: Oxidation of the C=C bond

Equipment

- Three-necked, round-bottomed flask (2 L)
- Magnetic plate stirrer
- Syringes (50 and 100 mL)
- Teflon-coated magnetic stirrer bar
- A cooling bath (−20 °C)
- Pressure-equalizing addition funnels (250 mL)
- Separating funnels (1 L and 2 L)
- Filter paper
- Filter funnel
- Erlenmeyer flask (1 L)
- Source of dry nitrogen or argon
- Glass chromatographic column (ca. 50 cm × 5 cm) fitted with a sintered glass disc (porosity 70–100 μm) at the base of the column
- Rubber septums
- A glass bubbler
- Thermometer (−40 → +40 °C)

Materials

- Activated 3-Å molecular sieves, 5 g   **hygroscopic**
- (E)-2-Hexen-1-ol, 11.8 mL, 100 mmol   **flammable**
- Dry dichloromethane,[a] 600 mL   **harmful by inhalation**
- Titanium tetraisopropoxide, 35.7 mL, 120 mmol   **fumes in air**
- (R,R)-(+)-Diethyl tartrate, 24 mL, 140 mmol   **flammable**
- t-Butyl hydroperoxide, 4.5 M in isooctane, 40 mL, 180 mmol   **flammable**
- Tartaric acid, 90 g, 600 mmol   **irritant**
- Dichloromethane for extraction, 400 mL   **harmful by inhalation**
- Diethyl ether, 300 mL   **flammable, explosive, harmful by inhalation**
- Sodium chloride saturated solution, 100 mL
- 15% Sodium hydroxide solution saturated with sodium chloride, 100 mL   **caustic**
- Celite
- Silica gel for chromatography (Merck, 0.015–0.04 mm)

Method

1. Equip a three-necked round-bottomed flask (2 L) with a magnetic stirrer bar, a pressure-equalizing addition funnel (250 mL) stoppered with a rubber septum, and two rubber septums. In the septums connected to the flask and the addition funnel, insert two syringe needles, one of which is connected to a source of dry nitrogen or argon, the other one to a bubbler (see Fig. 2.3.1). Dry using a flame or a high-temperature heating gun under a positive pressure of dry nitrogen or argon.

2. To the flask add activated 3-Å molecular sieves (5 g) and maintain the apparatus under an inert gas atmosphere. Add directly to the flask, dry dichloromethane (600 mL) and freshly distilled titanium tetraisopropoxide (35.7 mL, 120 mmol) via syringe or cannula. Cool down the mixture to between −30 and −20 °C[b] and stir it.

3. Add to the flask via syringe (R,R)-(+)-diethyl tartrate (24 mL, 140 mmol) and stir for 15 min.

4. Prepare a solution of (E)-2-hexen-1-ol (11.8 mL, 100 mmol) in dichloromethane (20 mL) and transfer it to the pressure-equalizing addition funnel. Add the solution dropwise over 5 min and stir for 10 min.

5. Transfer a solution of t-butyl hydroperoxide[c] (4.5 M in isooctane, 40 mL,

Protocol 1. *Continued*

   180 mmol) to the pressure-equalizing addition funnel, add to the reaction mixture slowly over a period of 10 min and continue stirring for 4–5 h.

6. Prepare a 15% aqueous solution (600 mL) of tartaric acid (90 g, 600 mmol) and add it to the mixture.[d] Then allow the mixture to warm to room temperature and continue stirring for 30 min. Filter the mixture through a pad of Celite.
7. Transfer the mixture to a separating funnel (2 L). Separate the organic phase and wash the remaining layer with two portions of dichloromethane (2 × 100 mL). Combine the organic layers and remove the solvent on a rotary evaporator to leave a wet oil.
8. Take out the oily residue in diethyl ether (300 mL), cool it to 0 °C, and treat it with an aqueous sodium hydroxide solution (1 M) saturated with NaCl (100 mL). Stir the resulting two-phase mixture vigorously for 30 min at 0 °C and transfer the mixture to a separating funnel (1 L).
9. Wash the ether solution with saturated aqueous sodium chloride solution (100 mL), dry over anhydrous magnesium sulfate, filter, and remove the solvent using a rotary evaporator to leave an oily residue.
10. Prepare a column for chromatography using silica gel (0.015–0.04 mm, Merck) and a mixture of hexane and ethyl acetate (15%). Dissolve the residue in the minimum volume of the solvent mixture, apply the solution at the top of the column, and elute the column with the solvent mixture. Evaporate the solvent on a rotary evaporator to yield the 2,3-epoxy alcohol **2** as a colourless oil, $[a]_D^{25} = -46.5°$ (c = 0.93, CHCl$_3$), 9.28 g, 80%.

[a] Dichloromethane is the solvent used in all asymmetric epoxidation reactions. Use only the methanol-free solvent dried over activated 3-Å molecular sieves. Avoid contamination by chelating solvents such as alcohols, esters, nitriles, amines, or ketones.
[b] Carbon tetrachloride–dry ice slurry.
[c] The syringe or cannula should never be inserted directly into the stock solution. Pour the estimated amount of the solution into a vessel, measure the amount needed, and transfer. Discard the remaining solution.
[d] In small-scale reactions the excess of *t*-butyl hydroperoxide can be ignored because it can be easily removed by azeotropic distillation with toluene or carbon tetrachloride (up to 0.5 mol) or in the product purification step. However, oxidant destruction may be desired to avoid peroxide hazard and is essential in large-scale reactions. If so, use a solution of tartaric acid including ferrous sulfate heptahydrate (30% w/v) to quench the reaction and to reduce the oxidant excess.

Various aspects of this reaction, including its mechanism,[7] early synthetic applications,[8–11] further synthetic applications of the epoxy alcohol product,[2] and the full scope and limitations,[12] have been reviewed. The reaction involves the rapid exchange of metal alkoxides with the allylic alcohol, the alkyl hydroperoxide, and the dialkyl tartrate to afford a complex **3** in which one enantioface of the olefin is preferentially epoxidized. The structure **4** represents the view along the axis of the double bond, distal peroxide and titanium.[7]

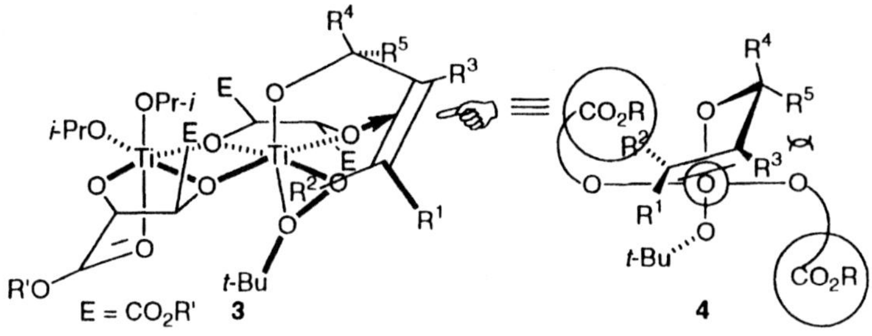

2: Oxidation of the C=C bond

[Structure diagrams 3 and 4 showing Ti-catalyst complexes with labels: OPr-i, i-PrO, E = CO_2R', t-Bu, R^1, R^2, R^3, R^4, R^5, R'O, CO_2R]

The model explains the following practical aspects:

- When R^2 is a large substituent, as for example in Z-olefins, the asymmetric epoxidation exhibits poor enantioselectivity and the reaction rate decreases (Scheme 2.3.3).
- When a substrate is a tertiary allylic alcohol ($R^4 \neq H$, $R^5 \neq H$), its reactivity is very poor.
- E-substituent (R^1) is located in an open quadrant, and the asymmetric epoxidation is practically insensitive to this kind of E-substituent. This is one of the most important features of this reaction, because regardless of the type of E-allylic alcohol used, the reaction proceeds with the expected selectivity (Scheme 2.3.3).

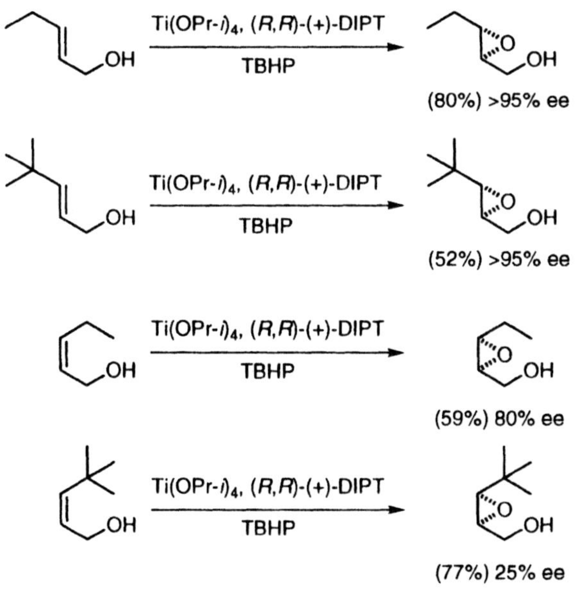

[Scheme showing four reactions, all with Ti(OPr-i)$_4$, (R,R)-(+)-DIPT / TBHP:

1. E-crotyl alcohol → epoxy alcohol (80%) >95% ee
2. E-4,4-dimethyl-2-pentenol → epoxy alcohol (52%) >95% ee
3. cyclohexenyl methanol → epoxy alcohol (59%) 80% ee
4. Z-4,4-dimethyl-2-pentenol → epoxy alcohol (77%) 25% ee]

Scheme 2.3.3

Asymmetric epoxidation of some allylic alcohols proceeds with diminished enantioselectivity. In many cases, however, the product epoxy alcohols are crystalline compounds that can be recrystallized to optical purity.[13] Non-crystalline epoxy alcohols can often be converted to crystalline compounds that also allow enhancement of enantiomeric purity through recrystallization (Scheme 2.3.4).[14] These derivatives can be prepared by catalytic asymmetric epoxidation and subsequent *in situ* derivatization.[6]

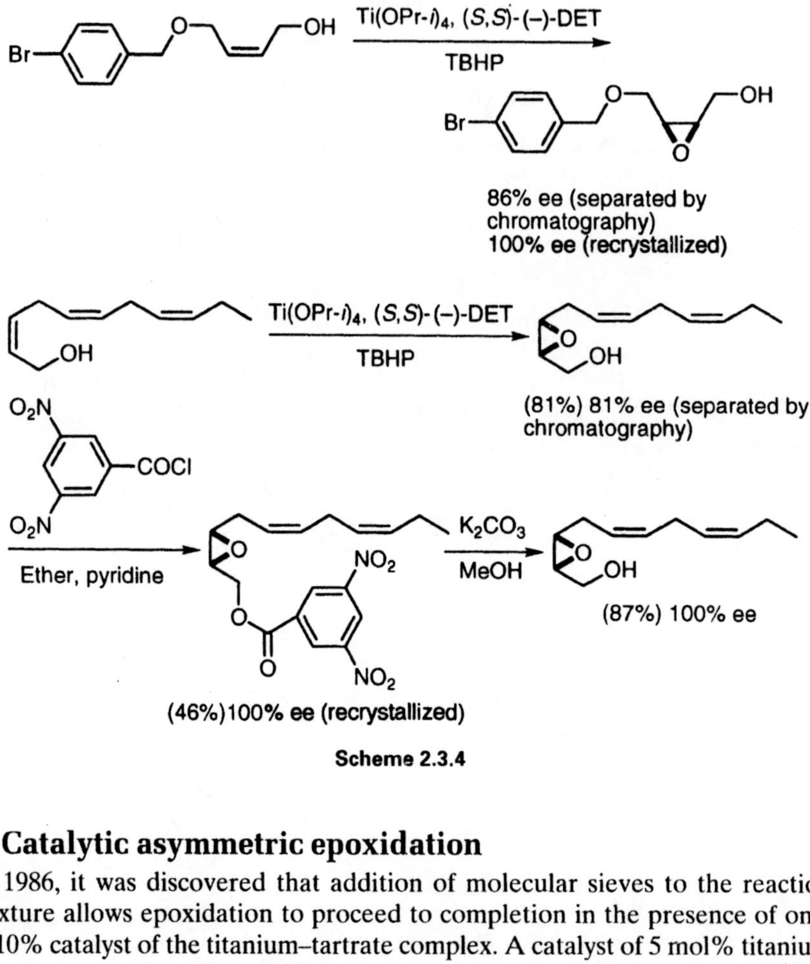

Scheme 2.3.4

3. Catalytic asymmetric epoxidation

In 1986, it was discovered that addition of molecular sieves to the reaction mixture allows epoxidation to proceed to completion in the presence of only 5–10% catalyst of the titanium–tartrate complex. A catalyst of 5 mol% titanium tetraisopropoxide and 6 mol% tartrate is recommended as the most widely usable system for a catalytic asymmetric epoxidation. Although the enantioselectivity is often reduced by 1–5% relative to stoichiometric reactions, it is the recommended option in most cases. Using the catalytic procedure, unstable and water-soluble epoxy alcohols are obtained in good to moderate yields (Scheme 2.3.5).[6]

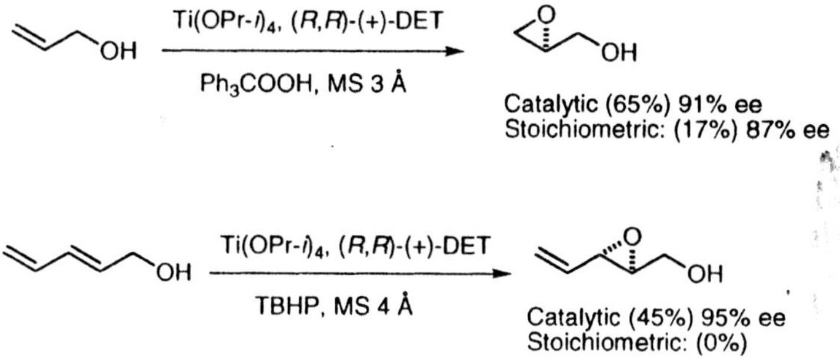

2: Oxidation of the C=C bond

Scheme 2.3.5

Allyl alcohol → Ti(OPr-i)$_4$, (R,R)-(+)-DET, Ph$_3$COOH, MS 3 Å → glycidol
Catalytic (65%) 91% ee
Stoichiometric: (17%) 87% ee

(E)-2-butenol → Ti(OPr-i)$_4$, (R,R)-(+)-DET, TBHP, MS 4 Å → epoxy alcohol
Catalytic (45%) 95% ee
Stoichiometric: (0%)

Protocol 2.
Synthesis of (2S,3S)-2,3-epoxycinnamyl alcohol [(2S-trans)-3-phenyloxiranemethanol] (Structure 5, Scheme 2.3.6)

Caution! The oxidant *t*-butyl hydroperoxide is susceptible to violent decomposition under the action of strong acids or certain metals. Titanium tetraisopropoxide decomposes very quickly when exposed to atmospheric moisture causing white fumes. Carry out all procedures in a well-ventilated hood and wear disposable gloves and safety goggles.

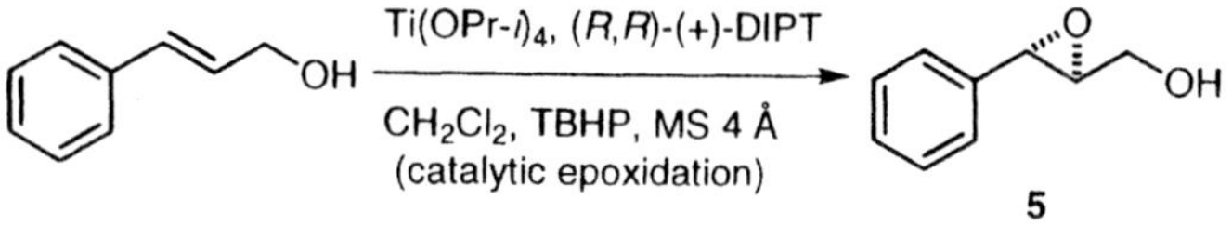

Cinnamyl alcohol → Ti(OPr-i)$_4$, (R,R)-(+)-DIPT, CH$_2$Cl$_2$, TBHP, MS 4 Å (catalytic epoxidation) → **5**
(89%) >98% ee

Scheme 2.3.6

This is a simple example of synthesis of an unstable epoxy alcohol by the catalytic asymmetric epoxidation of a suitable allylic alcohol.[a]

Equipment

- Three-necked, round-bottomed flask (5 L)
- Overhead mechanical stirrer
- Syringes (50 and 100 mL)
- A cooling bath (−20 °C)
- Pressure-equalizing addition funnel (250 mL)
- Separating funnel (1 L)
- Filter paper
- Filter funnel
- Erlenmeyer flask (5 L)
- Source of dry nitrogen or argon
- Rubber septums
- A glass bubbler
- Thermometer (−40 → +40 °C)
- High vacuum pump (0.2 mmHg)

Materials

- Activated 3-Å molecular sieves, 20 g — hygroscopic
- (E)-3-Phenylpropen-1-ol (cinnamyl alcohol), 50 g, 373 mmol
- Dry dichloromethane, 3.6 L — harmful by inhalation
- Titanium tetraisopropoxide, 5.5 mL, 5.3 g, 19 mmol — fumes in air
- (R,R)-(+)-Diisopropyl tartrate, 5.9 mL, 6.5 g, 28 mmol[b] — flammable

Protocol 2. *Continued*

- *t*-Butyl hydroperoxide, 7.7 M in dichloromethane, 97 mL, 746 mmol — flammable
- Dichloromethane for extraction, 400 mL — harmful by inhalation
- Diethyl ether, 400 mL — flammable, explosive, harmful by inhalation
- Toluene, 200 mL — flammable, harmful by inhalation, ingestion, and skin contact
- Sodium hydroxide saturated with sodium chloride, 30 mL — corrosive
- Petroleum ether/diethyl ether for recrystallization
- Celite

Method

1. Equip a three-necked round-bottomed flask (5 L) with an overhead mechanical stirrer fitted with a Teflon-coated stirrer paddle, a thermometer, and a pressure-equalizing addition funnel, with a nitrogen or argon inlet (see Fig. 2.3.2). Dry using a flame or a high-temperature heating gun under a positive pressure of dry nitrogen or argon.
2. Charge the flask with activated powdered 4-Å molecular sieves (20 g), dry dichloromethane (3.5 L), and (*R,R*)-(+)-diisopropyl tartrate (5.9 mL, 6.5 g, 28 mmol) and cool the mixture to −20°C. Then add titanium tetraisopropoxide (5.5 mL, 5.3 g, 19 mmol) and *t*-butyl hydroperoxide (7.7 M in dichloromethane, 97 mL, 746 mmol),[c] and stir the mixture for 1 h.
3. Prepare a solution of (*E*)-3-phenylpropen-1-ol (cinnamyl alcohol) (50 g, 373 mmol) in 70 mL of dichloromethane and transfer it to the pressure-equalizing addition funnel. Add the solution slowly over a period of 1 h.
4. After the addition is complete, continue stirring for 3 h at −20°C. Quench the reaction by adding at the same temperature a 10% aqueous solution of sodium hydroxide saturated with sodium chloride (30 mL). Add diethyl ether (400 mL) and allow the mixture to warm to 10°C. Add magnesium sulfate (30 g) and Celite (4 g) and continue stirring at 10°C for 15 min.
5. Allow the mixture to settle and filter the clear solution through a pad of Celite and wash with diethyl ether.
6. Add toluene (200 mL) and evaporate the mixture on a rotary evaporator for azeotropical removal of the excess of *t*-butyl hydroperoxide. Subject the residue to high vacuum (0.2 mmHg). The remaining yellow oil is recrystallized from petroleum ether/diethyl ether at −20°C yielding the epoxy alcohol (50 g, 89% yield) as slightly yellow crystals, m.p. 51.5–53°C, $[\alpha]_D^{25} = -49.6°$ (*c* = 2.4, CHCl$_3$).

[a] The stoichiometric asymmetric epoxidation gives yields of ~45%.
[b] The amount of tartrate ester must be carefully controlled, because a large amount of tartrate (>100% excess) will decrease the reaction rate [the titanium–tartrate (1:2) complex is inactive], while with too little tartrate (<10% excess) the enantioselectivity suffers.
[c] The syringe or cannula should never be inserted directly into the stock solution. Pour the estimated amount of the solution in a vessel, measure the amount needed, and transfer. Discard the remaining solution.

2: Oxidation of the C=C bond

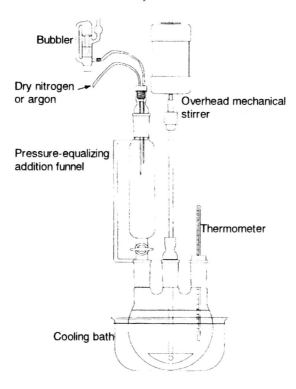

Fig. 2.3.2

One of the great advantages of the catalytic asymmetric epoxidation over the stoichiometric version is the ability of *in situ* derivatization of the product, improving the isolation.[6] Those derivatives can serve as versatile intermediates for further transformations (Scheme 2.3.8). The combination of asymmetric epoxidation and *in situ* titanium-mediated epoxide-opening reaction provides an access to diol derivatives without isolating the epoxy-alcohol (Scheme 2.3.7).

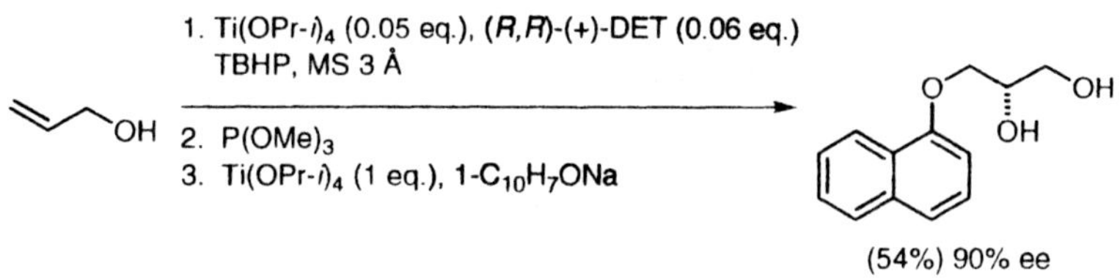

Scheme 2.3.7

Protocol 3.
(R)-2,3-Epoxypropyl p-nitrobenzoate (Structure 6, Scheme 2.3.8)

Caution! The oxidant t-butyl hydroperoxide is susceptible to violent decomposition under the action of strong acids or certain metals. Titanium tetraisopropoxide decomposes very quickly when exposed to atmospheric moisture causing white fumes. Carry out all the procedures in a well-ventilated hood and wear disposable gloves and safety goggles.

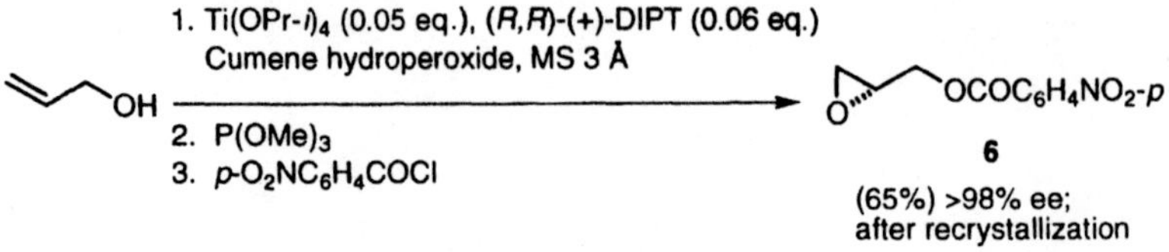

$$\text{allyl alcohol} \xrightarrow[\substack{2.\ P(OMe)_3 \\ 3.\ p\text{-}O_2NC_6H_4COCl}]{\substack{1.\ Ti(OPr\text{-}i)_4\ (0.05\ eq.),\ (R,R)\text{-}(+)\text{-}DIPT\ (0.06\ eq.) \\ \text{Cumene hydroperoxide, MS 3 Å}}} \text{6}\ \ OCOC_6H_4NO_2\text{-}p$$

(65%) >98% ee; after recrystallization

Scheme 2.3.8

This is an example of a tandem reaction of catalytic asymmetric epoxidation and *in situ* derivatization to obtain a crystalline derivative of a soluble and unstable epoxy alcohol.

Equipment

- Three-necked, round-bottomed flask (5 L)
- Overhead mechanical stirrer
- Syringes (50 and 100 mL)
- A cooling bath (−20°C)
- Pressure-equalizing addition funnel (250 mL)
- Separating funnel (5 L)
- Filter paper
- Filter funnel
- Erlenmeyer flask (5 L)
- Source of dry nitrogen or argon
- Rubber septums
- A glass bubbler
- Thermometer (−40 → +40°C)
- High-vacuum pump (0.2 mmHg)

Materials

- Activated 3-Å molecular sieves, 35 g — hygroscopic
- Allyl alcohol, 58.1 g, 1000 mmol[a]
- Dry dichloromethane, 2.5 L — harmful by inhalation
- Titanium tetraisopropoxide, 15 mL, 14 g, 50 mmol — fumes in air
- (R,R)-(+)-Diisopropyl tartrate, 12.5 mL, 13.9 g, 59.5 mmol[b] — flammable
- Cumene hydroperoxide, 80% technical grade, 360 mL, ca. 2000 mmol[c] — flammable, explosive
- Diethyl ether, 400 mL — flammable, harmful by inhalation
- Trimethyl phosphite, 180 mL, 189 g, 1500 mmol — flammable
- Triethylamine, 170 mL, 123 g, 1200 mmol — flammable
- p-Nitrobenzoyl chloride, 186 g, 1000 mmol — irritant
- Diethyl ether for recrystallization — flammable, explosive, harmful by inhalation
- 10% Aqueous solution of tartaric acid, 500 mL — irritant
- Saturated NaHCO$_3$ solution, 750 mL
- Saturated NaCl solution (brine), 500 mL
- Celite

Method

1. Assemble and flame-dry the glassware as in Protocol 2 (Fig. 2.3.2) under positive pressure of dry nitrogen or argon.

2: Oxidation of the C=C bond

2. Transfer into the three-necked round-bottomed flask (5 L), activated 3-Å molecular sieves (35 g), dry dichloromethane (1.9 L), (R,R)-(+)-diisopropyl tartrate (12.5 mL, 13.9 g, 59.5 mmol), and allyl alcohol (68 mL, 58.1 g, 1000 mmol).
3. With stirring, cool the mixture to −5 °C. Add titanium tetraisopropoxide (15 mL, 14 g, 50 mmol) and stir at −5 ± 2 °C for 30 min. Transfer via syringe cumene hydroperoxide (80% technical grade, 360 mL, ca. 2000 mmol) to the pressure-equalizing addition funnel.^{d} Add the solution slowly over a period of 30 min.
4. After the addition is complete, continue stirring for 6 h and cool the reaction mixture to −20 °C. Transfer trimethyl phosphite (180 mL, 189 g, 1500 mmol) to the funnel and add it carefully over a period of 1 h, taking care that the temperature does not rise above −20 °C.
5. Add triethylamine (170 mL, 123 g, 1200 mmol) and a solution of p-nitrobenzoyl chloride (186 g, 1000 mmol) in dichloromethane (250 mL). When the addition is complete, raise the temperature to 0 °C and continue stirring for a further 1 h. Filter the mixture through a pad of Celite, and transfer the filtrate to a separating funnel (5 L).
6. Wash the solution with 10% aqueous tartaric acid (2 × 250 mL), saturated $NaHCO_3$ (3 × 250 mL), and brine (2 × 250 mL). Dry the organic phase over sodium sulfate, filter through a small pad of silica gel, and concentrate first on a rotary evaporator, then at 0.2 mmHg at 60 °C, to remove any remaining cumene, 2-phenylpropan-2-ol, trimethyl phosphite, and trimethyl phosphate. The resultant residual oil solidifies on standing. Recrystallize twice from diethyl ether to give 136 g (61%) of the product, m.p. 59.5–60 °C, $[\alpha]_D^{25}$ = −38.8° (c = 3.02, $CHCl_3$) (92–95% ee).

^{a} Stored over activated 3-Å molecular sieves.
^{b} The amount of tartrate ester must be carefully controlled, because a large amount of tartrate (>100% excess) will decrease the reaction rate [the titanium–tartrate (1:2) complex is inactive], while with too little tartrate (<10% excess) the enantioselectivity suffers.
^{c} Dried over activated 3-Å molecular sieves.
^{d} The syringe or cannula should never be inserted directly into the stock solution. Pour the estimated amount of the solution into a vessel, measure the amount needed, and transfer. Discard the remaining solution.

---

4. Kinetic resolution of racemic secondary allylic alcohols

When a racemic secondary allylic alcohol (R^4 = H, R^5 ≠ H or R^4 ≠ H, R^5 = H in **3**) is epoxidized, one enantiomer with the group oriented towards the complex (R^5 ≠ H) reacts slower than the other enantiomer (R^5 = H). In practical terms, such a difference in the rate yields a method of kinetic resolution of the enantiomers of racemic secondary allylic alcohols.[15] One enantiomer reacts faster yielding the expected *erythro* epoxy alcohol, the remaining one becoming enriched in its optical purity. With a good relative rate it is possible with enough conversion to recover the unreacted olefin as a pure enantiomer (Scheme 2.3.9). Usually only the unreacted allylic alcohol is used and the epoxy alcohol is discarded. In some cases, however, with a large relative rate (>500) both the remaining olefin and the epoxide produced are obtained in high optical purity (Scheme 2.3.10).

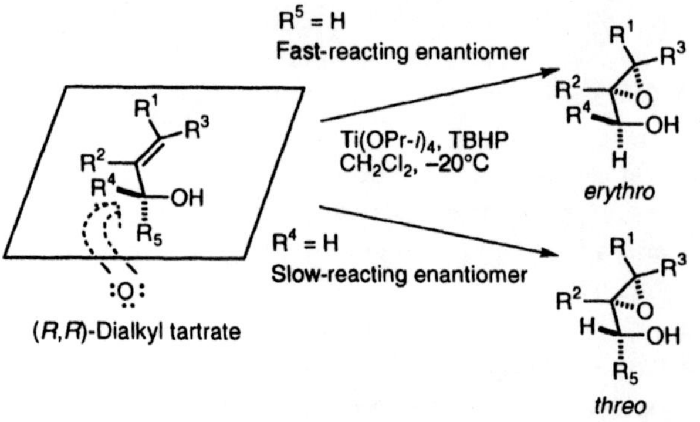

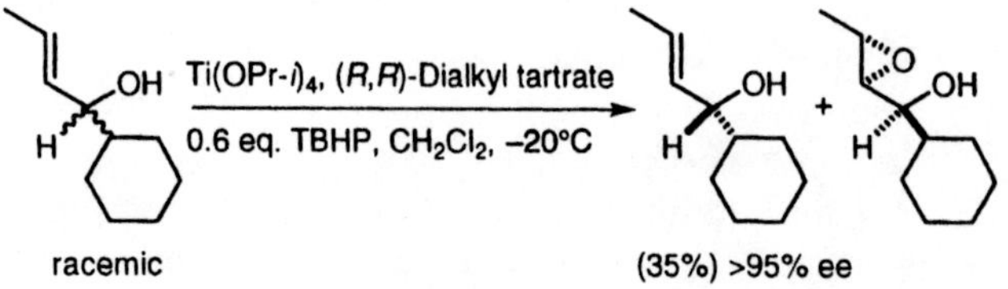

Scheme 2.3.9

Protocol 4.
(1*E*,3*R*)-1-Trimethylsilyl-1-octen-3-ol (Structure 7, Scheme 2.3.10) and (1*S*,2*S*,3*S*)-1-trimethylsilyl-1,2-epoxyoctan-3-ol (Structure 8, Scheme 2.3.10)[16]

Caution! The oxidant *t*-butyl hydroperoxide is susceptible to violent decomposition under the action of strong acids or certain metals. Titanium tetraisopropoxide decomposes very quickly when exposed to atmospheric moisture causing white fumes. Carry out all procedures in a well-ventilated hood and wear disposable gloves and safety goggles.

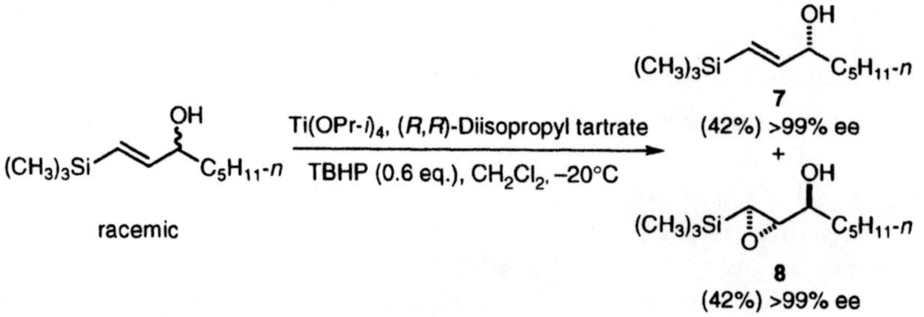

Scheme 2.3.10

2: Oxidation of the C=C bond

This is one example of a kinetic resolution procedure in which both the remaining allylic alcohol and the product formed can be obtained in high optical purity.

Equipment

- Three-necked, round-bottomed flask (1 L)
- Magnetic plate stirrer
- Syringes (50 and 100 mL)
- Teflon-coated magnetic stirrer bar
- A cooling bath (−20 °C)
- Pressure-equalizing addition funnel (250 mL)
- Separating funnel (1 L)
- Filter paper
- Filter funnel
- Erlenmeyer flask (1 L)
- Source of dry nitrogen or argon
- Glass chromatographic column (ca. 50 cm × 5 cm) fitted with a sintered glass disc (porosity 70–100 μm) at the base of the column
- Rubber septums
- A glass bubbler
- Thermometer (−40 → +40 °C)

Materials

- Activated 3-Å molecular sieves, 1 g   hygroscopic
- (E)-1-Trimethylsilyl-1-octen-3-ol, 3.12 g, 15.6 mmol   flammable
- Dry dichloromethane, 150 mL   harmful by inhalation
- Titanium tetraisopropoxide, 4.6 mL, 4.5 g, 15.6 mmol   fumes in air
- (R,R)-(+)-Diisopropyl tartrate, 3.94 mL, 4.4 g, 18.7 mmol[a]
- t-Butyl hydroperoxide, 3.5 M in dichloromethane, 4.46 mL, 15.6 mmol[b]   flammable
- Dimethyl sulfide, 3.94 mL, 5.1 g, 46.8 mmol   stench, flammable
- 10% Aqueous solution of tartaric acid, 280 mL   irritant
- Diethyl ether, 280 mL   flammable, explosive, harmful by inhalation
- Sodium chloride saturated solution, 100 mL
- Celite
- Sodium fluoride, 10.9 g, 260 mmol.
- Silica gel for chromatography (Merck, 0.015–0.04 mm)

Method

1. Fit and flame-dry the glass equipment as in Protocol 1 (Fig. 2.3.1) under a positive pressure of nitrogen or argon. Add to the flask, activated 3-Å molecular sieves (1 g), dichloromethane (140 mL) and titanium tetraisopropoxide (4.6 mL, 15.6 mmol) with stirring. Cool the mixture to −20 °C and add (R,R)-(+)-diisopropyl tartrate (3.94 mL, 18.7 mmol). Continue stirring for 10 min.

2. Prepare a solution of (E)-1-trimethylsilyl-1-octen-3-ol (3.12 g, 15.6 mmol) in 10 mL of dichloromethane and add it to the pressure-equalizing funnel. Add the solution slowly over a period of 5 min and stir for 15 min.

3. Transfer a solution of t-butyl hydroperoxide (3.5 M in dichloromethane, 4.46 mL, 15.6 mmol) to the pressure-equalizing addition funnel and add to the reaction mixture slowly over a period of 10 min.[c] Continue stirring for 7 h at −20 °C.

4. Transfer dimethyl sulfide (3.94 mL, 46.8 mmol) to the pressure-equalizing funnel and add slowly to the reaction mixture. Stir for 30 min at −20 °C.

Protocol 4. *Continued*

5. Add sequentially a 10% aqueous tartaric acid solution (280 mL), diethyl ether (280 mL), NaF (10.9 g), and Celite (6.2 g). Allow the mixture to warm up to room temperature and continue stirring for 30 min.
6. Filter through a pad of Celite, wash with diethyl ether, and concentrate on a rotary evaporator.
7. Purify using a column for chromatography of silica gel (0.015–0.04 mm) and a hexane–ethyl acetate mixture as eluent. Evaporate the eluates containing the products to afford 2.64 g (42%) of (1E,3R)-1-trimethylsilyl-1-octen-3-ol, $[\alpha]_D^{25} = -9.8°$ (c = 1.1, CHCl$_3$), and 2.75 g (42%) of (1S,2S,3S)-1-trimethylsilyl-1,2-epoxyoctan-3-ol, $[\alpha]_D^{25} = -7.5°$ (c = 1.04, CHCl$_3$).

[a] The choice of tartrate ester is critical in kinetic resolution since k_{rel} increases in the order DMT < DET < DIPT. In the catalytic procedure, dicyclohexyl tartrate (DCHT) and dicyclododecyl tartrate (DCDT) have shown greater enantioselectivity than DIPT, although DIPT is usually the first choice because it is most readily available and gives acceptable selectivity.
[b] Usually the kinetic resolution procedures are performed by using 0.6 equivalents of the peroxide. Thus, in order to ensure the desired conversion, the concentration of hydroperoxide should be measured accurately by iodometric titration.
[c] The syringe or cannula should never be inserted directly into the stock solution. Pour the estimated amount needed into a vessel, measure and transfer the amount needed and discard the remaining amount.

When prochiral or *meso* diallyl alcohols are used as substrates, only one of the four possible epoxides is formed. In these cases epoxidation occurs in preference at one of the two enantiotopic olefinic groups. The slowly formed stereoisomers react preferentially in a subsequent kinetic resolution step to give diepoxy alcohols, thus enhancing the optical purity of the desired epoxy alcohol (Scheme 2.3.11).[17]

Protocol 5.
(2R,3S)-1,2-Epoxy-4-penten-3-ol (Structure 9, Scheme 2.3.11)

Caution! The oxidant *t*-butyl hydroperoxide is susceptible to violent decomposition under the action of strong acids or certain metals. Titanium tetraisopropoxide decomposes very quickly when exposed to atmospheric moisture causing white fumes. Carry out all procedures in a well-ventilated hood and wear disposable gloves and safety goggles.

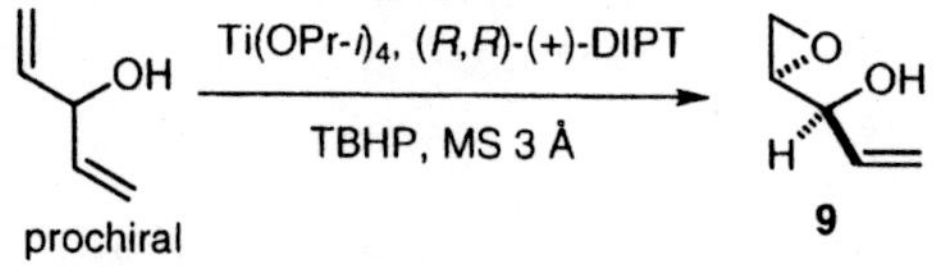

Scheme 2.3.11

2: Oxidation of the C=C bond

This is an example of catalytic asymmetric epoxidation of a prochiral divinylcarbinol.

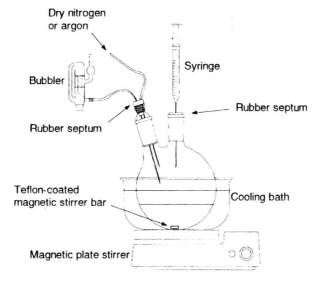

Fig. 2.3.3

Equipment

- Two-necked, round-bottomed flask (1 L)
- Magnetic plate stirrer
- Syringes (50 and 100 mL)
- Teflon-coated magnetic stirrer bar
- A cooling bath of dry ice–carbon tetrachloride ($-23\,°C$)
- Separating funnel (1 L)
- Filter paper
- Filter funnel

- Erlenmeyer flask (1 L)
- Source of dry nitrogen or argon
- Glass chromatographic column (ca. 50 cm × 5 cm) fitted with a sintered glass disc (porosity 70–100 μm) at the base of the column
- Rubber septums
- A glass bubbler
- Thermometer ($-40 \rightarrow +40\,°C$)

Materials

Activated 4-Å molecular sieves, 4.3 g	**hygroscopic**
Penta-1,4-dien-3-ol (divinyl carbinol), 20 g, 238 mmol	**flammable**
Dry dichloromethane, 250 mL	**harmful by inhalation**
Titanium tetraisopropoxide, 5 mL, 16.8 mmol	**fumes in air**
(R,R)-(+)-Diisopropyl tartrate, 4.5 mL, 21.5 mmol	
t-Butyl hydroperoxide, 3.0 M in isooctane, 160 mL, 480 mmol	**flammable**
Dimethyl sulfide, 3.94 mL, 46.8 mmol	**stench, flammable**
Saturated aqueous solution of sodium sulfate, 17 mL	**irritant**
Diethyl ether, 500 mL	**flammable, explosive, harmful by inhalation**
Sodium chloride saturated solution, 100 mL	
Celite	
Silica gel for chromatography (Merck, 0.015–0.04 mm)	

Method

1. Equip a two-necked round-bottomed flask (1 L) with a magnetic stirrer bar and two rubber septums, one of which is connected to a source of dry

Protocol 5. *Continued*

    nitrogen or argon, the other one to a bubbler (Fig. 2.3.3). Dry using a flame under positive pressure of inert gas. Add to the flask 4-Å molecular sieves (4.3 g) and dichloromethane (250 mL). Cool to $-23\,°C$ with stirring. Add titanium tetraisopropoxide (5 mL, 16.8 mmol) and (R,R)-(+)-diisopropyl tartrate (4.5 mL, 21.5 mmol) via syringe. Stir for a period of 10 min.

2. Add divinylcarbinol (20 g, 238 mmol) via cannula, and *t*-butyl hydroperoxide (3.0 M solution in isooctane, 160 mL, 480 mmol) in portions via syringe.

3. Remove all needles, place the reaction vessel in a $-15\,°C$ freezer and store for 118 h.

4. At the freezer temperature, with stirring, add a saturated aqueous solution of sodium sulfate (17 mL) and diethyl ether (250 mL). Allow the mixture to warm to room temperature and continue stirring for 2 h.

5. Filter the resulting slurry through a pad of Celite, washing with several portions of diethyl ether.

6. Transfer the Celite pad to an Erlenmeyer flask, and heat gently with diethyl ether. Filter the supernatant through a second fresh pad of Celite. Combine all the organic layers.

7. Remove carefully most of the solvent on a rotary evaporator with cooling to avoid undue loss of the somewhat volatile product.

8. Subject the resulting oil to flash chromatography on silica gel using a mixture of 3:1 pentane–diethyl ether followed by distillation at aspirator pressure to afford *ca.* 24 g of the desired epoxide, which should be about 50% pure by NMR.

9. Perform a second flash chromatography using a mixture of 3:1 pentane–diethyl ether. Remove the solvent carefully and distil under aspirator pressure to obtain 13.1 g (50%) of the product as a colourless oil, $[\alpha]_D^{25} = +48.8°$ ($c = 0.73$, $CHCl_3$).

Kinetic resolution can be extended to other secondary carbinol systems, which carry a group equivalent to olefin vicinal to the carbinol carbon (Scheme 2.3.12). Thus 2-furyl **10**[18] and 2-thienyl carbinols **11**[19] can be resolved efficiently in both stoichiometric[19] and catalytic manner.[18] α-Furfuryl toluenesulfonamide **12** is also resolved efficiently by using modified reaction conditions, although the stereochemistry is opposite to that observed for allylic alcohols.[20]

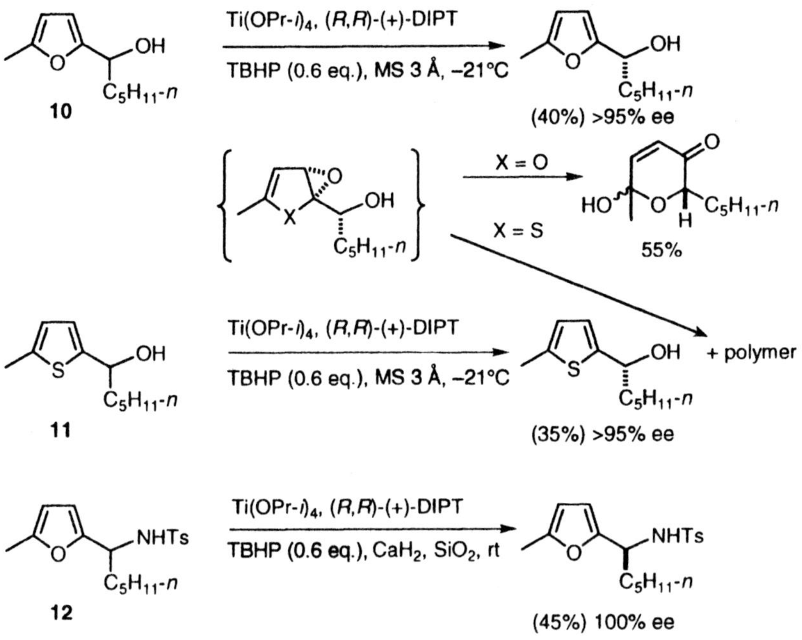

2: Oxidation of the C=C bond

Scheme 2.3.12

Protocol 6.
(*R*)-1-(2-Furyl)hexan-1-ol (Structure 13, Scheme 2.3.13)

Caution! The oxidant *t*-butyl hydroperoxide is susceptible to violent decomposition under the action of strong acids or certain metals. Titanium tetraisopropoxide decomposes very quickly when exposed to atmospheric moisture causing white fumes. Carry out all procedures in a well-ventilated hood and wear disposable gloves and safety goggles.

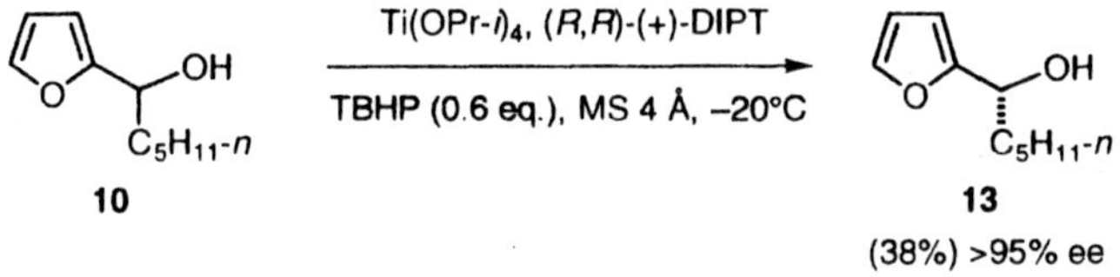

Scheme 2.3.13

This is an example of a catalytic kinetic resolution of a secondary furyl carbinol.

Equipment
- Three-necked, round-bottomed flask (500 mL)
- Magnetic plate stirrer
- Syringes (10 and 100 mL)
- Teflon-coated magnetic stirrer bar
- A cooling bath (−20 °C)
- Pressure-equalizing addition funnel (100 mL)
- Separating funnel (1 L)
- Filter paper

Protocol 6. Continued

- Filter funnel
- Erlenmeyer flask (1 L)
- Source of dry nitrogen or argon
- Glass chromatographic column (ca. 50 cm × 2.5 cm) fitted with a sintered glass disc (porosity 70–100 μm) at the base of the column
- Rubber septums
- A glass bubbler
- Thermometer ($-40 \rightarrow +40$ °C)

Materials

- Activated 4-Å molecular sieves, 5 g   *hygroscopic*
- 1-(2-Furyl)hexan-1-ol, 20.7 g, 123 mmol   *flammable*
- Dry dichloromethane, 120 mL   *harmful by inhalation*
- Titanium tetraisopropoxide, 7.35 mL, 24.7 mmol   *fumes in air*
- (R,R)-(+)-Diisopropyl tartrate, 6.23 mL, 29.6 mmol   *flammable*
- t-Butyl hydroperoxide, 4.35 M in dichloromethane, 17 mL, 74 mmol (0.6 eq.)   *flammable*
- Dimethyl sulfide, 5.4 mL, 74 mmol   *stench, flammable*
- 10% Aqueous tartaric acid solution, 5 mL
- Diethyl ether, 300 mL   *flammable, explosive, harmful by inhalation*
- Sodium fluoride, 30 g, 714 mmol
- Sodium chloride saturated solution, 100 mL
- 3 N Sodium hydroxide solution saturated with sodium chloride, 100 mL   *caustic*
- Celite
- Silica gel for chromatography (Merck, 0.015–0.04 mm)

Method

1. Equip a three-necked round-bottomed flask (500 mL) with a magnetic stirrer bar, a pressure-equalizing addition funnel (100 mL) stoppered with a rubber septum, and two rubber septums as in Protocol 1. In the septums connected to the flask and the addition funnel insert two syringe needles, one of which is connected to a source of dry nitrogen or argon, the other one to a bubbler (see Fig. 2.3.1). Dry using a flame or a high-temperature heating gun under a positive pressure of dry nitrogen or argon.

2. Add crushed 4-Å molecular sieves (5 g), dry dichloromethane (100 mL), and titanium tetraisopropoxide (7.35 mL, 24.7 mmol) via syringes. Cool down the mixture to -30 °C and stir. Add (R,R)-(+)-diisopropyl tartrate (6.23 mL, 29.6 mmol) via syringe and continue stirring for 10 min.

3. Prepare a solution of racemic 1-(2-furyl)hexan-1-ol (20.7 g, 123 mmol) in dichloromethane (20 mL) and transfer it to the pressure-equalizing addition funnel. Add the solution slowly over a period of 5 min and continue stirring for 30 min at -30 °C.

4. Transfer t-butyl hydroperoxide (4.35 M solution in dichloromethane, 17.0 mL, 74 mmol) to the pressure-equalizing addition funnel and add slowly to the reaction vessel over a period of 5 min. Stir the reaction mixture for 14 h at -21 °C.

5. Add dimethyl sulfide (5.4 mL, 74 mmol) slowly and continue stirring for 30 min at -21 °C.

6. Add a 10% aqueous tartaric acid solution (5 mL), diethyl ether (100 mL), and sodium fluoride (30 g, 714 mmol). Allow the mixture to warm to room temperature with vigorous stirring for a further 2 h. Filter the white precipitate through a pad of Celite and wash with diethyl ether (100 mL). Remove the solvent using a rotary evaporator to give an oil.

7. Dissolve the oil obtained in diethyl ether (100 mL), add aqueous 3 N sodium hydroxide solution saturated with sodium chloride (100 mL) and stir vigorously at 0°C for 30 min. Transfer the solution to a separating funnel (1 L).

8. Wash the organic solution with aqueous sodium chloride saturated solution (100 mL), dry over magnesium sulfate, and concentrate using a rotary evaporator to give an oil. Purify using a silica gel chromatographic column, and concentrate to yield 7.94 g of the product (38%), $[\alpha]_D^{25} = 13.8°$ (c = 1.07, $CHCl_3$).

References

1. Sheng, M. N.; Zajacek, J. *J. Org. Chem.* **1970**, *35*, 1839.
2. Sharpless, K. B.; Verhoeven, T. R. *Aldrichim. Acta* **1979**, *12*, 63.
3. Still, W. C. *J. Am. Chem. Soc.* **1979**, *101*, 2493.
4. Katsuki, T.; Sharpless, K. B. *J. Am. Chem. Soc.* **1980**, *102*, 5974.
5. *Merck Index* **12**, 1604.
6. Gao, Y.; Hanson, R. M.; Klunder, J. M.; Ko, S. Y.; Masamune, H.; Sharpless, K. B. *J. Am. Chem. Soc.* **1987**, *109*, 5765.
7. Finn, M. G.; Sharpless, K. B. In *Asymmetric Synthesis*; Morrison, J. D., ed.; Academic Press: Orlando, FL, **1985**, vol. 5, pp. 247–348.
8. Pfenninger, A. *Synthesis* **1986**, 89.
9. Rossiter, B. E. In *Asymmetric Synthesis*; Morrison, J. D., ed.; Academic Press: Orlando, FL, **1985**, vol. 5, pp. 193–246.
10. Katsuki, T. *J. Syn. Org. Chem. Jpn.* **1987**, *45*, 90.
11. Johnson, R.; Sharpless, K. B. In *Comprehensive Organic Synthesis*; Trost, B. M. ed.; Pergamon: Oxford, **1991**, vol. 7, Ch. 3, pp. 2–389.
12. Katsuki, T.; Martín, V. S. In *Organic Reactions*; Paquette, L. A., ed.; Wiley: New York, **1996**, vol. 48, Ch. 1, pp. 1–299.
13. Chong, J. M.; Wong, S. *J. Org. Chem.* **1987**, *52*, 2596.
14. Mori, K.; Ebata, T. *Tetrahedron* **1986**, *42*, 3471.
15. Martín, V. S.; Woodard, S.; Katsuki, T.; Yamada, Y.; Ikeda, M.; Sharpless, K. B. *J. Am. Chem. Soc.* **1981**, *103*, 6237.
16. Kitano, Y.; Matsumoto, T.; Sato, F. *J. Chem Soc., Chem. Commun.* **1986**, 1323–1325; Kitano, Y.; Matsumoto, T.; Sato, F. *Tetrahedron* **1988**, *44*, 4073–4086.
17. Schreiber, S. L.; Schreiber, T. S.; Smith, D. B. *J. Am. Chem. Soc.* **1987**, *109*, 1525.
18. Kusakabe, M.; Kitano, Y.; Kobayashi, Y.; Sato, F. *J. Org. Chem.* **1989**, *54*, 2085
19. Kitano, Y.; Kusakabe, M.; Kobayashi, Y.; Sato, F. *J. Org. Chem.* **1989**, *54*, 994.
20. Zhou, W.-S.; Lu, Z.-H.; Wang, Z.-M. *Tetrahedron Lett.* 1991, *32*, 1467.

I. Lantos

2.4 Enantioselective epoxidations of electron-deficient olefins

I. LANTOS

1. Phase-transfer catalysed epoxidation

Several reagents are effective for the epoxidation of electron-deficient olefins; most of them are hydroperoxides and function under basic conditions. Thus hydrogen peroxide,[1a] *t*-butyl hydroperoxide (TBHP),[1b] and potassium hypochlorite have all been used to carry out epoxidations of α,β-unsaturated carbonyl compounds. Mechanistically, the reaction is believed to involve nucleophilic addition of the hydroperoxide anion to the electron-deficient β-carbon followed by epoxide closure via elimination of hydroxide or alkoxide anion.[2] Since the reaction proceeds through a C_α—C_β partially single-bonded intermediate, conformational equilibration may occur to the thermodynamically most stable intermediate. Therefore, both *cis*- and *trans*-olefins often yield the same product.[3]

The first practical chiral phase-transfer catalysed epoxidation of α,β-unsaturated carbonyl compounds was reported in a series of publications[4] from Wynberg and colleagues using phase-transfer agents derived from the cinchona alkaloids quinine and quinidine. The alkaloids were alkylated by benzyl chloride giving benzylquinidinum (QUIBEC) or quininium chloride salts respectively. The enantioselectivities were low, of the order of 30–50%. The best results were achieved with **1b** reaching 55% ee (Scheme 2.4.1).

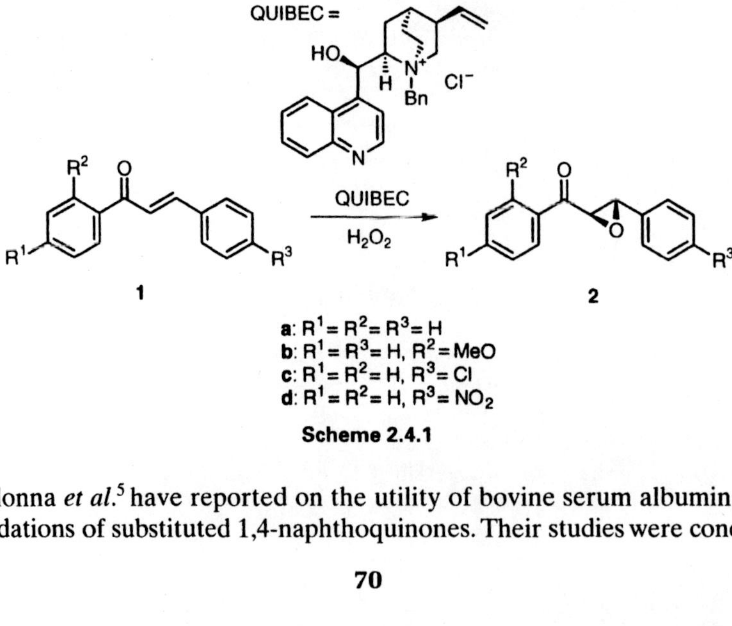

a: $R^1 = R^2 = R^3 = H$
b: $R^1 = R^3 = H$, $R^2 = MeO$
c: $R^1 = R^2 = H$, $R^3 = Cl$
d: $R^1 = R^2 = H$, $R^3 = NO_2$

Scheme 2.4.1

Colonna *et al.*[5] have reported on the utility of bovine serum albumin in the epoxidations of substituted 1,4-naphthoquinones. Their studies were conducted

2: Oxidation of the C=C bond

under heterogeneous phase-transfer conditions as well as in homogeneous aqueous solutions. t-BuO$_2$H or H$_2$O$_2$ was the oxidant and in some instances (i.e. with the addition of small amounts of organic cosolvents) workable levels of enantioselectivities were obtained (50–77% ee). The method never received wide application, however, because both the level and sense of the stereoselectivity were difficult to predict reliably.

The use of cyclodextrin was also examined by Colonna and Takahashi and their co-workers for the epoxidation of naphthoquinones[6] or chalcone.[7] Neither investigation afforded optical purities in excess of 41% ee.

A much improved practical method was reported by Julia and Vega[8a] in their first publication of polyamino acid-catalysed chiral epoxidations of chalcones **1a**. The catalysts were prepared from *N*-carboxyanhydride (NCA) precursors obtained by treating the amino acids with benzyl chloroformate and SOCl$_2$. Most of the polymerizations were carried out with n-BuNH$_2$ or H$_2$O$_2$ initiator but polymerization via the NCA-anions was also explored. The degree of polymerization was determined from the molar ratio of amino acid–NCA vs. the initiator.

Initially, three polyamino acids, poly-*S*-alanine, poly-*S*-β-benzyl-glutamate and poly-*S*-β-butyl-glutamate, were employed for the oxidation of **1a** in a triphasic media consisting of solvent–30% H$_2$O$_2$/aq. NaOH–polyamino acid. A 1:1 gram ratio of catalyst to substrate was found most efficient and of the several solvents toluene and CCl$_4$ were found most suitable. The study was extended[8e] to a much broader series of amino acids and some representative results are listed in Table 2.4.1.

Using poly-*S*-alanine, chemical yields were in the range 70–80% and optical yields were in the range 86–93%. The table indicates that: (1) amino acids Ala,

Table 2.4.1 Epoxidation of chalcone **1a** with various polyamino acids

Entry	Catalyst (MW)	Time (h)	Yield (%)	ee (%)
1	*S*-Ala (~1000)	28	75	93
2	*R*-Ala (~1000)	37	53	90
3	*R,S*-Ala (~1000)	24	5	0
4	*S*-Ala (~3000)	28	77	96
5	*S*-Val (~1000)	168	5.5	10
6	*S*-Val (~3000)	144	4	33
7	*S*-Lue (~1500)	28	60	84
8	*S*-Leu (~4500)	28	44	88
9	*S*-Isoleu (~1500)	72	76	95
10	*S*-Phe (~2000)	72	32	1
11	*S*-β-Bn-Asp	456	7.5	3
12	*S*-β-Bn-Glu	144	12	11.6
13	[*S*-Leu-*S*-Ala]$_{10}$	24	67	95
14	[*S*-Val-*S*-Ala]$_{10}$	96	39	88
15	*S*-Ala-*S*-Ala	144	41	2

Table 2.4.2 Poly-S-leucine and poly-R-leucine catalysed epoxidations of chalcones[8b,11,14a,b]

Entry	R¹	R²	Time (h)	Solvent	Yield (%)	ee (%)
1	p-NO$_2$-Ph	Ph	48	toluene	83	82
2	Ph	2-Thio	48	toluene	96	80
3	Ph	3-Thio	48	toluene	30	70
4	Ph	o-Me-Ph	48	toluene	54	50
5	p-Cl-Ph	Ph	48	toluene	47	66
6	Ph	p-Me-Ph	49	CCl$_4$	87	92
7	Ph	2-Naphthyl	40	CCl$_4$	83	99
8	o-(8-Ph)octyl-Ph	p-Me-Ph	40	hexane	76	94
9	o-(8-Ph)octyl-Ph	2-Naphthyl	24	hexane	76	94
10	Ph	Isobutyl	40	hexane	67	40
11	Ph	Cyclohexyl	40	hexane	70	65
12	Ph	n-Dodecyl	40	hexane	57	85
13	2-Pyridyl	Ph	16	CH$_2$Cl$_2$	84	72
14	4-Pyridyl	2-Naphthyl	16	CH$_2$Cl$_2$	67	96
15	2-Furyl	2-Naphthyl	50	CH$_2$Cl$_2$	75	96
16	Ph	i-Pr	168	CH$_2$Cl$_2$	60	62
17	2-Naphthyl	Cyclo-Pr	38	CH$_2$Cl$_2$	56	90
18	2-Quinolyl	Cyclo-Pr	18	CH$_2$Cl$_2$	94	79
19	PhCH=CH	Cyclo-Pr	42	CH$_2$Cl$_2$	73	74
20	PrCH=CH	Cyclo-Pr	19	CH$_2$Cl$_2$[a]	52	98
21	Bu	Ph	18	CH$_2$Cl$_2$	85	90
22	Cyclo-Pr	2-Naphthyl	28	CH$_2$Cl$_2$	73	98
23	Ph	Ph-alhynyl	96	CH$_2$Cl$_2$	57	90
24	Ph	t-Bu	15	toluene[b]	75	95
25	t-Bu	t-Bu	15	toluene[b]	100	95
26	Ph	Ot-Bu	15	toluene[b]	66	95

[a] NaBO$_3$ was the oxidant.
[b] Polystyrene-immobilized poly-S-Leu was the catalyst.

Leu, and isoLeu were the most efficient catalysts, and among these the higher homologues (entries 4 and 8) consistently afforded faster reaction times and higher ee than the shorter homopolymers; (2) as expected, the R-amino acids provided *identical results* but *of the opposite absolute stereochemistry*; (3) poly-Val was not an efficient catalyst; (4) the mixed polymers afforded no improvement over the homopolymeric materials (entries 13 and 14); (5) recycling the catalysts led to loss of catalytic activity,[8c] a phenomenon observed more with the lower molecular weight materials. It was also noted that the soluble S-Ala-S-Ala afforded racemic materials. The polyamino acids also functioned when anchored on a polystyrene matrix affording similar high yields and ee.

The effect of the chalcone substitution pattern was also examined by Julia et al.[8d] who prepared substituted chalcones (entries 1–5 in Table 2.4.2). They found that most of the compounds were well behaved under the standard procedure affording the epoxides in 55–95% isolated yields and 80–86% ee. The more electron-rich substrates had consistently lower chemical yields (entries 3

2: Oxidation of the C=C bond

and 4). Attempted application of the method to quinones, however, did not meet with much success.

Colonna et al.[8e] attributed the chiral selectivity to hydrogen-bonded interactions of the substrate with the α-helical polyamino acid catalysts. This interpretation was later supported by Carriere and co-workers[10] in their studies of chalcone epoxidations with acrylamide–methacrylamide copolymer-anchored polyamino acids.

The approach was later re-examined by Flisak and co-workers[11] with the intention of developing this epoxidation procedure into an efficient synthesis of glycidic acid esters **3** via a Bayer–Villiger transformation (Scheme 2.4.2).

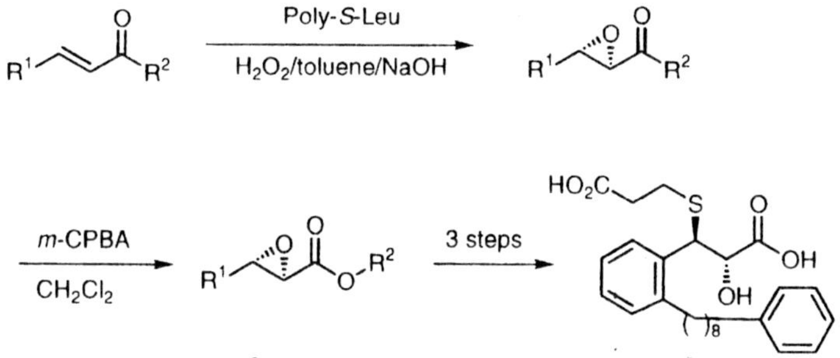

Scheme 2.4.2

The catalyst was prepared by the NCA approach but, unable to carry out solution-phase polymerization, these workers initiated solid-phase polymerization in a humidity chamber.

The catalyst obtained in this way had a molecular weight in the 20,000–30,000 range. On account of the higher molecular weight, these catalysts were found to be reusable with very little loss of ee. The epoxidations were carried out using poly-S-Leu catalyst in hexane and/or CCl_4 solvents. Additionally, in order to shorten the reaction time and inhibit the slow decomposition of the peroxide caused by minute amounts of Fe^{2+}, EDTA was added to the reaction. Representative results of the epoxidation experiments are listed in Table 2.4.2 (entries 6–9).

Protocol 1.
Solid-phase polymerization of NCAs and preparation of (2R-trans)-(2-naphthalenyl)-3-phenyloxiranylmethanone (entry 7 in Table 2.4.2)[11]

Caution! Carry out all procedures in a well-ventilated hood and wear disposable gloves and safety goggles.

Equipment
- Wide-mouthed bottle (1 L)
- Thermometer (0–100 °C)
- Addition funnel (1 L)
- Separatory funnel (4 L)

Protocol 1. Continued

- Humidity chamber or a room with at least 70% humidity
- Three-necked round-bottomed flask (12 L)
- Mechanical stirrer equipped with a shaft and Teflon-coated stirring paddle

Materials

- S-Leucine–NCA — hygroscopic, irritant
- Sodium hydroxide — corrosive, toxic
- Ethylenediaminetetraacetic acid — irritant
- Naphthalchalcone[a]
- Aqueous hydrogen peroxide (30%) — oxidant, irritant
- Hexanes — flammable
- Ethyl acetate — flammable
- Magnesium sulfate anhydrous

Method

1. Place S-leucine–NCA[9] (672 g, 4.27 mol) in a wide-mouthed bottle to a depth of about 6–8 cm.
2. Place the uncapped bottle in a humidity chamber maintained at 70% humidity and 25°C.
3. Follow the progress of the polymerization by the disappearance of the anhydride carbonyl bands at 1830 and 1780 cm^{-1}. Complete polymerization usually requires 7–9 days.
4. Charge a 12-L three-necked round-bottomed flask, equipped with a mechanical stirrer, pressure-equalizing addition funnel, and a thermometer, with 440 mL of deionized water.
5. Add 172.6 g (4.31 mol) of sodium hydroxide and initiate stirring and cool the solution in an ice bath until the internal temperature is stabilized at 10°C.
6. Sequentially charge the flask with 148.5 g of poly-S-leucine, 3.5 g (0.012 mol) of ethylenediaminetetraacetic acid, 2.7 L of hexanes, and 100.0 g (0.387 mol) of the naphthalchalcone.[a]
7. Allow the mixture to stir for 24 h at ambient temperature.
8. Add to the reaction via the addition funnel 780 mL of 30% hydrogen peroxide at such a rate that the internal temperature does not exceed 40°C. Once addition is complete, continue stirring the triphasic mixture at room temperature for 24 h.
9. Check for the presence of starting material. If still present as indicated by TLC, add an additional 200 mL of 30% hydrogen peroxide and stir the reaction for an additional 24 h.
10. When the reaction reaches completion, add 2.0 L of ethyl acetate and continue stirring for 1 h at room temperature.
11. Filter off the catalyst and wash with 500 mL of hot ethyl acetate.

2: Oxidation of the C=C bond

12. Transfer the combined filtrates to a separatory funnel, separate the layers and wash the organic layer with water (2 × 1 L), brine (1 L) and dry over magnesium sulfate.
13. Filter the solution and concentrate the filtrate by rotary evaporation to yield an off-white solid. The solid is dried at 60 °C under vacuum to yield 90.3–105.0 g (85–98%) of the crude epoxide product that has an optical purity[b] of 81–84% ee compared with the corresponding racemate.[c]

---

[a] The naphthalchalcone was prepared from benzaldehyde (Aldrich Chemical Company) and 2-acetonaphthone (Aldrich Chemical Company) according to Ref. 19. Use of commercial 25% sodium methoxide (Aldrich Chemical Company) in place of sodium metal works just as well. The product was precipitated from the reaction mixture by the addition of deionized water, collected by filtration, and dried at 50 °C under vacuum.
[b] The optical purity was determined via HPLC using a Daicel Chilracel OA column with a mobile phase of hexane–2-propanol (1:1) at a flow rate of 2.0 mL/min and a wavelength of 211 nm.
[c] The racemate was prepared by two-phase epoxidation of the naphthalchalcone with 30% H_2O_2 in toluene using a phase-transfer catalyst (Aliquat 336).

The nature of the substituents was dictated by fine-tuning the migratory behaviour of the substituted ketones during the ensuing Bayer–Villiger transformation as well as the ease of purification and optical purity enrichment of the products by crystallization. The epoxidation proceeded uniformly for all of the compounds in 70–85% chemical yield and enantioselectivities in the 85–96% range. It was even more gratifying to see that the ensuing Bayer–Villiger reaction proceeded selectively and the required glycidates **3** were delivered in 60–80% yield and 90–99% ee. The naphthyl-substituted substrates were especially easy to work with on account of the highly crystalline property engendered by the 2-naphthyl substituent. Thus, these substrates (entries 8 and 9, Table 2.4.2) were found to serve as excellent intermediates in the large-scale chemical synthesis of drug candidate **4** by Novack and co-workers.[12]

The process was also extended by Flisak[11b] to the alkyl aryl ketones (entries 10–12, Table 2.4.2). These compounds were difficult to purify by crystallization and were isolated by flash chromatography. As seen from the data, none of the α,β-unsubstituted aliphatic ketones reproduced the efficiency observed with the chalcones. Nevertheless, synthetically useful enantioselectivities could be obtained for some of the compounds. Bayer–Villiger oxidation, however, proceeded uneventfully affording high yields and no loss of optical purity.

In a series of reports, the method was applied to electron-rich polyoxygenated chalcones **5a–e** by Ferreira and Bezuidenhoudt and co-workers.[13] The epoxidations were carried out with poly-S-Ala or poly-R-Ala in triphasic mixtures identically to the previous experiments. The optical purities of the epoxides were assessed by chiral shift reagents and were in the range 40–84% (Scheme 2.4.3).

The stereochemical assignment of the epoxy ketones rested on CD correlations with **2a** whose absolute stereochemistry had been established previously.[4]

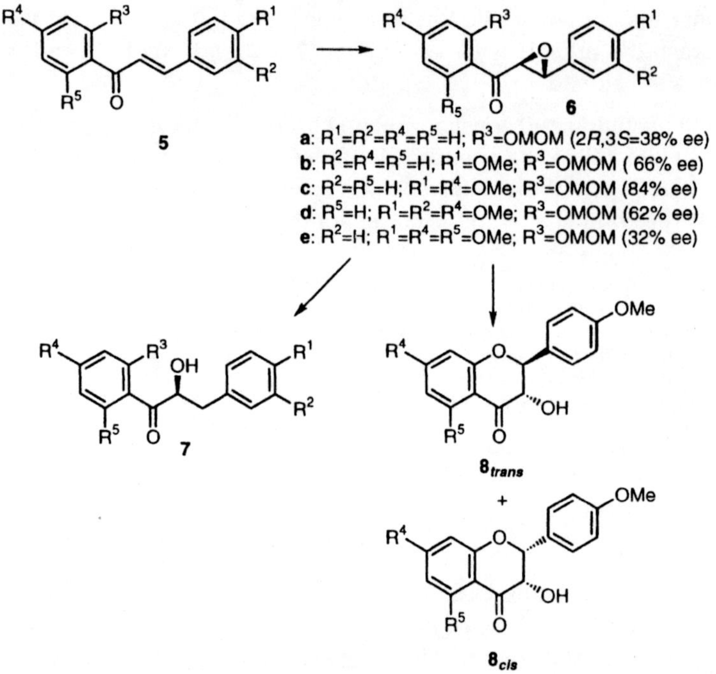

Scheme 2.4.3

a: $R^1=R^2=R^4=R^5=H$; $R^3=OMOM$ (2R,3S=38% ee)
b: $R^2=R^4=R^5=H$; $R^1=OMe$; $R^3=OMOM$ (66% ee)
c: $R^2=R^5=H$; $R^1=R^4=OMe$; $R^3=OMOM$ (84% ee)
d: $R^5=H$; $R^1=R^2=R^4=OMe$; $R^3=OMOM$ (62% ee)
e: $R^2=H$; $R^1=R^4=R^5=OMe$; $R^3=OMOM$ (32% ee)

The experiments led to the conclusion that *ortho*-substitution adversely affected the stereoselectivity of the reaction. Compounds **6a–e** served as precursors to α-hydroxy-dihydrochalcones **7**[13b] and flavanoids[13d] **8**$_{cis}$ and **8**$_{trans}$ as shown in Scheme 2.4.3.

The overall scope and potential of the epoxidation procedure was greatly expanded by Roberts and co-workers.[14] In a major departure from the previously employed substituted chalcone-type substrates, these workers extended the reaction to ketones with aliphatic substituents, C_2 symmetrical α,β-unsaturated ketones, dienes, and bis-dienes. The polyamino acid catalysts were prepared either by solid-state NCA polymerization or by polymerization in CH_2Cl_2 solutions with 1,3-diaminopropane initiator. In addition, an excellent procedure was developed for immobilized catalyst preparation.[15,16]

The oxidant was 30% H_2O_2 but very efficient epoxidations could also be achieved with sodium perborate or percarbonate containing minute amounts of phase-transfer agents. As seen from Table 2.4.2, and noted by the authors, successful epoxidations were observed for ketones containing the 2-, 4-pyridyl or 2-furyl substituents (entries 13–15), which is noteworthy considering the sensitive nature of these heterocycles towards oxidations. It is also of note that the reaction was carried out successfully on aliphatic ketones, as shown by the results on the *t*-butyl ketones (entries 24 and 25). Cyclopropyl aryl ketones undergo the oxidation efficiently but isopropyl substrates are not transformed.

2: Oxidation of the C=C bond

Highly insightful results were obtained from the epoxidations of dienyl or tetraenyl substances. As the structures **9, 10, 11,** and **12** indicate, all epoxidations afforded the 2R,3S structures. The high degree of selectivity is a very significant result as the formation of *meso*-structures is also possible, and occurred in a 4:1 ratio of related structures.[14a] Mechanistically, these epoxides give rise to the very interesting question of how stereoselective delivery of the oxygen occurs in molecules **9, 10,** and **11** onto opposite faces of the initial double bonds.

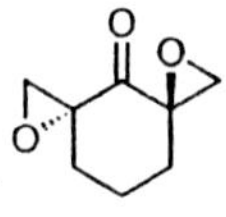

9

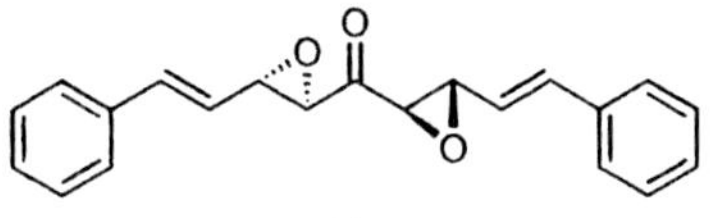

10

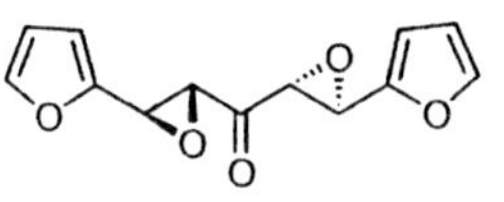

11

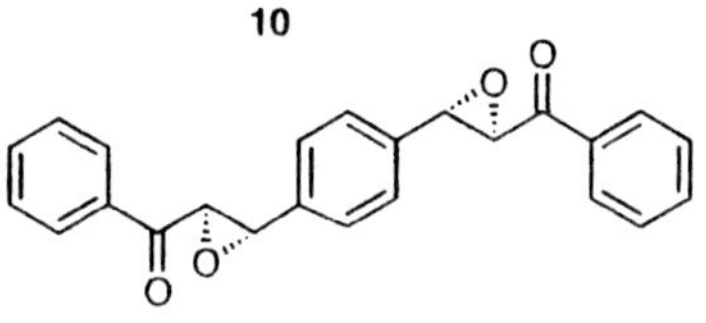

12

Most recently, Roberts and co-workers[15] have devised a convenient new non-aqueous protocol for the reaction. Urea–hydrogen peroxide complex is used as oxidant in THF solution containing a tertiary amine base. Reaction times are much reduced without loss of ee of products.*

2. Epoxidations promoted by organometallic agents

Several new approaches have emerged from the literature employing organometallic complexes as catalyst in asymmetric epoxidation of electron-deficient systems. Jacobsen *et al.*[17] applied the (salen)Mn complex to the epoxidation of *cis*-ethyl cinnamate (Scheme 2.4.4).

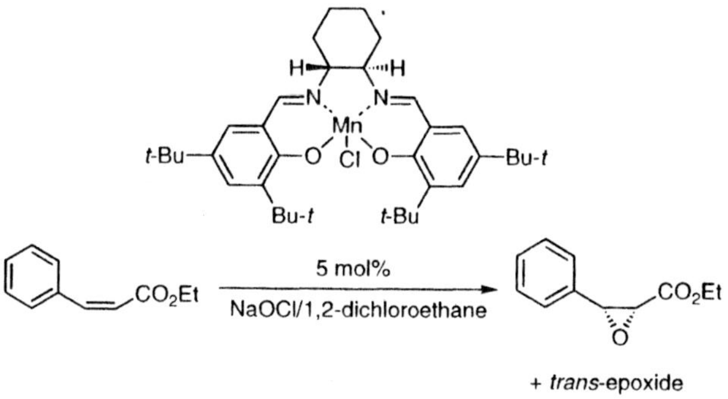

Scheme 2.4.4

*We are grateful to Professor Roberts and Dr Nugent for sending us a previously unpublished procedure, which unfortunately could not be included on account of space limitations.

I. Lantos

The yield of the mixture of *cis*- and *trans*-epoxides (4:1) was 82 to 98% depending on the additive (mostly pyridine *N*-oxides). Enantiopurities, however, have been very high (~93%) for both isomers.

Another very efficient and general process belonging to this category was reported by Shibasaki and co-workers[18] employing chiral lanthanoid complexes in the epoxidation of α,β-unsaturated ketones. The chemistry is shown in Scheme 2.4.5.

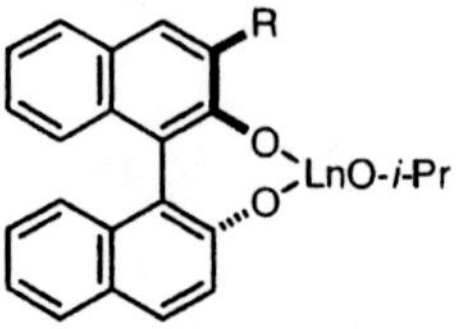

La-13: Ln = La, R = H
La-14: Ln = La, R = CH$_2$OH

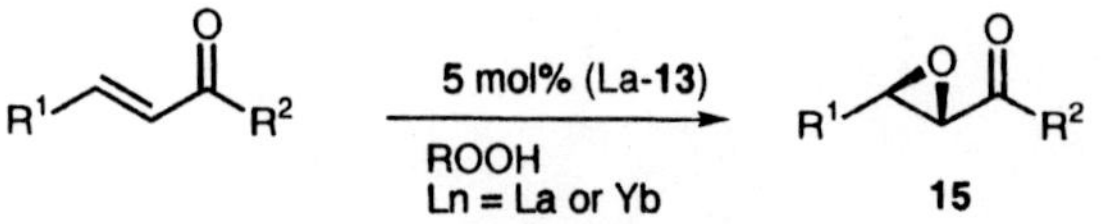

5 mol% (La-13)
ROOH
Ln = La or Yb

Scheme 2.4.5

The complexes were prepared from (*R*)-1,1'-bi-2-naphthol [(*R*)-BINOL] or (*R*)-3-hydroxymethyl-BINOL ligands, *t*-BuOOH or cumene hydroperoxide (CHP) served as the oxidant, and the reactions proceeded at room temperature. It was found that ketones with aryl substituents, i.e. either R^1 or R^2 aryl, were best epoxidized with La metal (ee in the range 83–94%), while those with aliphatic substituents [i.e. R^1 or R^2 being (CH$_2$)$_n$] were best converted with Yb metal. The absolute configuration of the products was α*S*,β*R*.

Protocol 2.
Catalytic asymmetric epoxidation of α,β-unsaturated ketones using lanthanoid–BINOL complexes (Structure 15: R^1 = *i*-Pr, R^2 = Ph in Scheme 2.4.5)

Caution! Carry out all procedures in a well-ventilated hood and wear disposable gloves and safety goggles.

Equipment
- Test-tube equipped with a magnetic stirring bar and three-way stopcock
- Inert gas supply
- Dry syringes and stainless-steel needles

Materials
- 4-Å Molecular sieves (MS)[a]
- Dry distilled THF[b]   flammable, liquid, irritant
- (*R*)-BINOL (FW 286.33); used as 0.1 M solution in THF   irritant
- Lanthanum triisopropoxide (FW 316.19)[c]; used as 0.2 M solution in THF   irritant, hygroscopic

2: Oxidation of the C=C bond

- Cumene hydroperoxide (FW 152.19)[d]; used as ca. 3 M solution in toluene   oxidant, toxic
- Ammonium chloride (FW 53.49)
- trans-4-Methyl-1-phenyl-2-penten-1-one (FW 174.24)

Method

1. Place in the test-tube MS 4-Å (80 mg), THF (1.8 mL), THF solution of (R)-BINOL (100 μL, 0.01 mmol), and La(O-i-Pr)$_3$ solution in THF (50 μL, 0.01 mmol).
2. Stir the mixture for 30 min at room temperature.
3. While maintaining room temperature, add trans-4-methyl-1-phenyl-2-penten-1-one (35 mg, 0.2 mmol) to the mixture and stir for an additional 5 min.
4. Add the toluene solution of CHP (ca. 100 μL, 0.3 mmol) in one portion and continue stirring the reaction mixture for 12 h.
5. Quench the reaction mixture with saturated aqueous NH$_4$Cl (1 mL) and extract with EtOAc (2 × 20 mL). Wash the combined extracts with brine, dry over anhydrous Na$_2$SO$_4$ and concentrate to afford an oily residue.
6. The trans-(2S,3R)-epoxy-4-methyl-1-phenylpentan-1-one is purified by flash chromatography (SiO$_2$, AcOEt–hexane 1:20). The yield is 35 mg (93%). The enantiopurity of this product is 86% ee as determined by chiral HPLC analysis (Daicel Chiralcel OD, i-PrOH–hexane (2:98), 1.0 mL/min flow rate, R$_t$ = 16.2 and 18.7 min for 2S,3R and 2R,3S respectively. The use of La-**14** complex affords 94% ee.

[a] MS 4-Å is used after drying at 180°C for 3 h under reduced pressure.
[b] Freshly distilled from benzophenone ketyl under an inert atmosphere.
[c] Obtained from Kojundo Chemical Co. Ltd., Japan.
[d] Pure CHP is obtained by the method given in Ref. 20.

References

1. (a) Weitz, E.; Scheffer, A. *Chem. Ber.* **1921**, *54*, 2327. (b) Yang, N. C.; Finnegan, R. A. *J. Am. Chem. Soc.* **1958**, *80*, 5845.
2. House, H. In *Modern Synthetic Reactions*, 2nd edn; W. A. Benjamin Inc.: Menlo Park, California, **1972**, p. 306.
3. Bauer, T. In *Stereoselective Synthesis*, vol. 8, Helmchen, G.; Hoffmann, R. W.; Mulzer, J.; Schaumann, E., ed., Georg Thieme Verlag: Stuttgart, **1996**, p. 4649.
4. (a) Helder, R.; Hummelen, J. C.; Laane, R. W. P. M.; Wiering, J. S.; Wynberg, H. *Tetrahedron Lett.* **1976**, 1831. (b) Wynberg, H.; Greijdanus, B. *J. Chem Soc., Chem. Commun.* **1978**, 427. (c) Marsman, B.; Wynberg, H. *J. Org. Chem.* **1979**, *44*, 2312.
5. (a) Colonna, S.; Gaggero, N.; Manfredi, A.; Spadoni, M.; Casella, L.; Carrera, G.; Pasta, P. *Tetrahedron* **1988**, *44*, 5169. (b) Colonna, S.; Manfredi, A.; Annunziata, R.; Spadoni, M. *Tetrahedron* **1987**, *43*, 2157.
6. Colonna, S.; Manfredi, A.; Annunziata, R.; Gaggero, N. *J. Org. Chem.* **1990**, *55*, 5862.

7. Hu, Y.; Harada, A.; Takahashi, S. *Synth. Commun.* **1988**, *18*, 1607.
8. (a) Julia, S.; Vega, J. C. *Angew. Chem., Int. Ed. Engl.* **1980**, *19*, 929. (b) Julia, S.; Guixer, J.; Masana, J.; Rocas, J.; Colonna, S.; Annuziata, R.; Molinary, H. *J. Chem. Soc., Perkin Trans. I* **1982**, 1317. (c) Banfi, S.; Colonna, S.; Molinary, H.; Julia, S.; Guixer, J. *Tetrahedron* **1984**, *40*, 5207. (d) Julia, S.; Guixer, J.; Masana, J.; Rocas, J.; Colonna, S.; Annuziata, R.; Molinari, H. *J. Chem. Soc., Perkin Trans. I* **1982**, 1317. (e) Colonna, S.; Molinary, H.; Banfi, S.; Julia, S.; Masana, J.; Alvarez, A. *Tetrahedron* **1983**, *39*, 1635.
9. Konopinska, D.; Sunion, J. Z. *Angew. Chem., Int. Ed. Engl.* **1967**, *6*, 248.
10. Boulahia, J.; Carriere, F.; Sekiguchi, H. *Macromol. Chem.* **1991**, *192*, 2969.
11. (a) Baures, P. W.; Eggleston, D. S.; Flisak, J. R.; Gombatz, K.; Lantos, I.; Mendelson, W.; Remich, J. *Tetrahedron Lett.* **1990**, *31*, 6501. (b) Flisak J. R. Asymmetric epoxidation of aliphatic and aromatic α,β-unsaturated naphthyl ketones and subsequent conversion to glycidic esters. Presented at the 48th Southwest Regional ACS Meeting, October 21–23, **1991**.
12. Flisak, J. R.; Gombatz, K. J.; Holmes, M. M.; Jarmas, A. A.; Lantos, I.; Mendelson, W. L.; Novack, V. J.; Remich, J. J.; Snyder, L. *J. Org. Chem.* **1993**, *58*, 6247.
13. (a) Bezuidenhoudt, B. C. B.; Swanepoel, A.; Augustyn, J. A. N.; Ferreira, D. *Tetrahedron Lett.* **1987**, *28*, 4857. (b) Augustyn, J. A. N.; Bezuidenhoudt, B. C. B.; Swanepoel, A.; Ferreira, D. *Tetrahedron* **1990**, *46*, 2651. (c) Augustyn, J. A. N.; Bezuidenhoudt, B. C. B.; Swanepoel, A.; Ferreira, D. *Tetrahedron* **1990**, *46*, 4429. (d) van Rensburg H., van Heerden, P. S.; Bezuidenhoudt, B. C. B.; Ferreira, D. *Chem. Commun.* **1996**, 2747.
14. (a) Lasterra-Sanchez, M. E.; Roberts, S. M. *J. Chem. Soc., Perkin Trans. I* **1995**, 1467. (b) Lasterra-Sanchez, M. E.; Felfer, U.; Mayon, P.; Roberts, S. M.; Thornton, S. R.; Todd, C. J. *J. Chem Soc., Perkin Trans. I* **1996**, 343. (c) Kroutil, W.; Mayon, P.; Lasterra-Sanchez, M. E.; Maddrell, S. J.; Roberts, S. M.; Thornton, S. R.; Todd, C. J.; Tuter, M. *Chem. Commun.* **1996**, 845. (d) Kroutil, W.; Lasterra-Sanchez, M. E.; Maddrell, S. J.; Mayon, P.; Morgan, P.; Roberts, S. M.; Thornton, S. R.; Todd, C. J.; Tuter, M. *J. Chem Soc., Perkins Trans. I* **1996**, 2837.
15. Bentley, P. A.; Bergeron, S.; Cappi, M. W.; Hibbs, D. E.; Hursthouse, M. B.; Nugent, T. C.; Pulido, R.; Roberts, S. M.; Wu, L. E. *Chem. Commun.* **1997**, 739.
16. Itsuno, S.; Sakakura, M.; Ito, K. *J. Org. Chem.* **1990**, *55*, 6047.
17. Jacobsen, E. N.; Deng, L.; Furukawa, Y.; Martinez, L. E. *Tetrahedron* **1994**, *50*, 4323.
18. Bougachi, M.; Watanabe, S.; Arai, T. Sasai, H.; Shibasaki, M. *J. Am. Chem. Soc.* **1997**, *119*, 2329.
19. Kohler, E. P., Chadwell, H. M. *Org. Synth. Coll. Vol. 1* **1941**, 78.
20. Perrin, D. D; Armarego, W. L., ed.; *Purification of Laboratory Chemicals*, 3rd edn; Pergamon Press: New York, **1988**.

2.5 Asymmetric dihydroxylation

H. BECKER and K. B. SHARPLESS

1. Introduction

In recent years, catalytic asymmetric dihydroxylation (the 'AD reaction') has become one of the most widely used practical accesses to chiral compounds.[1] The high regio- and stereoselectivity for a broad range of substrates are the most outstanding characteristics of this reaction.

The development of the OsO_4-mediated dihydroxylation of olefins towards a highly efficient synthetic method had already started in the 1930s. However, the stoichiometric osmylation developed by Criegee is not satisfactory regarding economy and ecology. In his pioneering work Criegee also observed the accelerating influence of tertiary amines on the reaction of osmium(VIII) oxide with olefins. It is therefore not surprising that catalytic methods were developed using cheap organic or inorganic cooxidants to reoxidize the resulting Os(VI) glycolates. The most widely used procedures use t-butyl hydroperoxide,[2] N-methylmorpholine N-oxide (Upjohn Process),[3] or $K_3Fe(CN)_6$ in the presence of K_2CO_3.[4]

The potential of the osmylation reaction to introduce one or two chiral centres in one step in a stereocontrolled way initiated the search for chiral amines to allow for the asymmetric dihydroxylation of a prochiral olefin.[5]

Derivatives of chinchona alkaloids, first introduced by Hentges and Sharpless, have proven to be (by far) the best system for catalytic asymmetric dihydroxylation. Derivatives of dihydroquinine (DHQ) and dihydroquinidine (DHQD) gave the most promising results. Those alkaloids are relatively cheap and, more important, their derivatives behave almost like enantiomers within this catalytic system. Therefore, both enantiomeric diols are accessible. Mechanistic insights allowed further development of the process[6] and development of other derivatives of chinchona alkaloids[7] increased the scope of the reaction.

A breakthrough was made by the discovery that ligands with two independent cinchona alkaloid units lead to dramatically increased enantioselectivities and rates (Scheme 2.5.1). Ligands with a phthalazine {(DHQD)$_2$PHAL and (DHQ)$_2$PHAL},[8] a diphenylpyrimidine {(DHQD)$_2$PYR and (DHQ)$_2$PYR},[9] and an anthraquinone core {(DHQD)$_2$AQN and (DHQ)$_2$AQN}[10] are now part of the catalytic system of choice for the asymmetric dihydroxylation of almost all types of olefins. Today, most olefins can be oxidized in an easy and convenient way to give diols with an ee >90%.

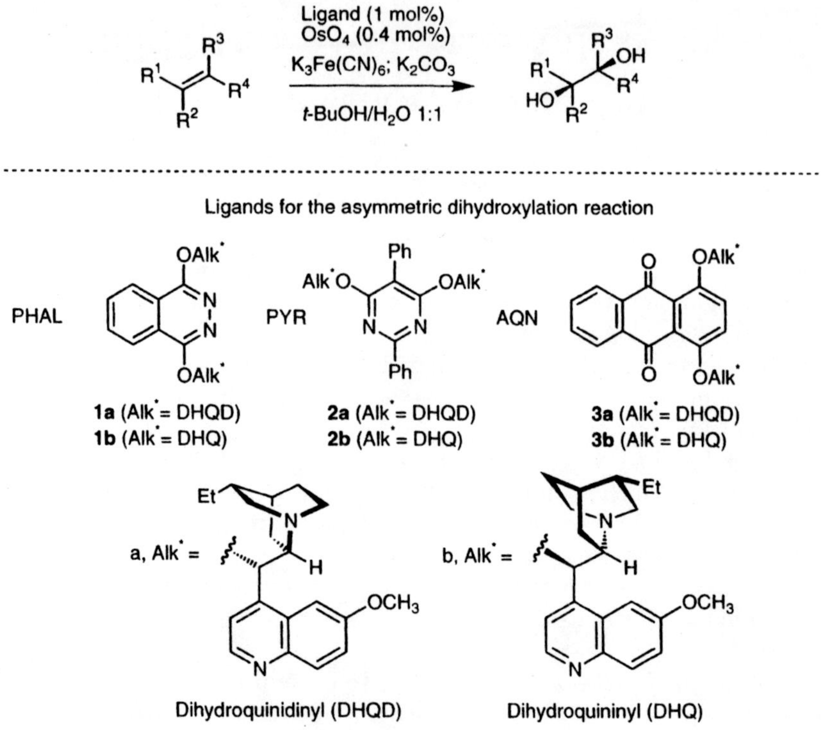

Scheme 2.5.1

2. Scope and limitations

The overall picture of the catalytic asymmetric dihydroxylation reaction using derivatives of chinchona alkaloids can be summarized as follows:

1. The reaction is stereospecific, resulting in 1,2-*cis*-addition of two OH groups to the olefin; those 1,2-diols provide access to synthetically very interesting 1,2-difunctionalized intermediates (a review of the synthetic applications of 1,2-diols can be found in Ref. 1). Amazingly, the reaction never makes mistakes, i.e. the product is always a diol derived from *cis*-addition, and side-products, such as epoxides or *trans*-diols, are never observed.
2. The reaction does not require any directing or sterically demanding group to yield high values for ee (e.g. *trans*-2-butene yields an ee of 72% with (DHQD)$_2$PHAL as ligand). Almost all olefins react in the process.
3. It typically proceeds with high chemoselectivity and enantioselectivity. Chemically different double bonds in polyunsaturated compounds can often be differentiated.[11]
4. The face selectivity is reliably predicted using a simple 'mnemonic device' (Scheme 2.5.2) and exceptions are very rare and are only observed in cases yielding low values for ee.[12]

2: Oxidation of the C=C bond

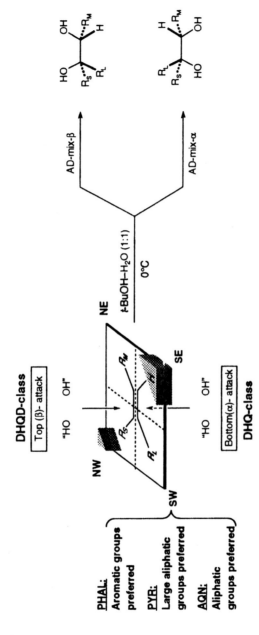

Scheme 2.5.2 The mnemonic device for predicting the face selectivity of the asymmetric dihydroxylation reaction.

The 'mnemonic device' can be applied to all olefins using a simple set of rules. The substituents of the olefin are fitted into the four quadrants of the 'mnemonic device' in such a way that the following requirements are fulfilled:

- The SE quadrant is sterically inaccessible, only hydrogen can be placed here (except for 1,2-*cis* and tetrasubstituted olefins).
- The NW quadrant allows larger substituents than the SE quadrant.
- The SW quadrant accepts the largest groups; in the case of aromatic groups this quadrant is very attractive (especially PHAL, DPP, and Ph-PHAL ligands). It also can be attractive to alkyl and haloalkyl groups (AQN ligands) or large groups (PYR ligands).
- The NE quadrant easily tolerates many different functional groups.

  If an olefin is placed into this plane according to this rule, the two OH groups are transferred from above (β-face) if a dihyroquinidine (DHQD) ligand is used; for a dihydroquinine (DHQ) ligand, the attack occurs from the bottom (α-face).

5. The experimental conditions are simple and the reaction can be easily scaled up (see Section 4); a single set of conditions is satisfactory for most olefin substitution patterns. Only slight modifications (amounts of ligand and osmium) may be applied to the standard procedure.
6. The AD reaction has a broad scope of substrates, tolerating the presence of most organic functional groups (e.g. sulfur(II)-containing groups,[13] acetylenes,[14] allylic halides).[15]
7. It usually shows a high catalytic turnover number, allowing low ligand and osmium loading and good chemical yields. The reaction conditions are very moderate, the reaction is usually run overnight and therefore allows efficient laboratory synthesis.
8. It makes use of inexpensive substrates to yield 'high value added' building blocks for synthesis.
9. It provides access to chemically very useful 1,2-difunctionalized intermediates which are set up for further manipulation.

3. Choice of ligand and reaction conditions

One standardized set of reaction conditions allows for the asymmetric dihydroxylation of most olefins with manifold substitution patterns. These conditions use 3 eq. of $K_3Fe(CN)_6$ in the presence of 3 eq. of K_2CO_3 for the reoxidation. *t*-Butyl alcohol–water 1:1 is used as a solvent, and the osmium is added as potassium osmate dihydrate. Large-scale reactions yielding up to 2 kg of product can also be performed using *N*-methylmorpholine *N*-oxide as a cooxidant.[16]

Source of osmium

Potassium osmate dihydrate $[K_2OsO_2(OH)_4]$ is the most convenient source of osmium for the AD reaction (in contrast to OsO_4, $K_2OsO_2(OH)_4$ is not

2: Oxidation of the C=C bond

volatile). However, for very small batches a solution of OsO_4 (*e.g.* in toluene) may be handled more conveniently.

Amounts of osmium and ligand

Usually, 0.4 mol% of osmium is enough to allow for appropriate reaction times (0.4 mol% of osmium gives better reproducible results than 0.2 mol%.[8]). If a slow reaction is observed, 1 mol% of osmium should be added; the amount of ligand does not have to be increased (this osmium load is generally recommended together with the use of PYR ligands[9]). Up to 5 mol% of osmium can be added, in this case 5 mol% of ligand should be used (not in the case of AQN ligands due to their lower solubility). In very slow cases, the reaction temperature should be brought to room temperature. It is not recommended to add more solvent; the concentration (not the total amount!) of the ligand in the organic phase determines the rate and enantioselectivity.

Addition of methanesulfonamide

For the AD-reaction of 1,2-disubstituted and trisubstituted olefins, 1 eq. of methanesulfonamide should be added to accelerate the reaction by facilitating the hydrolysis of the osmium(VI) glycolate. For tetrasubstituted olefins, 3 eq. of methanesulfonamide should be used. An extraction step (aqueous KOH) should be added in the work-up to remove the methanesulfonamide, except if a base-sensitive substrate is used.

Buffered conditions

Many moderately base-sensitive substrates (e.g. allylic halides, allylic tosylates, allylic mesylates) can be converted without significant side-reactions by adding 3 eq. of $NaHCO_3$ to the reaction system. Because the starting material and the product are mainly in the organic phase, these conditions do not usually hydrolyse the aforementioned compounds under those 'buffered conditions'. It is not possible to omit the use of K_2CO_3.

Solvents

In the standard procedure, *t*-butyl alcohol–water 1:1 is used. If the olefin is not soluble enough in *t*-butyl alcohol, cosolvents may be added. Of the many possibilities, *t*-butyl methyl ether has been found to be the best. Up to 50% of the *t*-butyl alcohol can be substituted by *t*-butyl methyl ether without significant impact on the enantiomeric excess. THF can also be used, but lower enantioselectivity is usually observed. Increasing the total volume of the organic phase should be avoided (*vide infra*).

Choice of the ligand

The characteristics of the most important ligands under investigation are listed in Table 2.5.1.

To give an overview of enantioselectivities to be expected, selected results

Table 2.5.1

Formula	Ref.	Acronym	Advantages	Disadvantages
(phthalazine with OAlk* groups)	8	PHAL	Commercially available, easy synthesis, available as ready-to-use AD-mix; most 'cost-effective' ligand	Moderate performance with terminal aliphatic olefins
(pyrimidine with Ph and OAlk* groups)	9	PYR	Commercially available, relatively easy synthesis; best ligand for branched aliphatic olefins	Low performance with aromatic olefins, moderate performance with allylic halides
(anthraquinone with OAlk* groups)	10	AQN	Commercially available; best ligand for aliphatic olefins, superior for allylic halides	Relatively demanding synthesis, moderate performance with aromatic olefins
(diphenylpyrazinopyridazine with OAlk*)	16	DPP	Superior with aromatic substituted olefins	Relatively demanding synthesis, not commercially available
(diphenyl-phthalazine with OAlk*)	16	DP-PHAL	Good 'all-round' ligand, best for certain *cis*-1,2-disubstituted olefins	Demanding synthesis, not commercially available
(phenanthryl with OAlk*)	7	PHN	Best first-generation ligand, commercially available	Moderate enantioselectivity
(indoline with AlkO carbamate)	17	IND	Good for most *cis*-1,2-disubstituted olefins	Not commercially available

are shown in Tables 2.5.2 and 2.5.3. For the sake of simplicity, only results for the dihydroquinidine ligands are shown. The ees obtained with the 'pseudo enantiomeric' dihydroquinine are slightly lower. A comprehensive overview can be found in Ref. 1.

2: Oxidation of the C=C bond

Table 2.5.2

Substrate	(DHQD)$_2$PHAL	(DHQD)$_2$PYR	(DHQD)$_2$DPP	(DHQD)$_2$AQN	Diol config.
1-butene (I-CH$_2$-CH=CH$_2$ type)	63	70	68	83	S
TsO-CH$_2$-CH=CH$_2$	40			83	S
F$_3$C-CH$_2$-CH=CH$_2$	63	64		81	R
C$_8$H$_{17}$-CH=CH$_2$	84	89	89	92	R
styrene (PhCH=CH$_2$)	88	96	89	86	R
t-Bu-CH=CH$_2$	64	92	59		R
Ph-C(CH$_3$)=CH$_2$ (α-methylstyrene)	97	80	99	89	R
H$_{11}$C$_5$-C(CH$_3$)=CH$_2$	78	76	78	85	R
Ph-C(=CH$_2$)-CH=CH$_2$	94	69	96	82	R
C$_4$H$_9$-CH=CH-C$_4$H$_9$	97	88	96	98	R, R
Cl-CH$_2$-CH=CH-CH$_2$-Cl	94			96	S, S
H$_{11}$C$_5$-CH=CH-CO$_2$Et	99			99	2S, 3S

Table 2.5.4 summarizes the experience gained up to now with various ligands and should be viewed as a rough guideline for the choice of the appropriate ligand and for the enantioselectivities to be expected. However, in more complex cases the guidelines are not unambiguous and at least the commercially available ligands PHAL, PYR, and AQN should be tested.

4. Protocols for the asymmetric dihydroxylation reaction

Two different protocols are described here. Protocol 1 describes a general AD reaction using individual reagents. Protocol 2 makes use of the 'ready to use' AD-mixes and is more easy to perform on a small scale. AD-mixes and the

Table 2.5.3

Olefin	DHQD-IND	(DHQD)₂PYR	(DHQD)₂DPP	(DHQD)₂AQN	Diol config.
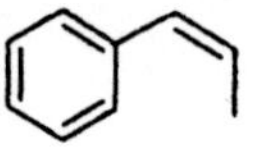 cis-stilbene	72		68	45	1R, 2S
ethyl cis-cinnamate	78				2R, 3R
cis-cyclohexylpropene	56				1R, 2S
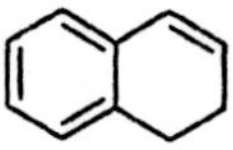 dihydronaphthalene	16	7	40	35	1R, 2S
7-MeO-2,2-dimethylchromene	32	67			3S, 4S
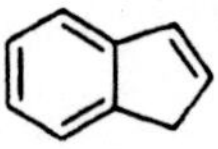 indene		35	20	63	1R, 2S

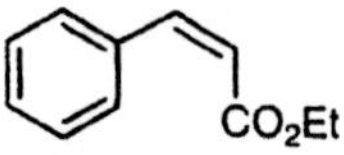

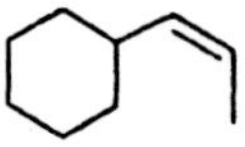

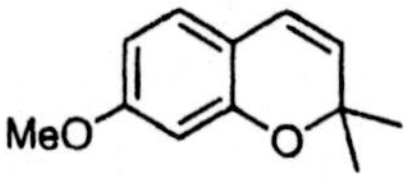

Table 2.5.4

Olefin class	 R—CH=CH₂	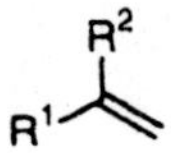 R¹R²C=CH₂	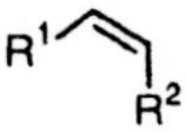 R¹HC=CHR² (cis)
Preferred ligand (expected ee range)	R = Aromatic DPP, PHAL (97–99%) R = Aliphatic AQN (80–92%) R = Branched PYR (90–96%)	R¹, R² = Aromatic DPP, PHAL (88–97%) R¹, R² = Aliphatic AQN (80–85%) R¹, R² = Branched PYR (70%)	Acyclic IND (50–80%) Cyclic PYR, DPP AQN (10–75%)

Table 2.5.4 (*continued*)

Olefin class	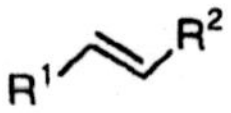 R¹HC=CHR² (trans)	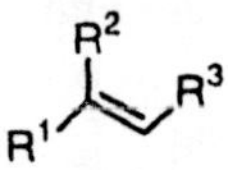 R¹R²C=CHR³	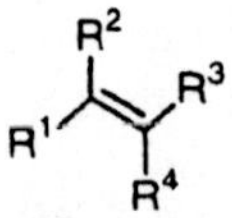 R¹R²C=CR³R⁴
Preferred ligand (expected ee range)	R¹, R² = Aromatic DPP, PHAL (97–99.8%) R¹, R² = Aliphatic AQN (75–98%)	PHAL DPP, AQN (95–99%)	PYR, PHAL (20–85%)

2: Oxidation of the C=C bond

ligands can be obtained from Aldrich Chemical Co.: AD-mix-α, No. 39,275–8; AD-mix-β, No. 39,276–6; (DHQD)$_2$PHAL, No. 39,273–1; (DHQ)$_2$PHAL, No. 39,272–3, (DHQD)$_2$PYR, No. 41,895–1; (DHQ)$_2$PYR, No. 41,897–8; (DHQD)$_2$AQN, No. 45,671–3; (DHQ)$_2$AQN, No. 45,670–5.

The alkaloids needed for the preparation of the ligands are also sold by Aldrich: dihydroquinidine, No. 35,934-3; dihydroquinine, No. 33,771-4.

Protocol 1.
A general AD reaction using individual reagents

Caution! All osmium compounds are toxic. The procedure must be undertaken in a well-ventilated hood and disposable vinyl or latex gloves and chemical-resistant safety goggles must be worn at all times.

The general procedure for the asymmetric dihydroxylation of olefins is given on a 10-mmol scale; in this procedure all reagents are added individually to allow for a maximum of flexibility. The general procedure given here usually affords good results for most olefins. Possible variations depend on the olefin used and are given as remarks.

Equipment

- One-necked round-bottomed flask (250 mL) with a Teflon-coated stirrer bar (egg-shaped)[a]
- Magnetic stirrer with a water–ethylene glycol bath equipped with a cryogenic cooler[b]
- Separatory funnel (250 mL)
- One-necked round-bottomed flask (500 mL)
- Chromatography column

Materials

- Potassium hexacyanoferrate(III) (FW 329.25), 9.88 g, 30 mmol — **toxic**
- Potassium carbonate (FW 138.21), 4.14 g, 30 mmol
- Sodium bicarbonate (FW 84.01), 2.52 g, 30 mmol (only for base-sensitive olefins[c])
- Ligand, 0.1 mmol (1 mol%) (78 mg in the case of (DHQD)$_2$PHAL or (DHQ)$_2$PHAL; 88 mg in the case of (DHQD)$_2$PYR or (DHQ)$_2$PYR; 86 mg in the case of (DHQD)$_2$AQN or (DHQ)$_2$AQN)
- Potassium osmate dihydrate [K$_2$OsO$_2$(OH)$_4$] (0.4 mol%) (FW 368.43), 14.7 mg, 0.04 mmol — **toxic**
- Methanesulfonamide (FW 95.12), 951 mg, 10 mmol [only added in the case of 1,2-disubstituted, trisubstituted and tetrasubstituted (30 mmol) olefins]: see Section 3 — **irritant**
- t-Butyl alcohol, 50 mL — **flammable, harmful**
- Sodium sulfite (FW 126.04), 8.0 g, 63 mmol — **harmful**
- Ethyl acetate, 200 mL — **flammable**
- Aqueous sodium hydroxide solution (2 M), 200 mL (only if methanesulfonamide is used) — **corrosive**
- Magnesium sulfate for drying
- Silica gel for flash chromatography — **harmful by inhalation**
- Hexane for chromatography — **flammable, irritant**
- Ethyl acetate for chromatography — **flammable**

Protocol 1. Continued

Method

1. Add potassium hexacyanoferrate(III) (9.88 g, 30 mmol), potassium carbonate (4.14 g, 30 mmol), ligand [1 mol%; 78 mg in the case of (DHQD)$_2$PHAL or (DHQ)$_2$PHAL; 88 mg in the case of (DHQD)$_2$PYR or (DHQ)$_2$PYR; 86 mg in the case of (DHQD)$_2$AQN or (DHQ)$_2$AQN], potassium osmate dihydrate [K$_2$OsO$_2$(OH)$_4$] (0.4 mol%; 14.7 mg, 0.04 mmol), methanesulfonamide (951 mg, 10 mmol; *only added in the case of 1,2-disubstituted and trisubstituted olefins*), and (*only for base-sensitive olefins*[c]) sodium bicarbonate (2.52 g, 30 mmol) into a 250-mL one-necked round-bottomed flask with a Teflon-coated stirrer bar. Add 50 mL of water and 50 mL of *t*-butyl alcohol[d] and stir at room temperature until all materials are dissolved.

2. Cool to 0 °C using a water–ethylene glycol bath equipped with a cryogenic cooler.[b] (Some of the dissolved salts precipitate.) Add the olefin[d] and stir vigorously for 6 to 24 h. Monitor the reaction by TLC (silica gel, hexane–ethyl acetate).[e]

3. Add sodium sulfite (8 g, 63 mmol), warm to room temperature and stir for 1 h. Transfer into a 250-mL separating funnel and add ethyl acetate (100 mL). Separate phases and extract the aqueous phase with ethyl acetate (2 × 50 mL).[f]

4. If methanesulfonamide is used, wash the combined organic phases with 200 mL of aqueous 2 M sodium hydroxide solution.

5. Combine the organic phases, dry over magnesium sulfate and remove the solvent using a rotary evaporator.

6. Purify the raw product by flash chromatography using hexane–ethyl acetate as eluent. The diols are usually obtained with yields of 65–90%.

[a] Magnetical stirring is sufficient for up to a total of 1 L of solvent mixture. For larger batches, mechanical stirring is better suited.
[b] If a cryogenic cooler is not available, a Styrofoam box or a low Dewar flask filled with water–ice is sufficient for cooling.
[c] Allylic halides, tosylates and mesylates can be dihydroxylated successfully using 'buffered' conditions.
[d] If the olefin is not soluble in *t*-butyl alcohol, cosolvents may be added (see the section on *Solvents*).
[e] Some olefins are not detected by their UV absorption; putting the TLC plate in a chamber filled with iodine–silica mixture detects most olefins.
[f] For more polar products, dichloromethane is better used for the extraction of the product. If dienes or trienes are oxidized, the resulting polyols are better extracted with chloroform–isopropanol 3:1.

Protocol 2.
An AD reaction using 'ready to use' AD-mix

Caution! All osmium compounds are toxic. The procedure must be undertaken in a well-ventilated hood and disposable vinyl or latex gloves and chemical-resistant safety goggles must be worn at all times.

2: Oxidation of the C=C bond

The general procedure for the asymmetric dihydroxylation of olefins is on a 1-mmol scale; this procedure uses the 'ready to use' AD-mixes and is easier to perform than the standard reaction (Protocol 1). Possible variations depend on the olefin used and are given as remarks.

Equipment

- One-necked round-bottomed flask (25 mL) with a Teflon-coated stirrer bar. A flask with a long neck is preferred because it can be mounted more easily in the cooling bath
- Magnetic stirrer with a water–ethylene glycol bath equipped with a cryogenic cooler[a]
- One-necked round-bottomed flask (100 mL)
- Separatory funnel (100 mL)
- Chromatography column

Materials

- AD-mix-α or AD-mix-β, 1.4 g   **toxic**
- Methanesulfonamide (FW 95.12), 95 mg, 1 mmol (only added in the case of 1,2-disubstituted and trisubstituted olefins)   **irritant**
- t-Butyl alcohol, 5 mL   **flammable, harmful**
- Sodium sulfite (FW 126.04), 0.8 g, 6.3 mmol   **harmful**
- Ethyl acetate, 30 mL   **flammable**
- Aqueous 2 M sodium hydroxide solution, 20 mL (only if methanesulfonamide is used)   **corrosive**
- Magnesium sulfate for drying
- Silica gel for flash column chromatography   **harmful by inhalation**
- Hexane for chromatography   **flammable, irritant**
- Ethyl acetate for chromatography   **flammable**

Method

1. Add AD-Mix-α or AD-mix-β (1.4 g), methanesulfonamide (95 mg, 1 mmol; *only added in the case of 1,2-disubstituted and trisubstituted olefins*), and (*only for base-sensitive olefins*[b]) sodium bicarbonate (252 mg, 3 mmol) into a 25-mL one-necked round-bottomed flask with a Teflon-coated stirrer bar. Add 5 mL of water and 5 mL of t-butyl alcohol[c] and stir at room temperature until all materials are dissolved.

2. Cool to 0°C using a water–ethylene glycol bath equipped with a cryogenic cooler.[a] (Some of the dissolved salts precipitate.) Add the olefin[c] and stir vigorously for 6 to 24 h. Monitor the reaction by TLC (silica, hexane–ethyl acetate).[d]

3. Add sodium sulfite (0.8 g, 6.3 mmol), warm to room temperature and stir for 1 h. Transfer into a 100-mL separating funnel along with ethyl acetate (20 mL). Separate the phases and extract the aqueous phase with ethyl acetate (2 × 5 mL).[e]

4. If methanesulfonamide is used, wash the combined organic phases with 20 mL of aqueous 2 M sodium hydroxide solution.

5. Combine the organic phases, dry over magnesium sulfate and remove the solvent using a rotary evaporator.

Protocol 2. Continued

6. Purify the raw product by flash chromatography using hexane–ethyl acetate as eluent. The diols are usually obtained with a yield of 65–90%.

[a] If a cryogenic cooler is not available, a Styrofoam box or a low Dewar flask filled with water–ice is sufficient for cooling overnight.
[b] Allylic halides, tosylates and mesylates can be dihydroxylated successfully using 'buffered' conditions.
[c] If the olefin is not soluble in *t*-butyl alcohol, cosolvents may be added (see the section on *Solvents*).
[d] Some olefins are not detected by their UV absorption; putting the TLC plate in a chamber filled with iodine–silica mixture detects most olefins.
[e] For more polar products, dichloromethane is better used for the extraction of the product. If dienes or trienes are oxidized, the resulting polyols are better extracted with chloroform–isopropanol 3:1.

---

5. Synthesis of the ligands

Protocol 3.
Preparation of 1,4-dichlorophthalazine (Structure 5, Scheme 2.5.3)

Caution! Carry out all procedures in a well-ventilated hood, and wear disposable vinyl or latex gloves and chemical-resistant safety goggles.

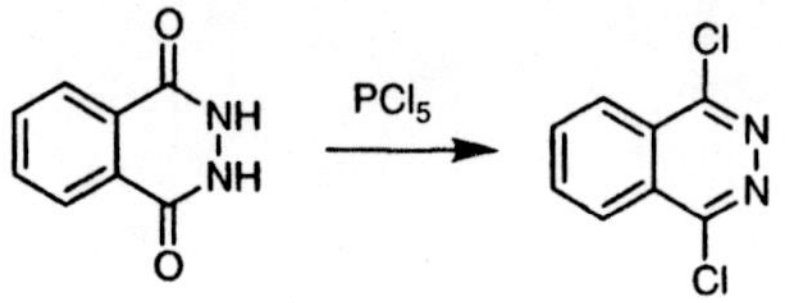

Scheme 2.5.3

Equipment

- One three-necked round-bottomed flask (1 L) with a mechanical stirrer and a reflux condenser which is fitted with a calcium chloride drying tube; the outlet is connected to a gas vent at the hood
- Thermostated hotplate stirrer with an oil bath and a thermometer
- Distillation apparatus
- Sintered glass funnel (diameter 12 cm)

Materials

- Phthalhydrazide **4** (FW 162.15), 81 g, 0.50 mol — toxic
- Phosphorus pentachloride (FW 208.24), 218.4 g, 1.05 mol — corrosive, moisture sensitive
- Dichloromethane, 1.2 L — harmful by inhalation
- THF, 750 mL — flammable, irritant
- Neutral alumina

Method

1. Equip an oven-dried 1-L three-necked round-bottomed flask with a condenser and a mechanical stirrer. Add phthalhydrazide (81 g, 0.50 mol) and

2: Oxidation of the C=C bond

phosphorus pentachloride (218.4 g, 1.05 mol) and heat slowly at 145 °C bath temperature (Caution! Strong evolution of HCl-gas!) Heat the mixture at this temperature for an additional 4 h. The mixture liquifies during the reaction. Replace the reflux condenser with a distilling apparatus and distil off the phosphorus oxychloride formed.

2. Crush the residual solid to a fine powder and dissolve in 1.2 L of dichloromethane. Filter off the residual solid and add 250 g of neutral alumina. Stir for an additional hour; filter the solution through an 8-cm deep and 12-cm wide pad of neutral alumina and wash with dichloromethane until the product is obtained (monitor by TLC: silica gel; dichloromethane, $R_f = 0.35$). Remove the solvent on a rotary evaporator.

3. Recrystallize the solid from 750 mL of THF to yield 58.2 g (0.292 mol, 58%) of 1,4-dichlorophthalazine (5) (FW 199.03, m.p. 162–163.5 °C) as white needles.[a]

The product is moisture sensitive and should be kept under nitrogen in the refrigerator.

[a] A second crop of 20 g of material can be obtained by concentrating the mother liquor.

Protocol 4
Preparation of 1,4-bis(dihydroquinidine)phthalazine [(DHQD)$_2$PHAL] (Structure 1a, Scheme 2.5.4)

Caution! Carry out all procedures in a well-ventilated hood, and wear disposable vinyl or latex gloves and chemical-resistant safety goggles.

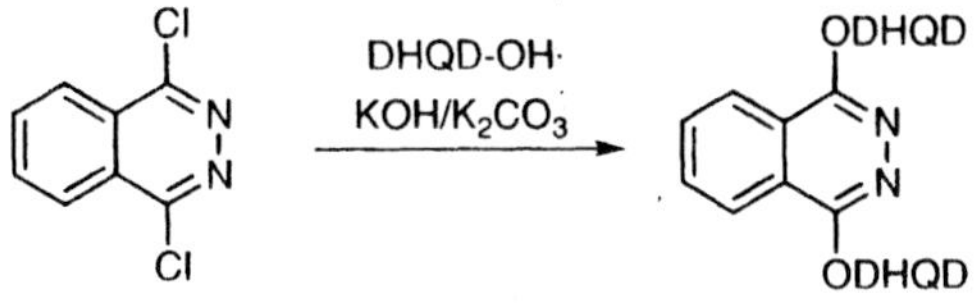

Scheme 2.5.4

Equipment

- Two-necked round-bottomed flask (1 L) equipped with a stopper, a Dean–Stark condensor and a Teflon-coated magnetic stirrer bar
- Dry nitrogen source with gas inlet
- Thermostated hotplate stirrer with an oil bath and a thermometer
- Separating funnels (500 mL and 1 L)
- Sintered (20 μm) glass funnel
- Erlenmeyer flasks (1 L and 2 L)
- Filter funnel
- One-necked round-bottomed flask (2 L)

Materials

- Dihydroquinidine (FW 326.44), 50.02 g, 153 mmol[a, b]   **irritant**
- 1,4-Dichlorophthalazine (FW 199.04), 15.55 g, 78 mmol[c]   **moisture sensitive, irritant**
- Potassium carbonate (FW 138.21), 31.75 g, 229 mmol

Protocol 4. Continued

- Potassium hydroxide 85% (FW 56.11), 15.11 g, 229 mmol[d]   corrosive
- Toluene, 500 mL   flammable
- Ethyl acetate, 800 mL   flammable
- Brine, 100 mL
- Magnesium sulfate, 100 g
- Ethanol, 850 mL   flammable
- Diethyl ether, 200 mL   irritant, flammable
- Sulfuric acid (FW 98.08, d = 1.85), 66.0 g (35.7 mL), 673 mmol   corrosive, toxic

Method

1. Add dihydroquinidine (50.02 g, 153 mmol), 1,4-dichlorophthalazine (15.55 g, 78 mmol), potassium carbonate (31.75 g, 229 mmol), and toluene (500 mL) in a 1-L two-necked round-bottomed flask equipped with a magnetic stirrer bar and Dean–Stark trap with condenser. Connect a nitrogen inlet to the condenser and flush the apparatus with nitrogen. Close the remaining opening of the flask with a stopper.

2. Heat and stir the reaction mixture to reflux for 2 h (135 °C bath temperature). Remove the heating bath (Care! Hot!) and cool down the mixture to a point where no boiling is observed; add potassium hydroxide pellets (85%, 15.11 g, 229 mmol) and heat to reflux for 12 h. The reaction can be monitored by TLC (silica gel, dichloromethane–MeOH 20:1 + 1% aqueous NH_3, R_f = 0.2).

3. Cool the solution to room temperature, add 100 mL of water, and transfer into a 1-L separatory funnel (use ethyl acetate to rinse the flask). Transfer the aqueous phase into a 500-mL separating funnel and extract with 200 mL of ethyl acetate. Combine the organic phases and wash with 100 mL of water and 100 mL of brine.

4. Transfer the solution into a 2-L Erlenmeyer flask and dry over 50 g of magnesium sulfate. Filter the solution into a 2-L one-necked flask and remove the solvent on a rotary evaporator to dryness. Dissolve the residual solid in 250 mL of ethanol (96%) and add a solution of 66.0 g (35.7 mL) of concentrated sulfuric acid in 500-mL of ethanol over 10 min. Refrigerate (−10 °C) overnight and collect the white precipitate by filtration through a sintered (20 μm) glass funnel. Wash with 100 mL of cold (−10 °C) ethanol and 200 mL of diethyl ether.

5. Add the solid into a 1-L separatory funnel and dissolve in 200 mL of water. Add saturated aqueous sodium bicarbonate solution until a pH of 9–10 is reached. Extract with ethyl acetate (3 × 200 mL) and transfer the combined organic phases into a 1-L Erlenmeyer flask. Add 50 g of magnesium sulfate, stir for 10 min and filter into a 1-L one-necked round-bottomed flask. Evaporate the solvent using a rotary evaporator and dry in vacuum to yield 52.30 g (67.1 mmol, 86%) of 1,4-bis(dihydroquinidine)phthalazine [(DHQD)$_2$PHAL, **1a**] (FW 779.00, m.p. 131–134 °C, $[\alpha]_{589}^{20}$ = −258°, c = 1.34, MeOH) as a colourless powder.

2: Oxidation of the C=C bond

Smaller reactions (1–20 mmol scale) are better carried out using another method to remove water. Replace step 1 in the Protocol with:

Add dihydroquinidine, 1,4-dichlorophthalazine, potassium carbonate, and toluene in a two-necked round-bottomed flask equipped with a magnetic stirrer bar and a pressure-equalizing dropping funnel with a condenser on top. Put a small pad of cotton wool into the dropping funnel and fill it with 20 g of activated 4-Å molecular sieves. Isolate the addition funnel with aluminium foil. Connect a nitrogen inlet to the condenser and flush the apparatus with nitrogen. Close the remaining opening of the flask with a stopper.

The water produced is absorbed by the molecular sieves.

[a] The alkaloid is dried in vacuum in a desiccator over P_4O_{10} or under vacuum at 50 °C before use.
[b] If the dihydroquinidine was obtained as the hydrochloride salt, the free base must be obtained first.[8]
[c] 1,4-Dichlorophthalazine may be obtained as a commercial product. However, the compound should be purified by filtration with dichloromethane over neutral alumina before use. The compound is moisture sensitive and should be stored in the refrigerator under nitrogen.
[d] 85% Potassium hydroxide (the rest is water) should be used directly in the form of pellets. The evaporation of water from the pellets produces a very fine powder of potassium hydroxide in toluene.

Protocol 5.
Preparation of 1,4-bis(dihydroquinine)phthalazine [(DHQ)$_2$PHAL] (Structure 1b, Scheme 2.5.5)

Caution! Carry out all procedures in a well-ventilated hood, and wear disposable vinyl or latex gloves and chemical-resistant safety goggles.

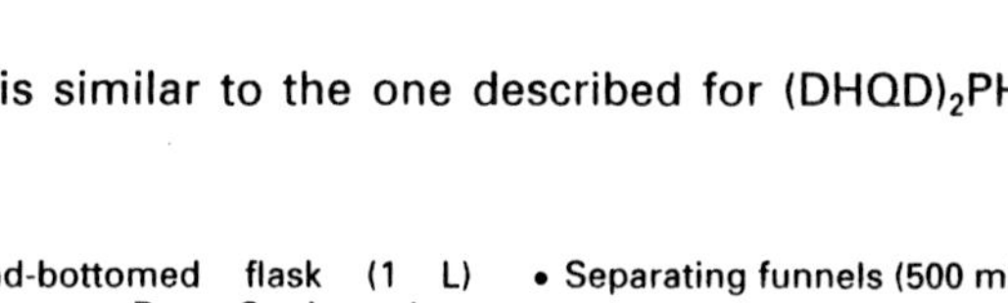

Scheme 2.5.5

This procedure is similar to the one described for (DHQD)$_2$PHAL (Protocol 4)

Equipment
- Two-necked round-bottomed flask (1 L) equipped with a stopper, a Dean–Stark condensor and a Teflon-coated magnetic stirrer bar
- Dry nitrogen source with gas inlet
- Thermostated hotplate stirrer with an oil bath and a thermometer
- Separating funnels (500 mL and 1 L)
- Sintered (20 μm) glass funnel
- Erlenmeyer flasks (1 L and 2 L)
- Filter funnel
- One-necked round-bottomed flask (2 L)

Materials
- Dihydroquinine (FW 326.44), 50.02 g, 153 mmol[a, b]   **irritant**
- 1,4-Dichlorophthalazine (FW 199.03), 15.55 g, 78 mmol[c]   **moisture sensitive, irritant**

Protocol 5. *Continued*

- Potassium carbonate (FW 138.21), 31.75 g, 229 mmol
- Potassium hydroxide 85% (FW 56.11), 15.11 g, 229 mmol[d]   **corrosive**
- Toluene, 500 mL   **flammable**
- Ethyl acetate, 1.2 L   **flammable**
- Brine, 100 mL
- Magnesium sulfate, 50 g
- Ethanol, 200 mL   **flammable**

Method

1. As step 1 in Protocol 4 except that dihydroquinine (50.02 g, 153 mmol) is used instead of dihydroquinidine.
2. Follow steps 2 and 3 of Protocol 4.
3. Transfer the solution into a 2-L Erlenmeyer flask and dry with 50 g of magnesium sulfate. Filter the solution into a 2-L one-necked flask and remove the solvent on a rotary evaporator to dryness.
4. Recrystallize the compound from 1 L of ethyl acetate. A second crop of crystals may be obtained by concentrating the mother liquor.

   An overall yield of 41.3 g (53.0 mmol, 68%) of 1,4-bis(dihydroquinine)phthalazine [(DHQ)$_2$PHAL, **1b**] (FW 779.00, m.p. 177–178°C, $[\alpha]_{589}^{20}$ = +347°, c = 0.79, MeOH) is obtained as a colourless powder.

[a] The alkaloid is dried in vacuum in a desiccator over P_4O_{10} or under vacuum at 50°C before use.
[b] If the dihydroquinine is obtained as the hydrochloride salt, the free base must be obtained first.[8]
[c] 1,4-Dichlorophthalazine may be obtained as a commercial product. However, the compound should be purified by filtration with dichloromethane over neutral alumina before use. The compound is moisture sensitive and should be stored in the refrigerator under nitrogen.
[d] 85% Potassium hydroxide (the rest is water) should be used directly in the form of pellets. The evaporation of water from the pellets produces a very fine powder of potassium hydroxide in toluene.

Protocol 6.
Preparation of AD-mix-α and AD-mix-β

Caution! All osmium compounds are toxic. The procedure must be undertaken in a well-ventilated hood and disposable vinyl or latex gloves and chemical-resistant safety goggles must be worn at all times.

Procedure for the preparation of 500 g of ready-to-use AD-mix-α or AD-mix-β containing 1 mol% of ligand and 0.4 mol% of osmium. This amount is sufficient to convert 0.35 mol of an olefin to the corresponding diol.

Equipment

- Blender (Waring, model 31BL92)
- Mortar with pestle (agate)

2: Oxidation of the C=C bond

Materials

- Potassium hexacyanoferrate(III) (FW 329.25), 350 g, 1.06 mol   **toxic**
- Potassium carbonate (FW 138.21), 147 g, 1.06 mol
- 1,4-Bis(dihydroquinine)phthalazine (**1b**) for AD-mix-α (FW 779.00), 2.76 g, 3.54 mmol

or

- 1,4-Bis(dihydroquinidine)phthalazine (**1a**) for AD-mix-β (FW 779.00), 2.76 g, 3.54 mmol
- Potassium osmate dihydrate [$K_2OsO_2(OH)_4$] (FW 368.43), 0.52 g, 1.41 mmol   **toxic**

Method

1. Grind potassium osmate dihydrate (0.52 g, 1.41 mmol) and the ligand 1,4-bis(dihydroquinine)phthalazine for AD-mix-α [or 1,4-bis(dihydroquinidine)-phthalazine for AD-mix-β] (2.76 g, 3.54 mmol) to a fine powder in a mortar.
2. Add this powder, $K_3Fe(CN)_6$ (350 g, 1.06 mol), and potassium carbonate (147 g, 1.06 mol) into a blender and mix for 30 min. The resulting orange powder should be kept dry and is ready for use.

Protocol 7
Preparation of 2,5-diphenyl-4,6-dichloropyrimidine (Structure 7, Scheme 2.5.6)

Caution! Carry out all procedures in a well-ventilated hood, and wear disposable vinyl or latex gloves and chemical-resistant safety goggles.

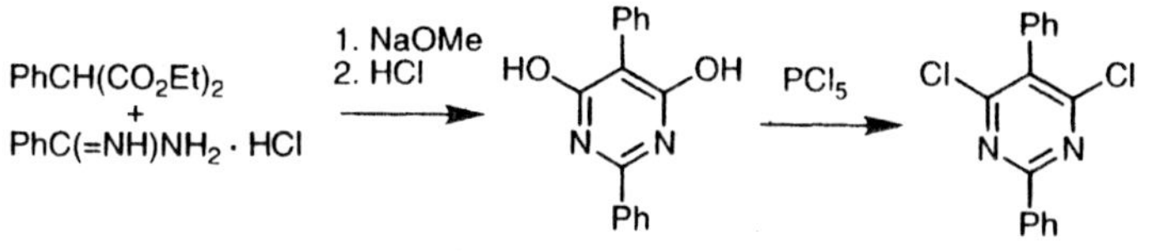

$$PhCH(CO_2Et)_2 + PhC(=NH)NH_2 \cdot HCl \xrightarrow{\text{1. NaOMe} \atop \text{2. HCl}} \mathbf{6} \xrightarrow{PCl_5} \mathbf{7}$$

Scheme 2.5.6

Equipment

- One three-necked round-bottomed flask (250 mL) with a Teflon-coated magnetic stirrer bar and a reflux condenser with a bubbler
- Thermostated hotplate stirrer with an oil bath and a thermometer
- Source of dry nitrogen
- Buchner filter funnel
- Distillation apparatus

- One three-necked round-bottomed flask (1 L) with a mechanical stirrer and a reflux condenser which is fitted with a calcium chloride drying tube; the outlet is connected to a gas vent at the hood
- Separatory funnel (250 mL)
- Sintered glass funnel, 3 cm wide

Materials

- Sodium (FW 22.99), 3.50 g, 150 mmol   **moisture sensitive!**
- Phenylmalonic acid diethyl ester (FW 236.27, d = 1.095), 11.8 g, 10.8 mL, 50 mmol   **may be harmful**
- Benzamidine hydrochloride (FW 156.61), 7.83 g, 50 mmol; dried over P_4O_{10} **irritant**
- Dry methanol (100 mL)   **flammable, harmful by inhalation**

Protocol 7. *Continued*

- Hydrochloric acid (37%), 11 mL — corrosive, irritant
- Ethanol, 50 mL — flammable
- Diethyl ether, 50 mL — irritant, flammable
- Phosphorus pentachloride (FW 208.22), 16.64 g, 94.3 mmol — corrosive
- N,N-Dimethylformamide — irritant, harmful by inhalation
- Dichloromethane, 300 mL — harmful by inhalation
- 2 M Sodium hydroxide solution, 50 mL — corrosive
- Magnesium sulfate
- Acetonitrile, 100 mL — flammable, toxic

Method

1. Equip an oven-dried 250-mL three-necked round-bottomed flask with a condenser and a mechanical stirrer bar. Flush the apparatus with nitrogen. Add dry methanol (100 mL) and then add sodium metal (3.50 g, 150 mmol) in small portions (allow the hydrogen produced to escape through a bubbler!).

2. After all of the sodium is dissolved, add phenylmalonic acid diethyl ester (11.8 g, 50 mmol) and benzamidine hydrochloride (7.83 g, 50 mmol; dried over P_4O_{10}); heat the mixture to reflux for 14 h. Cool the suspension to room temperature and remove the solvent on a rotary evaporator. Dissolve in 100 mL of warm (40 °C) water and add 11 mL of hydrochloric acid.

3. Collect the precipitate using a Buchner filter funnel and wash with 50 mL of water, 50 mL of ethanol and 50 mL of diethyl ether. Dry the compound in vacuo.

   2,5-Diphenyl-4,6-dihydroxypyrimidine (**6**) is obtained in 89% yield (11.87 g).

4. Equip an oven-dried 250-mL three-necked round-bottomed flask with a condenser and a mechanical stirrer. Add 2,5-diphenyl-4,6-dihydroxypyrimidine (**6**) (11.87 g, 44.9 mmol), phosphorus pentachloride (16.64 g. 94.3 mmol), and one drop of N,N-dimethylformamide and heat slowly to 145 °C bath temperature (Caution! Strong evolution of HCl gas!). Heat the mixture at this temperature for an additional 4 h. The mixture liquifies during the reaction. Replace the reflux condenser with a distilling apparatus and distil off the phosphorus oxychloride formed by applying a vacuum.

5. Dissolve the residue in 100 mL of dichloromethane, transfer into a 250-mL separatory funnel and wash with 50 mL of 2 M sodium hydroxide solution and 50 mL of brine, dry with magnesium sulfate. Concentrate the solution and take up the residue in dichloromethane. Filter the solution through a 3-cm wide and 6-cm deep plug of silica (sintered glass funnel) using 200 mL of dichloromethane. Concentrate the solution using a rotary evaporator and recrystallize from 100 mL of acetonitrile. 2,5-Diphenyl-4,6-dichloropyrimidine (**7**) is obtained in 67% yield (8.36 g, 30.2 mmol).

2: Oxidation of the C=C bond

Protocol 8
Synthesis of 2,5-diphenyl-4,6-bis(9-O-dihydroquinidinyl)pyrimidine [(DHQD)$_2$PYR] (Structure 2a, Scheme 2.5.7)

Caution! Carry out all procedures in a well-ventilated hood, and wear disposable vinyl or latex gloves and chemical-resistant safety goggles.

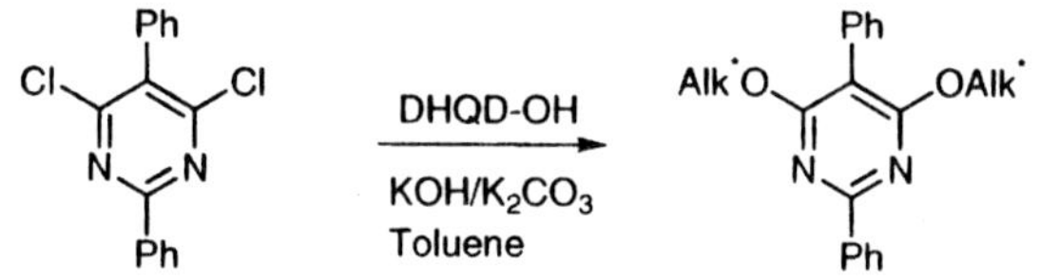

Scheme 2.5.7

Equipment

- Two-necked round-bottomed flask (50 mL) equipped with a stopper, pressure-equalizing funnel with condenser on top and a Teflon-coated magnetic stirrer bar
- Dry nitrogen source with gas inlet
- Thermostated hotplate stirrer with an oil bath and a thermometer
- Separatory funnel (250 mL)

Materials

- Dihydroquinidine (FW 326.44), 2.50 g, 7.66 mmol[a] — irritant
- 2,5-Diphenyl-4,6-dichloropyrimidine (FW 301.17), 1.15 g, 3.82 mmol
- Potassium carbonate (FW 138.21), 1.60 g, 11.5 mmol
- Potassium hydroxide 85% (FW 56.11), 730 mg, 13.0 mmol[b] — corrosive
- Toluene, 30 mL — flammable
- Dichloromethane, 150 mL — harmful by inhalation
- Magnesium sulfate for drying
- Acetonitrile, 100 mL — flammable, toxic

Method

1. Add dihydroquinidine (2.50 g, 7.66 mmol), 2,5-diphenyl-4,6-dichloropyrimidine (7) (1.15 g, 3.82 mmol), potassium carbonate (1.60 g, 11.5 mmol), and toluene (30 mL) in a 50-mL two-necked round-bottomed flask equipped with a magnetic stirrer bar and a pressure-equalizing dropping funnel with a condenser on top. Put a small pad of cotton wool into the dropping funnel and fill it with 30 g of activated 4-Å molecular sieves. Isolate the addition funnel with aluminium foil. Connect a nitrogen inlet to the condenser and flush the apparatus with nitrogen. Close the remaining opening of the flask with a stopper.

2. Heat and stir the reaction mixture to reflux for 2 h (135 °C bath temperature). Remove the heating bath (Care! Hot!) and cool down the mixture to a point where no boiling is observed; add potassium hydroxide pellets (85%, 730 mg, 13.0 mmol) and heat to reflux for 12 h. The reaction can be monitored by TLC (silica gel, dichloromethane–MeOH 20:1 + 1% aqueous NH$_3$).

Protocol 8. Continued

3. Cool the solution to room temperature, add into a 250-mL separatory funnel with 100 mL of water (use dichloromethane to rinse the flask). Separate the phases and extract the aqueous phase with dichloromethane (3 × 50 mL). Combine the organic phases, dry with magnesium sulfate and evaporate to dryness with a rotary evaporator.

4. Recrystallize from 100 mL of acetonitrile. 2,5-Diphenyl-4,6-bis(9-O-dihydroquinidinyl)pyrimidine (**2a**) [FW 881.14, m.p. 253–254°C, $[\alpha]_{589}^{25} = -389°$ ($c = 1.20$, MeOH)] is obtained in 77% yield (2.60 g, 2.95 mmol).

[a] The alkaloid is dried in vacuum in a desiccator over P_4O_{10} or under vacuum at 50°C before use.
[b] 85% Potassium hydroxide (the rest is water) should be used directly in the form of pellets. The evaporation of water from the pellets produces a very fine powder of potassium hydroxide in toluene.

Protocol 9
Synthesis of 2,5-diphenyl-4,6-bis(9-O-dihydroquininyl)pyrimidine [(DHQ)$_2$PYR] (Structure 2b, Scheme 2.5.8)

Caution! Carry out all procedures in a well-ventilated hood, and wear disposable vinyl or latex gloves and chemical-resistant safety goggles.

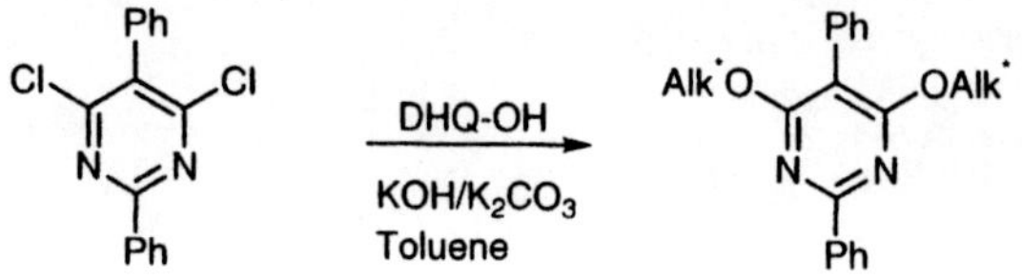

Scheme 2.5.8

Equipment

- The same equipment as Protocol 8, except that the size of the two-necked round-bottomed flask is not 50 mL but 25 mL and the size of the separating funnel is 100 mL

Materials

- Dihydroquinine (FW 326.44), 1.00 g, 3.06 mmol[a]   **toxic**
- 2,5-Diphenyl-4,6-dichloropyrimidine (FW 301.17), 461 mg, 1.53 mmol
- Potassium carbonate (FW 138.21), 640 mg, 4.60 mmol
- Potassium hydroxide 85% (FW 56.11), 350 mg, 5.3 mmol[b]   **corrosive**
- Toluene, 10 mL   **flammable**
- Dichloromethane, 60 mL   **harmful by inhalation**
- Magnesium sulfate for drying
- Ethyl acetate, 10 mL   **flammable**

Method

1. Add dihydroquinine (1.00 g, 3.06 mmol), 2,5-diphenyl-4,6-dichloropyrimidine (**7**) (461 mg, 1.53 mmol), potassium carbonate (640 mg, 4.60 mmol), and

toluene (10 mL) in a 25-mL two-necked round-bottomed flask equipped with a magnetic stirrer bar and a pressure-equalizing dropping funnel with a condenser on top. Put a small pad of cotton wool into the dropping funnel and fill it with 10 g of activated 4-Å molecular sieves. Isolate the addition funnel with aluminium foil. Connect a nitrogen inlet to the condenser and flush the apparatus with nitrogen. Close the remaining opening of the flask with a stopper.

2. As step 2 in Protocol 8 except that the amount of added potassium hydroxide pellets (85%) is 350 mg (5.3 mmol).

3. Cool the solution to room temperature, add into a 100-mL separatory funnel with 30 mL of water (use dichloromethane to rinse the flask). Separate the phases and extract the aqueous phase with dichloromethane (3 × 20 mL). Combine the organic phases, dry with magnesium sulfate and evaporate to dryness using a rotary evaporator.

4. Recrystallize from 10 mL of ethyl acetate. 2,5-Diphenyl-4,6-bis(9-O-dihydroquininyl)pyrimidine (**2b**) [FW 881.14, m.p. 247–248.5 °C, $[\alpha]_{589}^{25} = +456°$ (c = 1.25, MeOH)] is obtained in 73% yield (0.98 g, 1.11 mmol).

[a] The alkaloid is dried in vacuum in a desiccator over P_4O_{10} or under vacuum at 50 °C before use.
[b] 85% Potassium hydroxide (the rest is water) should be used directly in the form of pellets. The evaporation of water from the pellets produces a very fine powder of potassium hydroxide in toluene.

5. Determination of the enantiomeric excess (ee)

Many different methods have been used to determine the ee of chiral compounds.[18] Many of them work by converting an enantiomeric pair into diastereomes by derivatizing with a chiral enantiopure reagent. However, these methods bear the risk of kinetic resolution during derivatization and therefore may lead to incorrect determination of the ee. It is therefore recommended to use direct methods such as HPLC or GLC, using chiral stationary phases.

For 1,2-diols, we found that certain derivatives and HPLC analysis with chiral stationary phases usually give good results. For diols from aliphatic substituted olefins, bis(p-methoxy)benzoates usually show good resolution in HPLC. Enantiomeric diols from aromatic substituted olefins are usually better separated as their cyclic carbonates.

Protocol 10
Preparation of bis(*p*-methoxy)benzoates from 1,2-diols (Scheme 2.5.9)

Caution! Carry out all procedures in a well-ventilated hood, and wear disposable vinyl or latex gloves and chemical-resistant safety goggles.

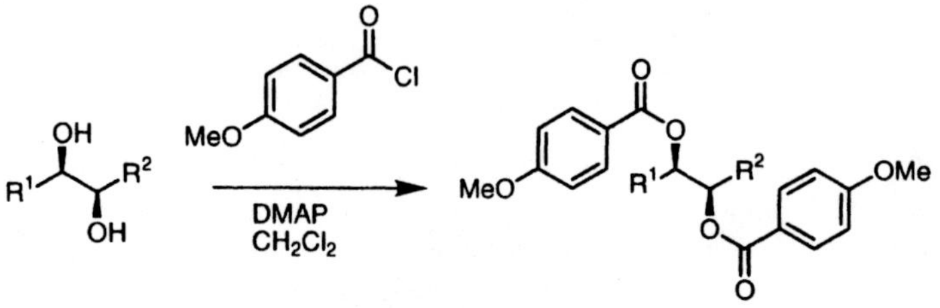

Scheme 2.5.9

Equipment

- Two-necked round-bottomed flask (10 mL) equipped with a stopper and a Teflon-coated magnetic stirrer bar
- Dry nitrogen source with gas inlet
- Two separatory funnels (100 mL)
- Chromatography column

Materials

- 1,2-Diol, 0.5 mmol
- 4-(*N*,*N*-Dimethylamino)pyridine (FW 122,17), 244 mg, 2 mmol — toxic, resorbed through skin
- *p*-Methoxybenzoyl chloride (FW 170.60), 341 mg, 2 mmol — corrosive
- Dichloromethane, 25 mL — harmful by inhalation
- Saturated aqueous NH_4Cl solution, 10 mL — harmful
- Concentrated aqueous $NaHCO_3$ solution, 10 mL
- Hexane for chromatography — flammable, irritant
- Ethyl acetate for chromatography — flammable

Method

1. Add 5 mL of dichloromethane, the diol (0.5 mmol), 4-(*N*,*N*-dimethylamino)pyridine (244 mg) and *p*-methoxybenzoyl chloride (341 mg) in a two-necked round-bottomed flask (10 mL) and stir for 60 min at room temperature under nitrogen. The reaction is monitored by TLC (hexane–ethyl acetate).

2. Add the reaction solution into a separatory funnel, dilute with 20 mL of dichloromethane and wash with 10 mL concentrated aqueous NH_4Cl solution and 10 mL of $NaHCO_3$ solution. Filter through cotton wool and remove the solvent with a rotary evaporator. Then filter it through a 1-cm diameter and 3-cm thick pad of silica gel using hexane–ethyl acetate as eluent. All fractions containing product are combined. The solvent is removed on a rotary evaporator.

The raw product (85% yield) is taken up in isopropanol–hexane. The enantiomeric excess is determined by HPLC. For most diols Chiralcel OD or OD-H columns have been found to give good results. Flow rates between 0.5 and 0.75 mL/min are used. The polarity of the eluent has to be adjusted to the diol used by changing the isopropanol content of the solvent mixture; 5% to 30% of isopropanol in hexane are typically used. Detection is by UV (254 nm).

Protocol 11
Preparation of cyclic carbonates from 1,2-diols (Scheme 2.5.10)

Caution! Carry out all procedures in a well-ventilated hood, and wear disposable vinyl or latex gloves and chemical-resistant safety goggles.

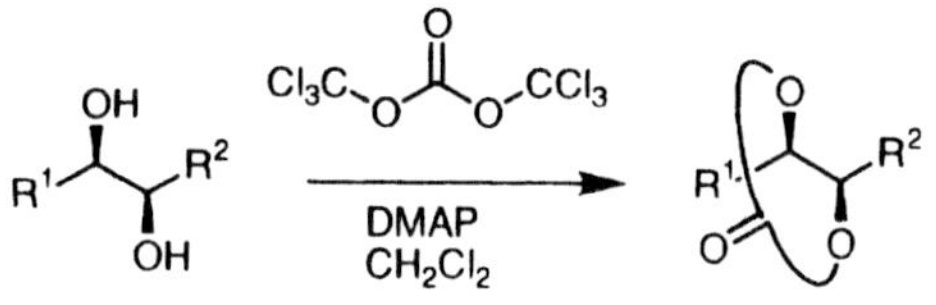

Scheme 2.5.10

Equipment
- Two-necked round-bottomed flask (10 mL) equipped with a stopper and a Teflon-coated magnetic stirrer bar
- Dry nitrogen source with gas inlet
- Two separatory funnels (100 mL)
- Chromatography column

Materials
- 1,2-Diol, 0.5 mmol
- 4-(N,N-Dimethylamino)pyridine (FW 122.17), 367 mg, 3 mmol — toxic, resorbed through skin
- Bis(trichlormethyl)carbonate (triphosgene) (FW 296.75), 742 mg, 2.5 mmol — toxic
- Dichloromethane, 25 mL — harmful by inhalation
- Saturated aqueous NH_4Cl solution, 10 mL — harmful
- Concentrated aqueous $NaHCO_3$ solution, 10 mL
- Hexane for chromatography — flammable, irritant
- Ethyl acetate for chromatography — flammable

Method

1. Add 5 mL of dichloromethane, 367 mg of 4-(N,N-dimethylamino)pyridine, and 742 mg of bis(trichlormethyl)carbonate (triphosgene) (FW 296.75) in a two-necked round-bottomed flask (10 mL) and stir for 10 min at room temperature under nitrogen. Add the diol (0.5 mmol) and stir for another 60 min. The reaction is monitored by TLC (hexane–ethyl acetate).

2. Follow step 2 in Protocol 10, except that Chiralcel OB-H or OD-H column is used in this protocol. The yield of carbonates is *ca.* 90%.

References

1. Kolb, H. C.; VanNieuwenhze, M. S.; Sharpless, K. B., *Chem. Rev.* **1994**, *94*, 2483–2547.
2. Sharpless, K. B.; Akashi, K. *J. Am. Chem. Soc.* **1976**, *98*, 1986.
3. VanRheenen, V.; Kelly, R. C.; Cha, D. Y. *Tetrahedron Lett.* **1976**, 1973.
4. Minato, M.; Yamamoto, K.; Tsuji, J. *J. Org. Chem.* **1990**, *55*, 766.
5. Various approaches towards asymmetric dihydroxylation are reviewed in Ref. 1.
6. Kwong, H. L.; Sorato, C.; Ogino, Y.; Chen, H.; Sharpless, K. B. *Tetrahedron Lett.* **1990**, *31*, 2999.
7. Sharpless, K. B.; Amberg, W.; Beller, M.; Chen, H.; Hartung, J.; Kawanami, Y.; Lubben, D.; Manoury, E.; Ogino, Y; Shibata, T.; Ukita, T. *J. Org. Chem.* **1991**, *56*, 4585.
8. Sharpless, K. B.; Amberg, W.; Bennani, Y. L.; Crispino, G. A.; Hartung, J.; Jeong, K.; Sung, K., Hoi, L.; Morikawa, K.; Wang, Z. M.; Xu, D.; Zhang, X.-L. *J. Org. Chem.* **1992**, *57*, 2768.
9. Crispino, G. A.; Jeong, K.-S.; Kolb, H.C .; Wang, Z.-M.; Xu, D.; Sharpless, K. B. *J. Org. Chem.* **1993**, *58*, 3785.
10. Becker, H.; Sharpless, K. B. *Angew. Chem., Int. Ed. Engl.* **1996**, *35*, 448.
11. Crispino, G. A.; Ho, P. T.; Sharpless, K. B. *Science* **1993**, *259*, 64; Becker, H.; Soler, M.; Sharpless, K. B. *Tetrahedron* **1995**, *51*, 1345; Xu, D.; Crispino, G. A.; Sharpless, K. B. *J. Am. Chem. Soc.* **1992**, *114*, 7570; Belley, M. L.; Hill, B.; Mitenko, H.; Scheigetz, J.; Zamboni, R. *Synlett* **1996**, 92.
12. Salvadori, P.; Superchi, S.; Minutolo, F. *J. Org. Chem.* **1996**, *61*, 4190.
13. Walsh, P. J.; Ho, P. T.; King, S. B.; Sharpless, K. B. *Tetrahedron Lett.* **1994**, *35*, 5129; Ohmori, K.; Nishiyama, S.; Yamamura, S. *Tetrahedron Lett.* **1995**, *36*, 6519.
14. Jeong, K.-S.; Sjö, P.; Sharpless, K. B. *Tetrahedron Lett.* **1992**, *33*, 3833.
15. For allylic halides, buffered conditions (addition of sodium bicarbonate) are used: Vanhessche, K. P. M; Wang, Z. M., Sharpless, K. B. *Tetrahedron Lett.* **1994**, *35*, 3469; Ahrgren, L.; Sutin, L. *Org. Process Res. Dev.* **1997**, *1*, 425; Wang, Z.-M.; Sharpless, K. B. *J. Org. Chem.* **1994**, *59*, 8302.
16. Becker, H.; King, S. B.; Taniguchi, M.; Vanhessche, K. P. M.; Sharpless, K. B. *J. Org. Chem.* **1995**, *60*, 3940.
17. Wang, L.; Sharpless, K. B. *J. Am. Chem. Soc.* **1992**, *114*, 7568.
18. Parker, D., *Chem. Rev.* **1991**, *91*, 1441.

2.6 Asymmetric aminohydroxylation

G. SCHLINGLOFF and K. B. SHARPLESS

1. Introduction

The β-aminoalcohol moiety is a widespread structural motif in natural products and synthetic drugs. Its generation in an enantioselective manner via metal catalysis represents a major challenge for synthetic organic chemists.

2: Oxidation of the C=C bond

Rapid progress has been made in catalytic transformations of olefins leading to epoxides or epoxide equivalents,[1,2] which can be reacted with nitrogen nucleophiles to give hydroxyamines. *Catalytic asymmetric aminohydroxylation* (AA) represents an even more elegant approach. Alkenes are therein converted into protected β-amino alcohols in a single step via a *syn*-cycloaddition catalysed by osmium salts and chiral quinuclidine-type ligands derived from cinchona alkaloids (Scheme 2.6.1). Three major aspects are addressed in this reaction: namely, chemo-, regio-, and enantioselection. The most important variable is the ultimate oxidant/nitrogen source for the generation of the active osmium(VIII) imido species responsible for aminohydroxylation. Several different reactants have been introduced thus far: the alkali metal salts of (1) *N*-halosulfonamides,[3,4] (2) *N*-halocarbamates,[5,6] and (3) *N*-haloamides.[7] The following account gives an overview of the different procedures that have been developed as of mid-1997 in our laboratories.

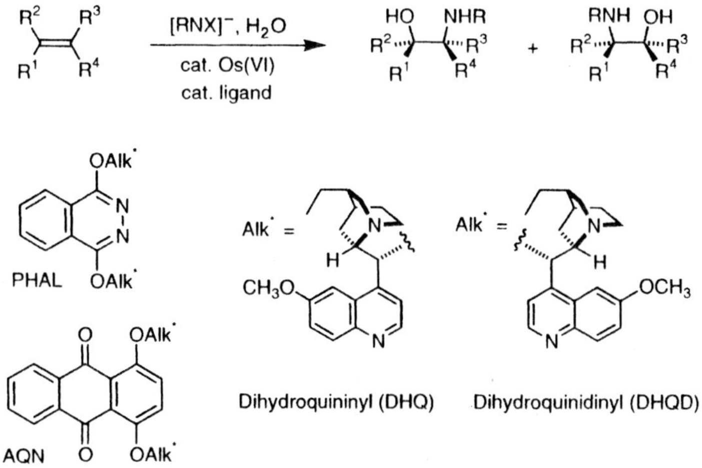

Scheme 2.6.1

2. Sulfonamide-based nitrogen sources

Chloramine-T (*N*-chloro-*N*-sodio-*p*-toluenesulfonamide) was used for the first time in 1976 as an oxidant/nitrogen source in the first catalytic aminohydroxylation process.[8] Limited success in terms of regioselectivity and reactivity was reported, however. These drawbacks were overcome when the reaction conditions were modified in such a way that significantly *more* water—crucial for the hydrolytic release of the aminoalcohol from the osmium(VI) intermediate —was added, also allowing enantioselective reactions[3,4] when working with chiral chinchona-derived ligands originally developed for the asymmetric dihydroxylation (AD) of olefins (see Section 2.5).[2] Both product enantiomers are available through the use of either quinine- or quinidine-type ligands. Remarkably, these ligands also have a beneficial effect on the regioselectivity in

Table 2.6.1 Examples of the sulfonamide-based AA using (DHQ)$_2$PHAL

Entry	Major product	R	ee (% yield)	Ref.
1	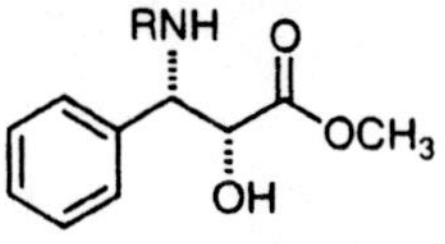 Ph-CH(NHR)-CH(OH)-C(O)OCH$_3$	—SO$_2$CH$_3$	95 (65)	4
2		—SO$_2$-p-tol	81 (64)	3
3	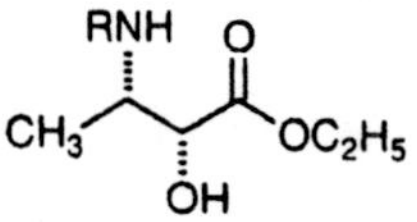 CH$_3$-CH(NHR)-CH(OH)-C(O)OC$_2$H$_5$	—SO$_2$-p-tol	74 (52)	4
4	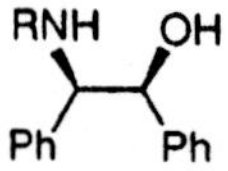 Ph-CH(NHR)-CH(OH)-Ph	—SO$_2$-p-tol	33 (48)	3
5	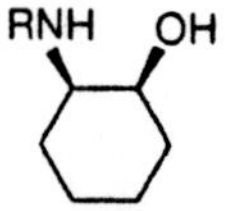 cyclohexane-1-NHR-2-OH	—SO$_2$CH$_3$	66 (49)	4
6		—SO$_2$-p-tol	45 (64)	3

most cases. Several different sulfonamides can be used as nitrogen sources (Table 2.6.1; yields in entries 1 and 3 refer to the overall AA yield).

Protocol 1.
Synthesis of (1*S*,2*S*)-2-[*N*-(*p*-toluenesulfonyl)amino]-1,2-diphenylethanol (Structure 1, Scheme 2.6.2)

Caution! Carry out all procedures in a well-ventilated hood, and wear disposable vinyl or latex gloves and safety glasses.

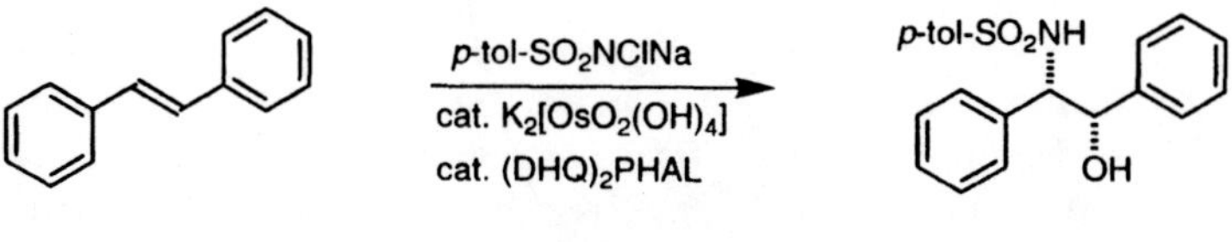

stilbene + p-tol-SO$_2$NClNa, cat. K$_2$[OsO$_2$(OH)$_4$], cat. (DHQ)$_2$PHAL → p-tol-SO$_2$NH-CH(Ph)-CH(OH)-Ph **1**

Scheme 2.6.2

Equipment

- Magnetic hotplate stirrer
- Water bath (1 L)
- Single-necked, round-bottomed flask (500 mL)
- Teflon-coated magnetic stirbar
- Sintered glass filter funnel (300 mL, medium porosity)

2: Oxidation of the C=C bond

Materials

- 1,4-Bis(dihydroquinine)phthalazine, (DHQ)$_2$PHAL, 2.20 g, 2.80 mmol — **harmful**
- t-Butanol, 150 mL — **flammable, irritant**
- Water, 350 mL
- *trans*-Stilbene, 10.5 g, 56 mmol — **harmful**
- Chloramine-T trihydrate, 48.4 g, 168 mmol — **oxidiser, irritant**
- Potassium osmate dihydrate (K$_2$[OsO$_2$(OH)$_4$]), 0.824 g, 2.24 mmol — **highly toxic**
- Ethyl acetate for recrystallization, 400 mL — **flammable, irritant**
- Hexane for recrystallization, 400 mL — **flammable, irritant**

Method

1. Dissolve with stirring 1,4-bis(dihydroquinine)phthalazine, (DHQ)$_2$PHAL, (2.20 g, 2.80 mmol) in *t*-butanol (100 mL) in a round-bottomed flask (500 mL) equipped with a magnetic stirbar, and add H$_2$O (100 mL).
2. To this solution, add the following reagents with stirring: *trans*-stilbene (10.5 g, 56 mmol), chloramine-T trihydrate (48.4 g, 168 mmol), and potassium osmate dihydrate (0.824 g, 2.24 mmol). You will obtain a brown suspension.
3. Immerse the reaction flask in a water bath kept at ambient temperature. Stir the dark green mixture over a period of 6 h.
4. Separate the greenish crystalline solid by filtration (sintered glass filter funnel, medium porosity), and wash the product with H$_2$O (2 × 100 mL) and ice-cold *t*-butanol–H$_2$O (100 mL, 1:1 v/v).
5. Dry the solid under high vacuum.
6. Take up the solid, 17.4 g, which contains product of 57% ee, in warm ethyl acetate (400 mL), filter while hot, and add hexane (300 mL) with stirring. Allow the mixture to cool to room temperature and collect the product on a sintered glass filter funnel, m.p. 166–167 °C, 10.08 g, 49%, $[\alpha]_D^{25} = -15.9°$ (c = 0.4, 95 vol% ethanol), 99% ee as determined by chiral HPLC on Daicel Chiralcel OD-H, 15% *i*-PrOH–hexane, 1 mL/min, 254 nm; retention times: 14.6 min (1*S*,2*S*), 23.3 min (1*R*,2*R*).

Our continuing efforts to replace sulfonamides with other nitrogen sources eventually led to the discovery of the asymmetric, carbamate-based aminohydroxylation.

3. Carbamate-based nitrogen sources

Since carbamoyl protecting groups for amines are extremely valuable from a synthetic standpoint, it was highly desirable to employ carbamate-based nitrogen sources for the aminohydroxylation. With the discovery of the 'water-rich' catalytic process, the addition of silver or mercury salts (in stoichiometric amounts!) to enable turnover in the aminohydroxylation became redundant. Good reactivity, combined with high enantioselection, is achieved when the *N*-chloro-*N*-sodio carbamates derived from ethyl, *t*-butyl, or benzyl carbamate are used (Table 2.6.2).[5,6]

Table 2.6.2 Examples of the carbamate-based AA using (DHQD)$_2$PHAL

Entry	Product	Regioselectivity	ee (% yield)	Ref.
1	BnOCONH-CH(Ph)-CH(OH)-COOCH$_3$	not determined	97 (65)	5
2	BnOCONH-CH(Ph)-CH$_2$OH	55:45	90 (50)	6
3	t-BuOCONH-CH(4-BnO-C$_6$H$_4$)-CH$_2$OH	83:17	96 (68)	6
4	BnOCONH-CH(3,5-BnO-4-CH$_3$O-C$_6$H$_2$)-CH$_2$OH	88:12	98 (68)	6
5	t-BuOCONH-CH(3-O$_2$N-C$_6$H$_4$)-CH$_2$OH	74:26	90 (59)	6
6	BnOCONH-CH(OC(O)OCH$_3$)-CH(OH)-COOCH$_3$	–	87 (55)	5

The reaction is typically run with excess oxidant (usually 3 eq with respect to the alkene) and a catalyst loading of 4 mol% Os and 5 mol% ligand. The oxidant itself is conveniently generated *in situ* by slow addition of *t*-butyl hypochlorite to a basic solution of the carbamate at room temperature. Strict control of the amount of base is crucial to the success of the AA reaction. Since the rate of solubility of powdered K$_2$[OsO$_2$(OH)$_4$] in the reaction medium may differ depending on supplier and/or age of the material, the outcome of the reaction has occasionally been unsatisfactory. Therefore, as reported in Proto-

2: Oxidation of the C=C bond

col 2, a modification of our old procedure should be used, which consists of addition of the potassium osmate as a solution in aqueous base.

The aminohydroxylation of aromatic olefins generally gives good results. Various styrenes were converted into the corresponding protected amino alcohols in a single step with good to high regioselectivity.[6] The undesired regioisomers were readily separated from the main products via silica gel chromatography. Also, electron-deficient alkenes such as fumarates or acrylates are excellent substrates for the carbamate AA.[5]

Protocol 2.
Synthesis of methyl (2S)-3-[N-(benzyloxycarbonyl)amino]-2-hydroxypropanoate (Structure 2, Scheme 2.6.3)

Caution! Cary out all procedures in a well-ventilated hood, and wear disposable vinyl or latex gloves and safety glasses.

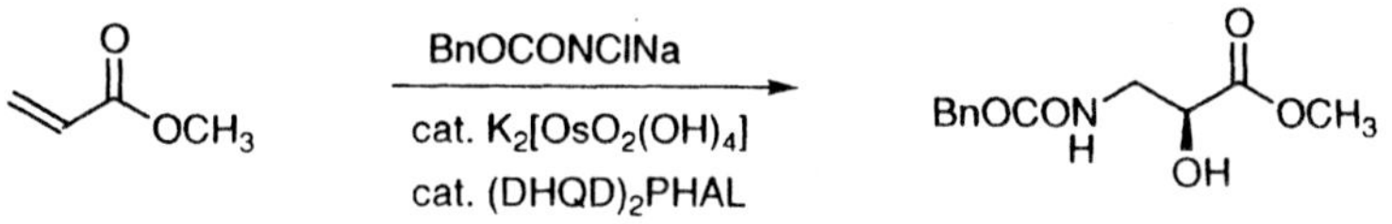

Scheme 2.6.3

Equipment

- Magnetic stirring plate
- Water bath (250 mL)
- Single-necked round-bottomed flasks (25 mL and 10 mL)
- Vial (2 mL)
- Two Teflon-coated magnetic stirbars
- Separation funnel (125 mL)
- Glass syringe (500 µL)
- Pasteur pipettes

Materials

Benzyl carbamate, 469 mg, 3.1 mmol	irritant
Acetonitrile, 7.5 mL	flammable, irritant
Water, 7.5 mL	
Sodium hydroxide, 122 mg, ca. 3.05 mmol	corrosive
t-Butyl hypochlorite, 356 µL, ca. 3.05 mmol	oxidizer, corrosive
Methyl acrylate, 90 µL, 1 mmol	toxic
Potassium osmate dihydrate (K$_2$[OsO$_2$(OH)$_4$]). 14.7 mg, 0.04 mmol	highly toxic
1,4-Bis(dihydroquinidine)phthalazine, (DHQD)$_2$PHAL, 39 mg, 0.05 mmol	harmful
Sodium sulfite, 2 g	very toxic by inhalation
Ethyl acetate, 75 mL	flammable, irritant
Saturated, aqueous sodium chloride solution, 10 mL	
Anhydrous sodium sulfate, 4 g	
Silica gel for flash chromatography	harmful by inhalation
Hexane for chromatography, ca. 0.5 L	flammable, irritant
Ethyl acetate for chromatography, ca. 0.5 L	flammable, irritant

Protocol 2. *Continued*

Method

1. Prepare a solution of sodium hydroxide[a] (122 mg, *ca.* 3.05 mmol) in water (7.5 mL) in a single-necked round-bottomed flask (25 mL) equipped with a magnetic stirbar.
2. With a Pasteur pipette, transfer *ca.* 0.5 mL of this solution into a vial (2 mL), and dissolve potassium osmate dihydrate (14.7 mg, 0.04 mmol) with gentle swirling.
3. Add acetonitrile (4 mL) and benzyl carbamate[b] (469 mg, 3.1 mmol) to the reaction flask and stir vigorously until *ca.* 90% of the carbamate is dissolved (10 min). Immerse the flask in a water cooling bath. Turn out the light in your fume hood. Add freshly prepared *t*-butyl hypochlorite[c] (346 μL, *ca.* 3.05 mmol) dropwise via a syringe (500 μL) with stirring over a period of 2 min to give a colourless solution.
4. In a single-necked, round-bottomed flask (10 mL) equipped with a magnetic stirbar, dissolve 1,4-bis(dihydroquinidine)phthalazine[b], (DHQD)$_2$PHAL (39 mg, 0.05 mmol), and methyl acrylate (90 μL, 1 mmol) in acetonitrile (3.5 mL) with stirring.
5. Transfer this solution with a Pasteur pipette into the reaction flask.
6. Add the pink K$_2$[OsO$_2$(OH)$_4$] solution to the reaction flask with a Pasteur pipette and stir vigorously at room temperature. The initial intense green colour of the reaction mixture turns bright yellow after about 2 min.
7. After 20 min, add sodium sulfite (2 g), and continue stirring for 45 min.
8. Transfer the mixture to a separatory funnel, separate the organic layer and wash the aqueous layer with ethyl acetate (4 × 25 mL). Combine the organic extracts and wash with saturated, aqueous sodium chloride solution (10 mL). Dry the organic extracts over anhydrous sodium sulfate (*ca.* 2 g). Filter the mixture and evaporate the filtrate under reduced pressure.
9. Prepare a column for flash chromatography using silica gel and hexane/ethyl acetate 1:1 v/v as the eluent (R_f = 0.25). Evaporate the eluates on a rotary evaporator to give the aminoalcohol (2*S*)-**2**, containing 4% regioisomer, as a viscous oil, 229 mg, 91%, $[\alpha]_D^{25}$ = +15.6° (*c* = 1.3, methanol), 90% ee as determined by HPLC on Daicel Chiralcel OB-H, *i*-PrOH–hexane 10:90 v/v, 0.7 mL/min, 210 nm; retention times: 62.7 min (2*R*), 65.5 min (2*S*).

[a] Use finely powdered material stored in a desiccator, and weigh quickly into the reaction flask. Do *not* employ more than the indicated amount of base.
[b] Use finely powdered material.
[c] Prepare according to Ref. 9.

2: Oxidation of the C=C bond

4. Acetamide-based nitrogen sources

Whereas alkali metal salts of N-chloro carboxamides are unstable and readily undergo Hofmann rearrangement, the corresponding N-bromo derivatives are convenient oxidants for AA reactions run at 4°C.[7] The standard substrates are converted smoothly into protected aminoalcohols with N-bromo alkali (K^+ or Li^+) salts of acetamide. When working with styrene derivatives and alkaloid ligands that contain an anthraquinone spacer (AQN ligands), the regiochemistry is reversed compared to the carbamate procedure with phthalazine (PHAL) ligands, see Table 2.6.3.

In contrast to the other AA protocols presented in Sections 2 and 3, 1.1 equivalents of the oxidant/nitrogen source suffice for complete conversion,

Table 2.6.3 Examples of the acetamide-based AA using (DHQD)$_2$PHAL (entries 1–3) or (DHQD)$_2$AQN (entries 4–6)

Entry	Major product	Regioselectivity	ee (% yield)
1	CH$_3$CONH, O-i-Pr, OH, CH$_3$O- (aryl)	>20:1	99 (71)
2	CH$_3$CONH, OH, (two phenyls)	–	93 (50)
3	CH$_3$CON(H), OC$_2$H$_5$, OH	>20:1	90 (46)
4	OH, NHCOCH$_3$ (phenyl)	13:1	88 (36)
5	O$_2$N-(aryl), OH, NHCOCH$_3$	4:1	94 (54)
6	OH, NHCOCH$_3$, CH$_3$O-(aryl)	9:1	86 (58)

and for isopropyl cinnamate as the olefin, the catalyst loading can be lowered significantly without affecting yield or enantioselectivity.

Protocol 3.
Synthesis of (2*R*,3*S*)-3-amino-2-hydroxy-3-phenylpropionic acid hydrochloride (Structure 3, Scheme 2.6.4)

Caution! Carry out all procedures in a well-ventilated hood, and wear disposable vinyl or latex gloves and safety glasses.

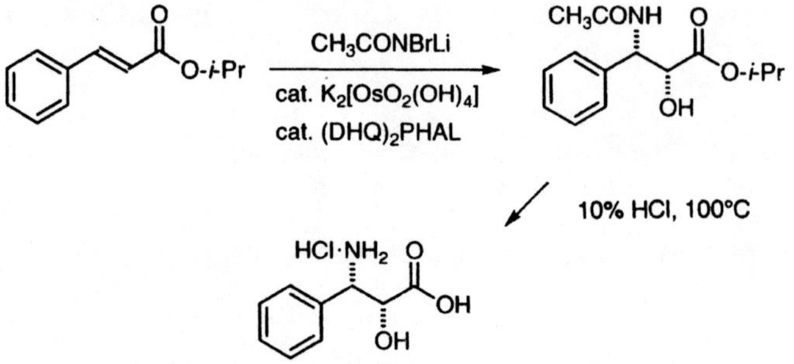

(2*R*,3*S*)-3

Scheme 2.6.4

Equipment

- Stainless-steel reaction vessel (40 L)
- Mechanical stirrer with stainless-steel stirring shaft (1 m) and four-blade rotor (6 cm)
- Immersion cooler with cold-finger probe
- Low-temperature thermometer
- Erlenmeyer flask (2 L)
- Teflon-coated magnetic stirbar (4 cm)
- Erlenmeyer flask (1 L)
- PVC [poly(vinyl chloride)] tubing (1 m × 10 mm)
- Two Erlenmeyer flasks (4 L)
- Glass filter (15 cm) with folded filter paper
- Single-necked round-bottomed flask (2 L)
- Sintered glass filter funnel (2 L, medium porosity)
- Recovery flask (4 L)
- High-vacuum pump
- Oil bath (0.5 L)
- Single-necked round-bottomed flask (1 L)
- Reflux condensor
- Sintered glass filter funnel (1 L, medium porosity)

Materials

- 1,4-Bis(dihydroquinine)phthalazine, (DHQ)$_2$PHAL, 9.738 g, 12.5 mmol — **harmful**
- *t*-Butanol, 7.5 L — **flammable, irritant**
- Lithium hydroxide monohydrate 56.12 g, 1.337 mol — **corrosive**
- Water, 11.25 L
- Potassium osmate dihydrate (K$_2$[OSO$_2$(OH)$_4$]), 6.909 g, 18.75 mmol — **highly toxic**
- Isopropyl cinnamate 237.8 g (96% pure by GC), *ca*. 1.250 mol
- *N*-Bromoacetamide[a] 195.6 g (97% pure by titration), 1.375 mol — **toxic**
- Sodium sulfite, 150 g — **very toxic by inhalation**
- Sodium chloride, 1.5 kg
- Ethyl acetate, 22.55 L — **flammable, irritant**
- Anhydrous sodium sulfate, 1.3 kg
- Chloroform, 0.4 L — **toxic**
- Silica gel for filtration — **harmful by inhalation**

2: Oxidation of the C=C bond

- Hexane, 0.3 L   flammable, irritant
- Diethyl ether, 0.6 L   flammable, irritant
- *t*-Butyl methyl ether, 0.1 L   flammable, irritant
- 10% Hydrochloric acid, 1.5 L   corrosive, toxic

Method

1. In a stainless-steel reaction vessel (40 L) equipped with a mechanical stirrer, a thermometer and an immersion cooler with a temperature controller set to 4 °C, dissolve 1,4-bis(dihydroquinine)phthalazine, $(DHQ)_2PHAL$ (9.738 g, 12.5 mmol), in *t*-butanol (7 L) with stirring (150 r.p.m.).
2. Add water (10 L) and start cooling.
3. In an Erlenmeyer flask (2 L), dissolve lithium hydroxide monohydrate (56.12 g, 1.337 mmol) in water (1.25 L) with stirring. Add $K_2[OsO_2(OH)_4]$ (6.909 g, 18.75 mmol) and continue stirring until you obtain a pink solution.
4. Pour the osmate solution into the reaction vessel.
5. Add isopropyl cinnamate (237.8 g, *ca.* 1.25 mol). Rinse the residual olefin with *t*-butanol (0.5 L). Add *N*-bromoacetamide (195.6 g, *ca.* 1.375 mol) quickly in a single portion at 4 °C.[a,b] Continue stirring the green reaction mixture at this temperature for 3.5 h.
6. At this point, a thin-layer chromatogram shows the absence of starting material (R_f = 0.85) and the mixture is red. Add sodium sulfite (150 g), remove the immersion cooler and stir for 12 h.
7. Add 1.5 kg NaCl and stir for 15 min. Stop stirring and allow phase separation. Siphon the upper organic layer through PVC tubing (length 1 m, diameter 10 mm) into two Erlenmeyer flasks (4 L) which contain anhydrous sodium sulfate (300 g each).
8. Filter the dried solution into a single-necked, round-bottomed flask (2 L) and concentrate the mixture on a rotary evaporator set to 60 °C.
9. Extract the remaining aqueous layer with ethyl acetate (1 × 6 L and 3 × 4 L). Decant the organic extract into the Erlenmeyer flasks.
10. Concentrate the organic extracts until you obtain a dark red oil. Add a mixture of warm ethyl acetate–chloroform (1.6 L, 3:1 v/v). Filter through a sintered glass filter funnel containing a 5-cm layer of SiO_2, covered with a 1-cm layer of Na_2SO_4. Wash with ethyl acetate (3 L). Evaporate the filtrate in a round-bottomed single-necked flask (2 L) to a volume of *ca.* 0.8 L, add hexane (250 mL) and a magnetic stirbar. Cool with stirring to 4 °C with an ice–water bath.
11. After 1 h, filter the suspension through a sintered glass filter funnel (1 L) and wash with ethyl acetate–hexane (250 mL, 3:2 v/v). Dry the filter cake under high vacuum for 4 h to yield 233 g of the AA product.
12. Concentrate the mother liquor and triturate with diethyl ether/*t*-butyl methyl ether (200 mL, 1:1 v/v). Filtration after 2 h and drying yields another

Protocol 3. *Continued*

27 g of product. The combined crystalline, white solid (260 g) is of 99% ee as determined by HPLC on Daicel Chiralcel OD-H, *i*-PrOH–hexane 40:60 v/v, 0.5 mL/min, 254 nm; retention times: 8.2 min (2S,3R), 12.6 min (2R,3S).

13. Heat the combined material in a 2-L round-bottomed flask immersed in an oil bath and equipped with a reflux condenser and a magnetic stirbar in 10% HCl (1.5 L, 4 h, 100°C).

14. Concentrate the mixture to *ca.* 20% of the initial volume on a rotary evaporator set to 60°C.

15. Filter the product through a sintered glass funnel (1 L), wash with cold diethyl ether (0.5 L), and dry the white crystals of hydrochloride (2R,3S)-3 for 12 h under high vacuum at 40°C; m.p. 224–226°C, $[\alpha]_D^{25} = -14.8$ (c = 0.55, 6 M HCl), 209.5 g, 77% (based on 96% GC-pure cinnamate).

a Prepared according to Ref. 10; titrate according to Ref. 11.
b To avoid hydrolysis of the ester, add the oxidant without delay.

References

1. Katsuki, T. *Coord. Chem. Rev.* **1995**, *140*, 189.
2. Kolb, H. C.; VanNieuwenhze, M. S.; Sharpless, K. B. *Chem. Rev.* **1994**, *94*, 2483.
3. Li, G.; Chang, H.-T.; Sharpless, K. B. *Angew. Chem., Int. Ed. Engl.* **1996**, *35*, 451.
4. Rudolph, J.; Sennhenn, P. C.; Vlaar, C. P.; Sharpless, K. B. *Angew. Chem.* **1996**, *35*, 2810.
5. Li, G.; Angert, H. H.; Sharpless, K. B. *Angew. Chem.* **1996**, *35*, 2813.
6. Laxma Reddy, K.; Sharpless, K. B. *J. Am. Chem. Soc.* **1998**, *120*, 1207.
7. Bruncko, M.; Schlingloff, G.; Sharpless, K. B. *Angew. Chem., Int. Ed. Engl.* **1997**, *36*, 1483–1486.
8. Sharpless, K. B.; Chong, A. O.; Oshima, K. *J. Org. Chem.* **1976**, *41*, 177.
9. Mintz, M. J.; Walling, C. *Org. Synth. Coll. Vol. V*; Wiley: New York, **1983**; pp. 183–187.
10. Oliveto, E. P.; Gerold, C. *Org. Synth. Coll. Vol. IV*; Wiley: New York, **1968**; pp. 104–105.
11. Bachand, C.; Driguez, H.; Paton, J. M.; Touchard, D.; Lessard, J. *J. Org. Chem.* **1974**, *39*, 3136.

2: Oxidation of the C=C bond

2.7 Asymmetric aziridination

M. M. FAUL and D. A. EVANS

1. Introduction

The synthesis of functionalized aziridines is currently of great interest due to the increasing importance of these compounds as synthetic intermediates.[1] Their ability to act as carbon electrophiles and their presence in a growing number of natural products[2] has led to substantial interest in the synthesis of these heterocycles. A variety of methods to prepare chiral non-racemic aziridines are available[3] which employ amino alcohols,[4] epoxides,[5] or vicinal diols[6] as starting materials. Two approaches to the asymmetric transition metal-catalysed synthesis of aziridines have been examined. Recent advances in the addition of carbenes to imines, catalysed by $Cu(OTf)_2$ or $Rh_2(OAc)_4$, have resulted in a highly enantioselective synthesis of aziridines through the intermediacy of a chiral sulfur ylid.[7] The second approach involves the transition metal-catalysed addition of a nitrogenous functional group to an olefin in analogy to the cyclopropanation[8] and epoxidation[9] reactions. While the latter functional group transfer reactions have been studied extensively, and highly enantioselective versions of each developed,[10] the catalytic transfer of nitrogen to olefins has received relatively little attention.

Only a few methods of olefin aziridination have demonstrated reasonable efficiency and generality. The formation of aziridines by addition of nitrenes, generated either thermally or photochemically, to olefins is a well-known reaction,[11] although its utility is limited by low yields and competing hydrogen abstraction and insertion processes. An exception to this is ethoxycarbonyl nitrene, produced by decomposition of ethyl azido formate, which reacts with a number of olefins to afford good yields of the corresponding aziridines.[12]

The first report of a stoichiometric asymmetric aziridination reaction was published by Atkinson and Tughan in 1987. Reaction of α,β-unsaturated ester **1** with $Pb(OAc)_4$ and the chiral *N*-aminoquinazoline **2** afforded the aziridine **3** in 64% yield and excellent diastereoselectivity (Scheme 2.7.1).[13] At that time no report of a transition metal-catalysed aziridination reaction employing a

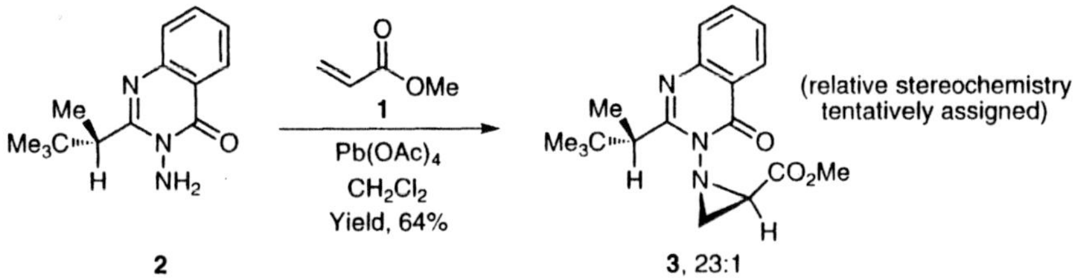

Scheme 2.7.1

chiral catalyst had been published, although racemic variants had been explored.

The first metal-catalysed nitrogen-atom transfer process was reported in 1967 by Kwart and Kahn,[14] who demonstrated that copper powder promoted the decomposition of benzenesulfonyl azide when heated with cyclohexene to afford the cyclohexene aziridine in 15% yield, along with 20% of products derived from C—H bond insertion.

In 1983, Groves and Takahashi reported the first example of activation and stoichiometric transfer of nitrogen from a nitridomanganese(V) porphyrin complex **5** to an olefin (Scheme 2.7.2).[15] Thus, treatment of **5**, isolated in 80% yield by irradiation of **4**, with trifluoroacetic anhydride (TFAA) in the presence of cyclooctene (11 eq.) afforded the *N*-(trifluoroacetyl)-protected aziridine **7** in 82–94% yield. Reaction of the nitrido complex **5** with other olefins was not reported. Recently, the generality of this methodology has been extended to reactions with enolsilanes[16] and carbohydrate glycals.[17]

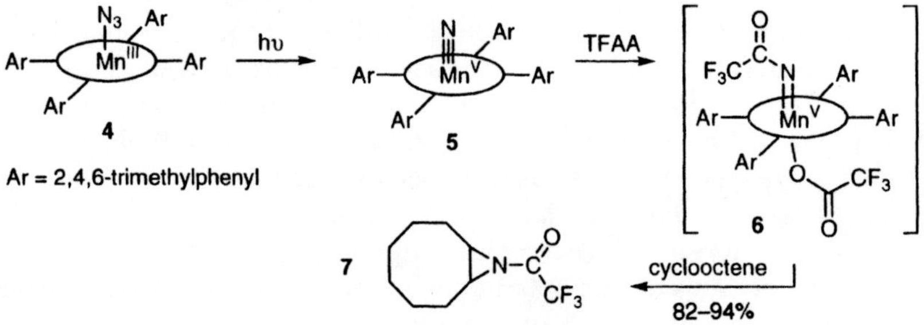

Scheme 2.7.2

In 1984, Mansuy and co-workers reported the aziridination of a number of olefins using [*N*-(*p*-toluenesulfonyl)imino]phenyliodinane, PhI=NTs **8**,[18] as the nitrogen atom source and Fe(III)- and Mn(III)-porphyrins as catalysts (Scheme 2.7.3).[19] Application of this reaction as a general method of aziridine synthesis, however, was limited. Although the Mn(TPP)Cl-catalysed reaction of styrene (**9**) with PhI=NTs gave aziridine **10** in 80% yield, much lower yields were obtained with other olefins (1-hexene, 23%; *trans*-stilbene, 36%). In addition, yields of aziridines in reactions with aliphatic alkenes were low. For example, reaction with cyclohexene (**11**) gave none of the desired aziridine **12** but afforded a 70% yield of the allylic sulfonamide **13**. Other evidence for catalytic imido-group transfer for a limited number of olefinic substrates has been reported.[20]

In 1991, Evans and co-workers made the observation that low-valent copper complexes (Cu(I) and Cu(II)) could catalyse the aziridination of various types of olefins with PhI=NTs.[21] In fact, copper was found to be superior to other metal complexes such as Mn(TPP)Cl, Fe(TPP)Cl, Rh$_2$(OAc)$_4$, and Co(acac)$_2$. Under the optimized conditions (MeCN, 5–10 mol% catalyst, 1 eq.

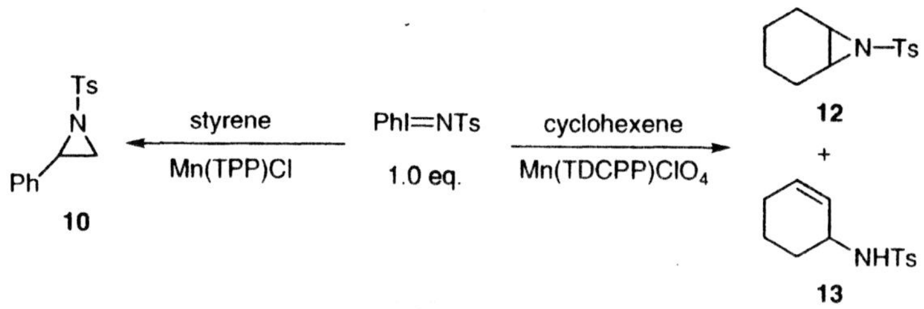

2: Oxidation of the C=C bond

Scheme 2.7.3

PhI=NTs, 5 eq. of olefin), both aromatic and aliphatic substituted olefins afforded the corresponding N-tosylaziridines in yields ranging from 55 to 95%. The cationic Cu(I) salts Cu(MeCN)$_4$ClO$_4$ and CuOTf·1/2PhH were more efficient than the Cu(I) halides.[22,23] Cu(MeCN)$_4$ClO$_4$ was preferred over CuOTf·1/2PhH since it is more easily prepared and less air sensitive. The use of Cu(II) salts was particularly attractive since these complexes are both commercially available and air stable. Their results on olefin aziridination using copper complexes suggested that an analogy between the transition metal-catalysed cyclopropanation and aziridination reactions could be drawn. This result was in contrast to results from the previous studies that employed as catalysts metal complexes that were commonly used in the epoxidation reaction. With the existence of a general method for transition metal-catalysed olefin aziridination, asymmetric variants of this reaction using (i) chiral bis(oxazoline) ligands (Scheme 2.7.4), and (ii) chiral bis(imine) ligands were explored.

As a consequence of the widespread interest in the development of practical asymmetric aziridination processes, a great deal of aziridine-related reaction methodology has appeared recently. A review by Osborn and Sweeney in 1989 provides a good general overview of the complementary methods available for asymmetric aziridine synthesis.[24]

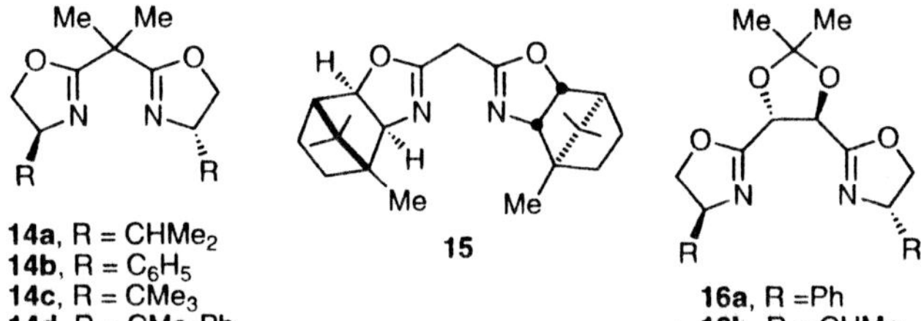

14a, R = CHMe$_2$
14b, R = C$_6$H$_5$
14c, R = CMe$_3$
14d, R = CMe$_2$Ph
14e, R = CMePh$_2$

16a, R = Ph
16b, R = CHMe$_2$
16c, R = CMe$_3$

Scheme 2.7.4

2. Asymmetric aziridination of olefins using chiral bis(oxazoline) ligands

The first example of an asymmetric transition metal-catalysed aziridination reaction was reported by Evans *et al.* in 1991.[25] Bis(oxazoline) ligands **14a–e**, that had previously been successfully employed in the asymmetric cyclopropanation reaction, were examined. Using bis(oxazoline) ligand **14c** and CuOTf·1/2PhH as catalyst, reaction of PhI=NTs **8** with styrene **9** afforded (*R*)-aziridine **17** in 63% enantioselectivity and 94% yield (Scheme 2.7.5). Examination of other bis(oxazoline) ligands **14a,b,d,e** indicated that reaction enantioselectivity was directly related to the steric bulk of the ligand substituent R, and inversely dependent on the solvent polarity and donicity.[26] Thus, for reactions in styrene as the solvent, enantioselectivity increased as the ligand substituent R increased from $CHMe_2$ **14a** (26% ee) to CMe_3 **14c** (63% ee). However, a further increase in the steric demand of the ligand substituent decreased the enantioselectivity of CMe_2Ph, $CMePh_2$ **14d,e** (30% ee, 0°C). Increase in the solvent polarity resulted in a decrease in enantioselectivity (C_6H_6, 57% ee; CH_2Cl_2, 36% ee; MeCN, 6% ee), with the highest selectivity being obtained in neat styrene (63% ee). This solvent effect was attributed initially to partial displacement of the ligand from the metal ion in the more polar solvents. However, investigation of other substrates demonstrated that both the direction and magnitude of the solvent effect was substrate dependent. These results were in contrast to those reported by Lowenthal and Masamune, who demonstrated that the complex derived from *ent*-**15** and CuOTf·1/2PhH afforded aziridine **17** in 88% enantioselectivity and 91% yield.[27] This laboratory has evaluated carefully the Masamune claim and has not been able to reproduce the reported results.[28] Aziridination of styrene using the tartrate-derived bis(oxazolines) **16** has also been examined.[29] Although yields of **17** were high (65–80%), the enantioselectivity decreased as the size of the ligand substituent R increased [**16a**, 12% (*R*); **16b**, 2% (*S*) and **16c**, <5% (*S*)].

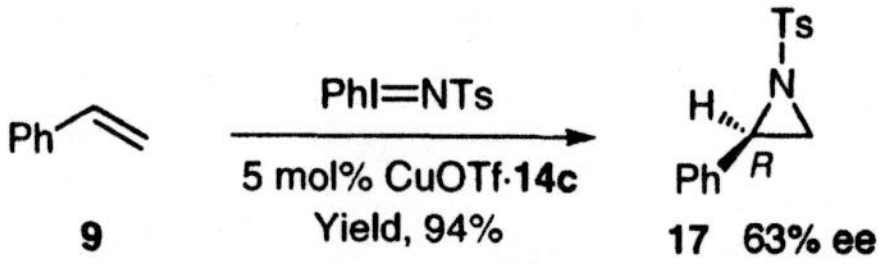

Scheme 2.7.5

In contrast to styrene, aziridination of *trans*-β-methylstyrene (**18**) with $Cu(MeCN)_4ClO_4$·**14c** as catalyst showed increased enantioselectivity for aziridine **19** with increasing solvent polarity and donor strength [benzene (15% ee, 23°C); CH_2Cl_2 (33% ee, 0°C); MeCN (54% ee, 25°C)] (Scheme 2.7.6). Interestingly, the configuration of the phenyl-bearing stereocentre in **19** was opposite to that observed in reaction with styrene. Furthermore, with solvent and temperature held constant (MeCN, 0°C), a nearly linear correlation between enantioselectivity and the *A* value of the ligand substituent R was observed [**14a** (R = $CHMe_2$) 21% ee; **14b**, (R = C_6H_5) 36% ee; and **14c**, (R = CMe_3) 54% ee]. However, as with styrene, the beneficial effect of increased steric bulk was limited and reaction enantioselectivity decreased when bis(oxazoline) ligand **14d** (R = CMe_2Ph) was employed (38% ee). Steric control can be

2: Oxidation of the C=C bond

further improved for cases using the *t*-butyl ligand **14c** by reducing the reaction temperature, although the rate of the reaction was significantly retarded (62% yield, 70% ee, at −30 °C). CuOTf·1/2PhH and Cu(OTf)$_2$ afforded similar selectivities under the reaction conditions.

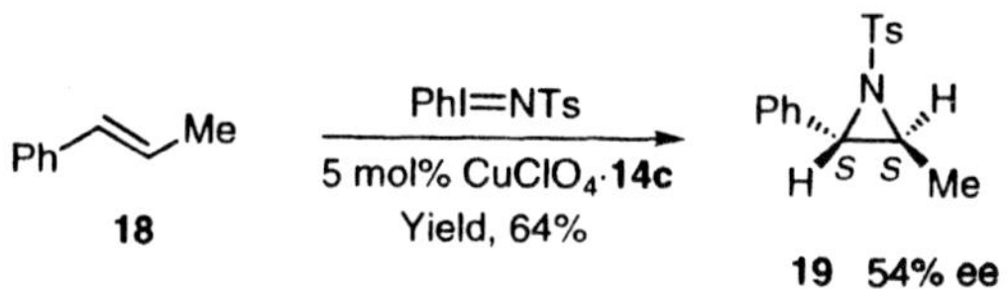

Ph–CH=CH–Me **18** $\xrightarrow{\text{PhI=NTs} \\ \text{5 mol% CuClO}_4\cdot\text{14c} \\ \text{Yield, 64%}}$ Ph–[N-Ts aziridine]–Me **19** 54% ee

Scheme 2.7.6

Although complete stereospecificity was observed in reactions with *trans*-β-methylstyrene, aziridination of *cis*-β-methylstyrene **20** afforded a 70% yield of a 62:38 mixture of *trans*- and *cis*-aziridines **19** and **21** (Scheme 2.7.7).[30] Isomerization of recovered olefin was also observed. The isolated *trans*-aziridine had an enantiomeric excess of 60%, similar to that obtained for *trans*-β-methylstyrene under similar conditions, but the *cis*-aziridine was isolated in only 8% enantioselectivity. These results provided a sharp contrast to aziridination reactions catalysed by ligand-free copper salts, in which very little *trans*-aziridine was produced (<7%) and no olefin isomerization was observed. Similar results were observed with Cu[bis(imine)] complexes *vide infra*.

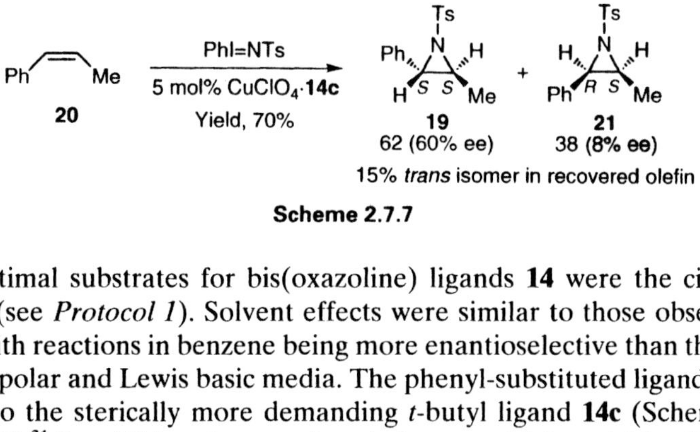

Ph–CH=CH–Me **20** $\xrightarrow{\text{PhI=NTs} \\ \text{5 mol% CuClO}_4\cdot\text{14c} \\ \text{Yield, 70%}}$ **19** (62, 60% ee) + **21** (38, 8% ee)

15% *trans* isomer in recovered olefin

Scheme 2.7.7

The optimal substrates for bis(oxazoline) ligands **14** were the cinnamate esters **22** (see *Protocol 1*). Solvent effects were similar to those observed for styrene with reactions in benzene being more enantioselective than those conducted in polar and Lewis basic media. The phenyl-substituted ligand **14b** was superior to the sterically more demanding *t*-butyl ligand **14c** (Scheme 2.7.9, Table 2.7.1).[31] In addition, for reactions in benzene higher yields of aziridine were obtained when 4-Å molecular sieves were employed. Interestingly, sieves had no effect on reactions run in other solvents, nor did they greatly affect the yields in reactions with other olefins. The nature of the ester group had no effect on the course of the reaction, phenyl cinnamate and *t*-butyl cinnamate gave essentially the same results as the methyl ester.

Although *cis*-olefins were not good substrates for the copper-catalysed aziridination reaction, the *cis*-aziridine was obtained in 84% yield as an 87:13 ratio of *cis*:*trans* isomers from the *trans*-aziridine by treatment with 10 mol%

t-BuOK in *t*-BuOH/THF (Scheme 2.7.8). No loss of optical activity was observed during this process.[28]

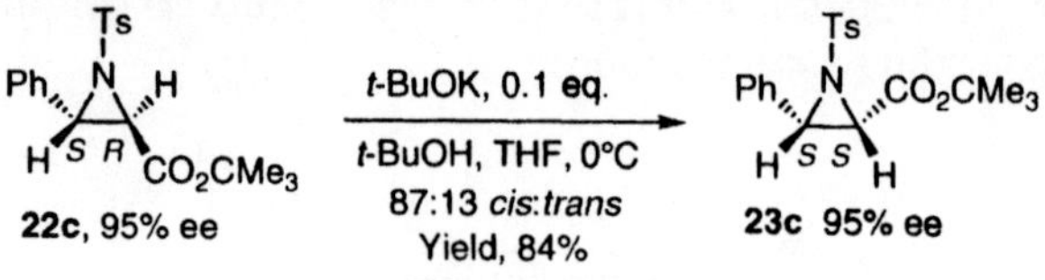
Ts
Ph
N
H
H
S R
CO2CMe3
22c, 95% ee
t-BuOK, 0.1 eq.
t-BuOH, THF, 0°C
87:13 cis:trans
Yield, 84%
Ts
Ph
N
CO2CMe3
H
S S
H
23c 95% ee

Scheme 2.7.8

Table 2.7.1 Enantioselective aziridination of cinnamate Esters **22** (Scheme 2.7.9)

Entry	R	Ligand	Solvent	Yield[a] (%)	ee (%)	Product
1	Me	**14c**	MeCN	16	19	**23a**
2	Me	**14c**	C_6H_6	16	0	**23a**
3	Me	**14b**	MeCN	21	70	**23a**
4	Me	**14b**	C_6H_6	20	94	**23a**
5	Me	**14b**	C_6H_6[b]	63	94	**23a**
6	Ph	**14b**	C_6H_6[b]	64	97	**23b**
7	CMe_3	**14b**	C_6H_6[b]	60	96	**23c**

[a] Values represent isolated yields of aziridine based on olefin.
[b] 4-Å molecular sieves were used.

Protocol 1. Synthesis of (2*R*,3*S*)-*N*-*p*-toluenesulfonyl-2-carbomethoxy-3-phenylaziridine (Structure 23a, Scheme 2.7.9)

Caution! Carry out all procedures in a well-ventilated hood, and wear disposable vinyl or latex gloves and chemical-resistant safety goggles.

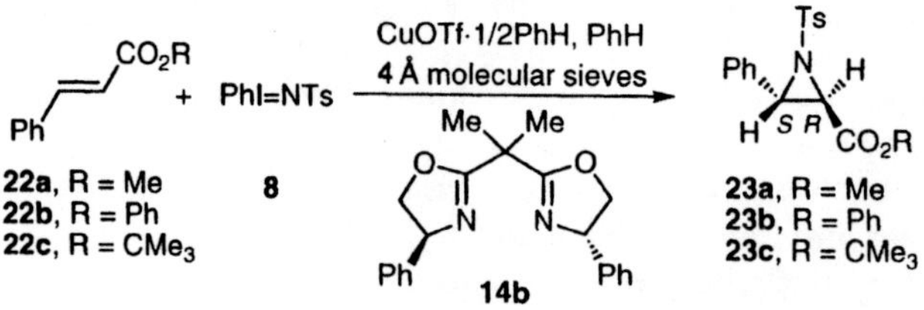
CO2R
Ph
+
PhI=NTs
22a, R = Me
22b, R = Ph
22c, R = CMe3
8
CuOTf·1/2PhH, PhH
4 Å molecular sieves
Me
Me
O
O
N
N
Ph
Ph
14b
Ts
Ph
N
H
H
S R
CO2R
23a, R = Me
23b, R = Ph
23c, R = CMe3

Scheme 2.7.9

This reaction is representative of the optimal procedure for aziridination of cinnamate esters. This method cannot be reliably extrapolated to other olefinic substrates.

Equipment

- Single-necked pear-shaped flask (10 mL)
- Teflon-coated magnetic stirrer bar
- Source of dry N_2 (preferably from a nitrogen line)
- All-glass syringe with a needle-lock luer (10 mL)

2: Oxidation of the C=C bond

- Single-necked, round-bottomed flask (50 mL)
- Septum, B14/20
- Cannula (12 in, 18 gauge)
- Sintered glass filtered funnel (porosity 10–20 μm)
- Single-necked, round-bottomed flask (250 mL)
- Column for chromatography (35 cm × 20 mm)
- Standard 5-mm NMR tube

Materials

(S)-2,2'-(1-Methylethylidene)bis(4,5-dihydro-4-phenyloxazole) **14b**, 50.1 mg, 0.150 mmol	irritant
Benzene,[a] 6.5 mL	toxic, cancer-suspect agent
CuOTf·1/2PhH,[b] 31.5 mg, 0.125 mmol	air sensitive
[N-(p-Toluenesulfonyl)imino]phenyliodinane, PhI=NTs **8**, 1.87 g, 5.00 mmol	
trans-Methyl cinnamate, 0.405 g, 2.50 mmol	
4-Å Molecular sieves,[c] 2.5 g	
Hexane for chromatography	flammable, irritant
Ethyl acetate for chromatography	flammable
EM reagents silica gel 60 (230–400 mesh)	harmful by inhalation
d_6-Benzene, 10 mL	toxic, cancer-suspect agent
Tris[3-(heptafluoropropylhydroxymethylene)-(+)-camphorato]europium(III), Eu(hfc)$_3$, 10 mg	hygroscopic

Method

1. Dry all glassware and stirrer bars in an electric oven (105°C) for 1 h prior to use.

2. In a dry-box, transfer into a pear-shaped round-bottomed flask (10 mL), (S)-2,2'-(1-methylethylidene)bis(4,5-dihydro-4-phenyloxazole) **14b** (50.1 mg, 0.150 mmol), CuOTf·1/2PhH (31.5 mg, 0.125 mmol) and a magnetic stirrer.[d] Remove the flask from the dry box and maintain under a positive pressure of dry nitrogen. Using an all-glass syringe (10 mL) with a Luer needle-lock, add benzene (6 mL) to the solids and stir the resulting mixture at room temperature for 30 min.

3. Add into a single-necked round-bottomed flask (50 mL), capped with a rubber septum, trans-methyl cinnamate (0.405 g, 2.5 mmol), PhI=NTs **8** (1.87 g, 5.00 mmol) and activated 4-Å molecular sieves (2.5 g).

4. Transfer by cannula, under a positive pressure of nitrogen, the CuOTf·1/2PhH·**14b** benzene solution into the round-bottomed flask (50 mL) prepared in step 3. Rinse the flask that contained the CuOTf·1/2PhH·**14b** with benzene (0.5 mL). Stir the reaction at room temperature for 24 h. PhI=NTs is insoluble in benzene. The reaction is complete when all the PhI=NTs has been drawn into solution.

5. Quench the reaction by diluting with 50% hexane–ethyl acetate (25 mL) and filter through a short plug of silica (2 × 2.5 cm) in a sintered glass filter funnel. Rinse the silica with additional portions of 50% hexane–ethyl acetate (2 × 25 mL). Transfer the filtrate into a round-bottomed flask (250 mL) and evaporate under reduced pressure.

6. Prepare a silica gel column using 80:20 hexane–ethyl acetate as eluent, and load the column by dissolving the residue in the minimum volume of 50:50

Protocol 1. *Continued*

hexane–ethyl acetate. Elute the column with 80:20 hexane–ethyl acetate and the desired aziridine elutes with an R_f of 0.21. Evaporate the eluates to yield 0.53 g (63%) of aziridine **23a** as a white crystalline solid.

7. The enantiomeric excess of (2*R*,3*S*)-*N*-*p*-toluenesulfonyl-2-carbomethoxy-3-phenylaziridine (**23a**) is determined by ^{1}H NMR (500 MHz) using the chiral shift reagent Eu(hfc)$_3$. To a solution of 0.5 mL of aziridine **23a** (0.08 M) in d_6-benzene in an NMR tube, add aliquots (0.1 mL) of the shift reagent (0.07 M) also in d_6-benzene. Record the NMR spectrum after each addition. Observe a baseline separation for the diagnostic resonances (typically seen after introduction of 5–15 mol% of the shift reagent). Employ a 1.00-s pulse delay and zero fill to 64K FIDs to enhance the digital resolution of the spectra. Monitor the resonances for the Ar-*H* and Ar-*CH$_3$* of the *N*-tosyl group and the methyl of the carboxylate. Calculate the enantiomeric excess from the integrated area of the two corresponding diastereomeric resonances. After the measurement is complete, add racemic aziridine (2–5%) to the sample and record a final spectrum to verify assignment of the diastereomeric resonances.[e]

[a] Benzene was distilled from calcium hydride under nitrogen prior to use.
[b] CuOTf·1/2PhH was prepared by the published method and handled in a dry box under an atmosphere of nitrogen. This reagent is grey/white in colour, it readily oxidizes when exposed to air to a green solid which will be inactive in the reaction.
[c] Powdered 4-Å molecular sieves were activated at 300°C for 24 h and stored in a dry box prior to use. Molecular sieves were employed only in aziridination reactions involving benzene.
[d] A reaction carried out under identical conditions using Cu(OTf)$_2$ 45.2 mg (0.125 mmol) as the copper source gave 0.309 g (37%) of aziridine **23a** in 95% ee.
[e] The enantiomeric excess of **23b** is determined using Eu(tfc)$_3$. However, the enantiomeric excess of **23c** was measured using Eu(hfc)$_3$.

---

3. Asymmetric aziridination of olefins using chiral bis(imine) ligands

Chiral bis(imine) ligands (Scheme 2.7.10) have been employed successfully in the catalytic asymmetric epoxidation reaction. Their use in the asymmetric aziridination reaction has been examined by a number of groups. The first report of the aziridination of styrene and *cis*-β-methylstyrene catalysed by complexes **24a,b** was published by Burrows and co-workers.[32] Although the aziridines were isolated in good yields (44–46%), no enantioselectivity was observed. Related complexes (M = Cr^{III}, Mn^{III}, Fe^{III}, Co^{II}, Cu^{II}, Rh^{III} and Pd^{II}) were also examined but all failed to promote aziridination.

Katsuki and co-workers examined the use of bis(imine) complexes in the aziridination of styrene.[33] The optimal catalyst was complex **25** which afforded aziridine **17** in 94% enantioselectivity and 76% yield. This result represents the highest enantioselectivity reported to date for aziridination of styrene. However, extension of these conditions to other olefins was unsuccessful. The

2: Oxidation of the C=C bond

only other substrate reported was indene, which was aziridinated in 50% enantioselectivity and 10% yield under identical reaction conditions.

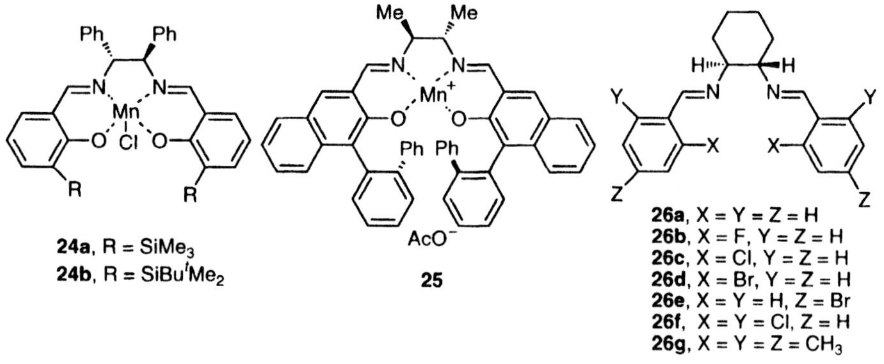

24a, R = SiMe$_3$
24b, R = SiBu^{t}Me$_2$

25

26a, X = Y = Z = H
26b, X = F, Y = Z = H
26c, X = Cl, Y = Z = H
26d, X = Br, Y = Z = H
26e, X = Y = H, Z = Br
26f, X = Y = Cl, Z = H
26g, X = Y = Z = CH$_3$

Scheme 2.7.10

Table 2.7.2 Asymmetric aziridination of 6-cyano-2,2-dimethylchromene (**27**) catalysed by CuOTf·1/2PhH in the presence of ligands 26a–g (Scheme 2.7.11)

Ligand	26a	26b	26c	26d	26e	26f	26g
ee (%)	50	64	72	81	42	>98	92
Turnover	10	3.6	3.6	8.2	3.6	16	6.1

The most extensive work using the bis(imine)-based ligand system has been reported by Jacobsen and co-workers, who determined that Cu(I) salts were the optimal catalysts for the bis(imine) ligands and that multiple open coordination sites on the copper were crucial for successful aziridination.[34] The optimal ligand was determined by examining the enantioselectivity and catalytic turnover in the aziridination of 6-cyano-2,2-dimethylchromene **27** with **26**, benzylidene derivatives of 1,2-diaminocyclohexane (Scheme 2.7.11, Table 2.7.2). Although the parent bis(benzylidenamino)cyclohexane **26a**, in association with CuOTf·1/2PhH, exhibited moderate catalytic turnover and enantioselectivity for **28** (10 turnovers, 50% ee), significant improvement was observed with the chloro-substituted derivative **26f**. Comparison of sterically similar ligands **26f** and **26g** indicated that the electronic properties have an effect on both the catalyst lifetime and selectivity.

Based on the results with 6-cyano-2,2-dimethylchromene, aziridination of a number of *cis*- and *trans*-olefins with CuOTf·1/2PhH·**26f** was examined (Table 2.7.3). In fact, it was determined that the olefins aziridinated with good enantioselectivity were those that were also successful in the corresponding Mn(salen)-catalysed epoxidation reaction. In both processes, *trans*-olefins such as stilbene were poor substrates with regard to both selectivity and rate, whereas 2,2-dimethylchromene derivatives undergo reaction with high selectivity. In addition, this oxidation reaction was found to be non-stereospecific,

Table 2.7.3 Asymmetric aziridination of alkenes catalysed by (S,S)-26f/CuOTf·1/2PhH

Entry	Olefin	Yield (%)	ee (%)	Aziridine confing.
1	1,2-Dihydronaphthalene	70	87	(1R,2S)-(+)
2	Indene	50	58	(1R,2S)-(+)
3	cis-β-Methylstyrene	79 (cis:trans 3:1)	67 (cis) 81 (trans)	(1R,2S)-(−) (1S,2S)-(−)
4	trans-Stilbene	nd	30	nd

with acyclic olefins affording a mixture of the *cis* and *trans* products (entries 3 and 4). Aziridination of styrene **9** with CuOTf·1/2PhH·**26f** afforded aziridine **17** in 66% enantioselectivity and 79% yield. The poor enantioselectivity observed with this substrate was thought to be due to this non-stereospecific pathway, since the *trans* pathway for terminal olefins constitutes a mechanism for enantiomeric leakage.[33a]

Protocol 2.
Synthesis of (3R,4R)-(+)-(N-p-toluenesulfonyl)-6-cyano-2,2-dimethylchromane(1,2)imine (Structure 28, Scheme 2.7.11)

Caution! Carry out all procedures in a well-ventilated hood, and wear disposable vinyl or latex gloves and chemical-resistant safety goggles.

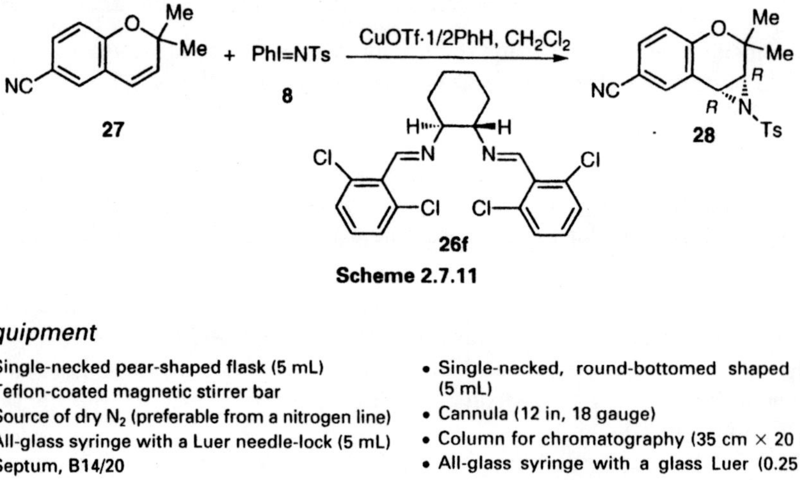

Scheme 2.7.11

Equipment

- Single-necked pear-shaped flask (5 mL)
- Teflon-coated magnetic stirrer bar
- Source of dry N_2 (preferable from a nitrogen line)
- All-glass syringe with a Luer needle-lock (5 mL)
- Septum, B14/20

- Single-necked, round-bottomed shaped flask (5 mL)
- Cannula (12 in, 18 gauge)
- Column for chromatography (35 cm × 20 mm)
- All-glass syringe with a glass Luer (0.25 mL)

Materials

- (1S,2S)-Bis(2,6-dichlorobenzylidenamino)cyclohexane **26f**, 23.5 mg, 0.055 mmol
- Dichloromethane,[a] 3.0 mL   harmful by inhalation
- CuOTf·1/2PhH,[b] 12.6 mg, 0.05 mmol   air sensitive
- [N-(p-Toluenesulfonyl)imino]phenyliodinane, PhI=NTs **8**, 279.9 mg, 0.75 mmol

2: Oxidation of the C=C bond

- 6-Cyano-2,2-dimethylchromene **27**, 92.6 mg, 0.50 mmol
- 4-Å Molecular sieves,[c] 400 mg
- Hexadecane, 60 μL
- Hexane for chromatography — flammable, irritant
- Ethyl acetate for chromatography — flammable
- Wolem 32064 silica gel, 5 g — harmful by inhalation

Method

1. Dry all glassware and stirrer bars in an electric oven (105 °C) for 1 h prior to use.

2. In a dry box, transfer into a pear-shaped round-bottomed flask (5 mL) CuOTf·1/2PhH (12.6 mg, 0.05 mmol) and a magnetic stirrer. Remove the flask from the dry-box and maintain under a positive pressure of dry nitrogen. Using an all-glass syringe with a Luer needle-lock (5 mL) add dichloromethane (1.5 mL) into the flask, followed by (1S,2S)-bis[(2,6-dichlorobenzylidene)diamino]cyclohexane (**26f**) (23.5 mg, 0.055 mmol). Stir the resulting bright yellow mixture at room temperature for 15 min.

3. Transfer by cannula, under a positive pressure of nitrogen, the CuOTf·1/2PhH·**26f** solution into a single-necked round-bottomed flask (5 mL) capped with a rubber septum. Dilute with additional dichloromethane to maintain a reaction volume of 1.5 mL. Add to this solution, 4-Å molecular sieves (400 mg), 6-cyano-2,2-dimethylchromene (92.6 mg, 0.5 mmol) and hexadecane (60 μL) as an internal standard. Remove an aliquot of the reaction and determine by GLC the ratio of olefin to hexadecane.[d] Cool the solution to −78 °C and add PhI=NTs **8** (279.9 mg, 0.75 mmol) against a positive nitrogen counterflow.

4. Stir the reaction at −78 °C and monitor by GLC until complete. When reaction progress has ceased (2–3 days), remove the solvent under reduced pressure on a rotary evaporator.

5. Prepare a silica gel column using 80:20 hexane–ethyl acetate as eluent, and load the column by dissolving the residue in the minimum volume of 50:50 hexane–ethyl acetate. Elute the column with 80:20 hexane–ethyl acetate. Evaporate the eluates to yield 133 mg (75%) of aziridine **28** as a crystalline solid.

6. The enantiomeric excess of aziridine **28** determined by HPLC was 98%.[e]

[a] Dichloromethane was distilled from calcium hydride under nitrogen prior to use.
[b] CuOTf·1/2PhH was prepared by the published method and handled in a dry box under an atmosphere of nitrogen. This reagent is grey/white in colour, it readily oxidizes when exposed to air to a green solid which will be inactive in the reaction.
[c] Powdered 4-Å molecular sieves, activated at 300 °C for 24 h, are stored in a dry box prior to use.
[d] The reaction is monitored by GLC analyses on a Hewlett Packard 5890 Series II instrument equipped with an FID detector and a Hewlett Packard 3396A integrator using a J & W Scientific 30 m × 0.32 mm i.d. DB-5 capillary column.
[e] Chiral HPLC analyses are performed using a Hewlett Packard 1050 series and a Spectra-Physics Isochrome high-performance liquid chromatography system on a commercial Whelk-O column (Regis) (10% 2-propanol:hexane).

Jacobsen and co-workers also completed a detailed study of the mechanism of the aziridination reaction. Although a number of mechanisms are possible, the evidence obtained to date supports a redox pathway for olefin aziridination catalysed by the [bis(imine)] copper complexes.[35]

4. Conclusions

The preceding discussion provides an overview of the status of asymmetric olefin aziridination with chiral metal complexes. It has been the collective experience of a number of groups that chiral copper(I), copper(II), and manganese(III) complexes are evolving as the catalysts of choice for enantioselective olefin aziridination. At this point, additional work is needed to make this reaction truly general; nevertheless, considerable progress has been made towards this goal.

References

1. (a) Deyrup, J. A. In *The Chemistry of Heterocyclic Compounds*; Hassner, A., ed.; John Wiley and Sons: New York, **1983**; *42*, Part 1, pp. 1–214. (b) Padwa, A.; Woolhouse, A. D. In *Comprehensive Heterocyclic Chemistry;* Lwowski, W., ed.; Pergamon Press: Oxford, **1984**; 7, pp. 47–93. (c) Dermer, O. C.; Ham, G. E. *Ethyleneimine and other Aziridines*; Academic Press: New York and London, **1969**.
2. (a) Iyer, V. N.; Szybalski, W. *Science* **1964**, *145*, 55. (b) Akhtar, M. H.; Begleiter, A.; Johnson, D.; Lown, J. W.; McLaughlin, L.; Sim, S.-K. *Can. J. Chem.* **1975**, *53*, 2891. (c) Connors, T. A.; Melzack, D. H. *Int. J. Cancer* **1971**, *7*, 86.
3. (a) Kemp, J. E. G. In *Comprehensive Organic Synthesis VII*; Trost, B. M.; Fleming, I., ed.; Pergamon Press: Oxford, **1991**, pp. 469–513. (b) Tanner, D. *Angew. Chem., Int. Ed. Engl.* **1994**, *33*, 599.
4. (a) Bates, G. S.; Varelas, M. A. *Can. J. Chem.* **1980**, *58*, 2562. (b) Yahiro, N. *Chem. Lett.* **1982**, 1479. (c) Pfister, J. R. *Synthesis* **1984**, 969. (d) Kelly, J. W.; Eskew, N. L.; Evans, S. A. *J. Org. Chem.* **1986**, *51*, 95.
5. (a) Sinou, D.; Emziane, M. *Tetrahedron Lett.* **1986**, *27*, 4423. (b) Godfrey, J. D. J.; Gordon, E. M.; Von Langen, D. J. *Tetrahedron Lett.* **1987**, *28*, 1603. (c) Legters, J.; Thijs, L.; Zwanenburg, B. *Tetrahedron Lett.* **1989**, *30*, 4881.
6. (a) Lohray, B. B.; Ahuja, J. R. *J. Chem. Soc., Chem. Commun.* **1991**, 95. (b) Lohray, B. B.; Gao, Y.; Sharpless, K. B. *Tetrahedron Lett.* **1989**, *30*, 2623. (c) Chang, H.-T.; Sharpless, K. B. *Tetrahedron Lett.* **1996**, *37*, 3219.
7. Aggrawal, V. K.; Thompson, A.; Jones, R. V. H.; Standen, M. C. H. *J. Org. Chem.* **1996**, *61*, 8368.
8. Doyle, M. P. *Chem. Rev.* **1986**, *86*, 919.
9. Jørgensen, K. A. *Chem. Rev.* **1989**, *89*, 431.
10. Cyclopropanations: (a) Doyle, M. P. In *Catalytic Asymmetric Synthesis*; Ojima, I., ed.; VCH Publishers, Inc.: New York, **1993**; pp. 63–99. Epoxidations: (b) Johnson, R. A.; Sharpless, K. B. In *Catalytic Asymmetric Synthesis*; Ojima, I. ed.; VCH Publishers, Inc.: New York, **1993**; pp. 101–158. (c) Jacobsen, E. N. In *Catalytic Asymmetric Synthesis*; Ojima, I., ed.; VCH Publishers, Inc.: New York, **1993**; pp. 159–202.

11. (a) Lwowski, W. In *Nitrenes*; Lwowski, W., ed.; Interscience: New York, **1970**; pp. 185–224. (b) Edwards, O. E. In *Nitrenes*; Lwowski, W., ed.; Interscience: New York, **1970**; pp. 225–243. (c) Lwowski, W. In *Acyl Azides and Nitrenes: Reactivity and Utility*; Scriven, E. F. V., ed.; Academic Press: Orlando, FL, 1985; pp. 205–246.
12. (a) Pellacani, L.; Persia, F.; Tardella, P. A. *Tetrahedron Lett.* **1980**, *21*, 4927. (b) Lociuro, S.; Pellacani, L.; Tardella, P. A. *Tetrahedron Lett.* **1983**, *24*, 593. (c) Loreto, M. A.; Pellacani, L.; Tardella, P. A. *Tetrahedron Lett.* **1989**, *30*, 2975. (d) Fioravanti, S.; Loreto, M. A.; Pellacani, L.; Tardella, P. A. *Tetrahedron* **1991**, *47*, 5877.
13. Atkinson, R. S.; Tughan, G. *J. Chem Soc., Perkin Trans. 1* **1987**, 2803. These reactions were run using racemic aminoquinazolinones.
14. Kwart, H.; Kahn, A. A. *J. Am. Chem. Soc.* **1967**, *89*, 1951.
15. Groves, J. T.; Takahashi, T. *J. Am. Chem. Soc.* **1983**, *105*, 2073.
16. Du Bois, J.; Hong, J.; Carreira, E. M.; Day, M. W. *J. Am. Chem. Soc.* **1996**, *118*, 915.
17. Du Bois, J.; Tomooka, C. S.; Hong, J.; Carreira, E. M. *J. Am. Chem. Soc.* **1997**, *119*, 3179.
18. (a) Yamada, Y.; Yamamoto, T.; Okawara, M. *Chem. Lett.* **1975**, 361. (b) Besenyei, G.; Németh, S.; Simándi, L. *Tetrahedron Lett.* **1993**, *34*, 6105.
19. (a) Mansuy, D.; Mahy, J.-P.; Dureault, A.; Bedi, G.; Battioni, P. *J. Chem. Soc., Chem. Commun.* **1984**, 1161. (b) Mahy, J.-P.; Bedi, G.; Battioni, P.; Mansuy, D. *Tetrahedron Lett.* **1988**, *29*, 1927. (c) Mahy, J.-P.; Bedi, G.; Battioni, P.; Mansuy, D. *J. Chem. Soc., Perkin. Trans. 2.* **1988**, 1517.
20. (a) Sharpless, K. B.; Chong, A. O.; Oshima, K. *J. Org. Chem.* **1976**, *41*, 177. (b) Breslow, R.; Gellman, S. H. *J. Chem. Soc., Chem. Commun.* **1982**, 1400. (c) Breslow, R.; Gellman, S. H. *J. Am. Chem. Soc.* **1983**, *105*, 6728. (d) Barton, D. H. R.; Hay-Motherwell, R. S.; Motherwell, W. B. *J. Chem. Soc., Perkin. Trans.1.* **1983**, 445. (e) Svastits, E. W.; Dawson, J. H.; Breslow, R.; Gellman, S. H. *J. Am. Chem. Soc.* **1985**, *107*, 6427. (f) Harlan, E. W.; Holm, R. H. *J. Am. Chem. Soc.* **1990**, *112*, 186.
21. (a) Evans, D. A.; Faul, M. M.; Bilodeau, M. T. *J. Org. Chem.* **1991**, *56*, 6744. (b) Evans, D. A.; Faul, M. M.; Bilodeau, M. T. *J. Am. Chem. Soc.* **1994**, *116*, 2742.
22. Kubas, G. J. *Inorg Synth.* **1979**, *19*, 90.
23. Cohen, T.; Ruffner, R. J.; Shull, D. W.; Fogel, E. R.; Flack, J. R. *Organic Syntheses, Collective Volume VI*; Wiley: New York, **1988**; pp. 737–744.
24. Osborn, H. M.; Sweeney, J. *Tetrahedron: Asymmetry* **1989**, *45*, 2875–2886.
25. Evans, D. A.; Woerpel, K. A.; Hinman, M. M.; Faul, M. M. *J. Am. Chem. Soc.* **1991**, *113*, 726.
26. Evans, D. A.; Faul, M. M.; Bilodeau, M. T.; Anderson, B. A.; Barnes, D. M. *J. Am. Chem. Soc.* **1993**, *115*, 5328.
27. Lowenthal, R. E.; Masamune, S. *Tetrahedron Lett.* **1991**, *32*, 7373.
28. Bilodeau, M. T., Ph. D. thesis, Harvard University, **1993**.
29. (a) Harm, A. M.; Knight, J. G.; Stemp, G. *Synlett* **1996**, 677. (b) Harm, A. M.; Knight, J. G.; Stemp, G. *Tetrahedron Lett.* **1996**, *37*, 6189.
30. Evans, D. A.; Anderson, B. A. Unpublished results, Harvard University, **1993**.
31. The asymmetric synthesis of aziridines by addition of carbenes to imines catalysed by [CuPF$_6$(CH$_3$CN)$_4$]·**14b** has been reported: Hansen, K. B.; Finney, N. S.; Jacobsen, E. N. *Angew. Chem., Int. Ed. Engl.* **1995**, *34*, 676.

32. O'Connor, K. J.; Wey, S.-J.; Burrows, C. J. *Tetrahedron Lett.* **1992**, *33*, 1001.
33. (a) Noda, K.; Hosoya, N.; Irie, R.; Ito, Y.; Katsuki, T. *Synlett* **1993**, 469. (b) Nishikori, H.; Katsuki, T. *Tetrahedron Lett.* **1996**, *37*, 9245.
34. Li, Z.; Conser, K. R.; Jacobsen, E. N. *J. Am. Chem. Soc.* **1993**, *115*, 5326.
35. (a) Zhang, W.; Lee, N. H.; Jacobsen, E. N. *J. Am. Chem. Soc.* **1994**, *116*, 425. (b) Li, Z.; Quan, R. W.; Jacobsen, E. N. *J. Am. Chem. Soc.* **1995**, *117*, 5889.

2.8 Asymmetric hydroxylations of enolates and enol derivatives

P. ZHOU, B.-C. CHEN, and F. A. DAVIS

1. Introduction

A feature common to many biologically relevant molecules is the α-hydroxy carbonyl array.[1,2] Compounds containing this moiety are also important auxiliaries and building blocks for the asymmetric synthesis of many natural products, including antitumour agents, antibiotics, pheromones, and sugars. Of the many methods devised for their synthesis, the α-hydroxylation of enolates and enol derivatives is one of the simplest and most frequently used methods (Scheme 2.8.1). In this procedure, a preformed enolate or its derivative is treated with an aprotic oxidizing reagent to give the corresponding α-hydroxy carbonyl product. With chiral enolates and/or asymmetric oxidizing reagents, optically active examples are available and, in many cases, high asymmetric induction can be achieved by appropriate choice of the oxidizing reagent and reaction conditions.

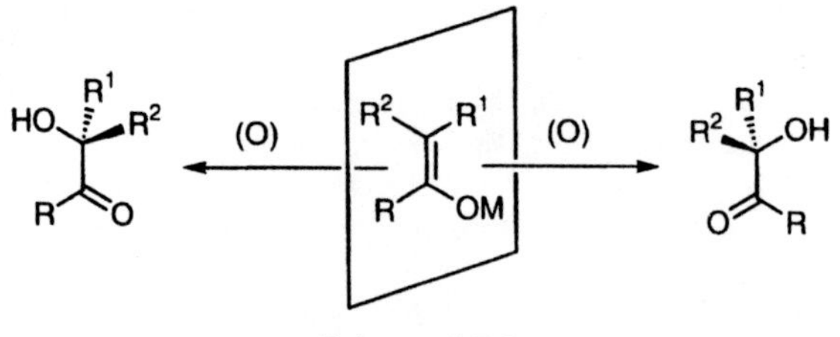

Scheme 2.8.1

2. Asymmetric α-hydroxylation of enolates

Enolates are versatile intermediates in organic synthesis and are generally prepared *in situ* from enolizable carbonyl substrates and a suitable base. A variety of bases are available for such purpose, among which the most common are the metal alkoxides and amides. The first oxidizing reagent explored for the α-hydroxylation of enolates was molecular oxygen. This reaction ini-

2: Oxidation of the C=C bond

tially gives an unstable α-hydroperoxy intermediate which usually undergoes decomposition resulting in the formation of a complex mixture of products.[2] In some cases, the intermediate α-hydroperoxy carbonyl compounds can be isolated and subsequently reduced to the α-hydroxy carbonyl products.[3] However, autoxidation of enolates with oxygen in the presence of an *in situ* reducing reagent such as trialkyl phosphite is the preferred method for the preparation of α-hydroxy carbonyl compounds. An example is found in the synthesis of the natural product clausenamide, **2** (Scheme 2.8.2).[4] In this reaction the lithium enolate of lactam **1** is treated with molecular oxygen in the presence of triethyl phosphite to give clausenamide **2** in 50% yield.

Protocol 1.
Synthesis of (3*S*,4*R*,5*R*,7*S*)-3-hydroxy-5-(α-hydroxybenzyl)-1-methyl-4-phenylpyrrolidin-2-one (Structure 2, Scheme 2.8.2)

Caution! Hexamethylphosphoric acid triamide and trimethyl phosphite are both potential carcinogens. Carry out all procedures in a well-ventilated hood, and wear disposable vinyl or latex gloves and chemical-resistant safety goggles.

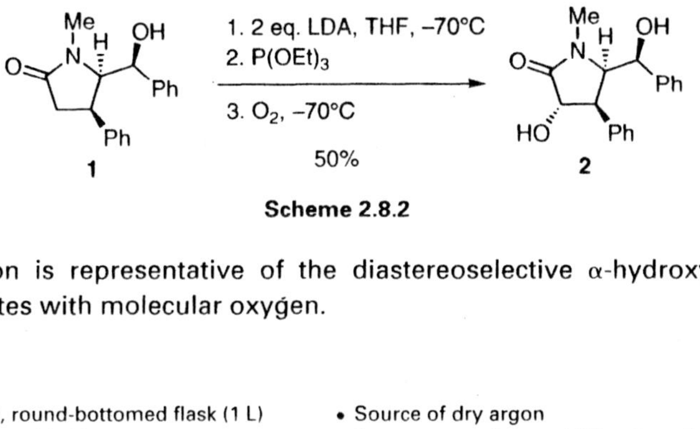

Scheme 2.8.2

1. 2 eq. LDA, THF, −70°C
2. P(OEt)$_3$
3. O$_2$, −70°C

50%

This reaction is representative of the diastereoselective α-hydroxylation of chiral enolates with molecular oxygen.

Equipment

- Three-necked, round-bottomed flask (1 L)
- Separating funnel (3 L)
- Thermostated hotplate stirrers
- Cooling baths
- Teflon-coated magnetic stirrer bar
- Source of dry argon
- Pressure-equalizing addition funnels (200 mL)
- Sintered glass filter funnels
- Three-necked round-bottomed flask (500 mL)
- Erlenmeyer flasks

Materials

- Dry tetrahydrofuran, 575 mL — flammable, irritant
- Diisopropylamine, 22.1 mL, 157.7 mmol — flammable, corrosive
- *n*-Butyllithium, 1.5 M in hexane, 103 mL, 154.5 mmol — pyrophoric
- (4*R*,5*R*,7*S*)-5-(α-Hydroxybenzyl)-1-methyl-4-phenylpyrrolidin-2-one, 17.7 g, 62.8 mmol — unknown toxicity, treat as toxic
- Hexamethylphosphoric acid triamide (HMPA), 130 mL — highly toxic, cancer-suspect agent
- Trimethyl phosphite, 5.3 mL, 44.9 mmol — flammable, cancer-suspect agent
- Dry oxygen, as needed — non-flammable gas oxidizer
- Hydrochloric acid, 0.5 M, 600 mL — highly toxic
- Magnesium sulfate, anhydrous, as needed

Protocol 1. *Continued*

• Diethyl ether, 100 mL	flammable, toxic
• Pentane, as needed	flammable, irritant
• 2-Propanol, as needed	flammable, irritant
• Ethyl acetate, 2 L	flammable, irritant
• Methanol, as needed	flammable, highly toxic
• Molybdophosphoric acid, as needed	corrosive, oxidizer
• Sulfuric acid, as needed	highly toxic oxidizer
• Phosphorus pentaoxide, as needed	highly toxic, corrosive

Method

1. Equip a 1-L three-necked round-bottomed flask with an argon bubbler, a 200-mL dropping funnel, a rubber septum and a magnetic stirring bar. The apparatus is maintained under a positive pressure of argon.

2. (4R,5R,7S)-5-(α-Hydroxybenzyl)-1-methyl-4-phenylpyrrolidin-2-one (17.7 g, 62.8 mmol) in anhydrous tetrahydrofuran (490 mL) and absolute HMPA (130 mL) is added to the reaction vessel, cooled to $-70°C$ and stirred.

3. A 500-mL three-necked round-bottomed flask is equipped with an argon bubbler, a 200-mL dropping funnel, a rubber septum and a magnetic stirring bar. The apparatus is maintained under a positive pressure of argon and diisopropylamine (22.1 mL) in THF (80 mL) is added. The solution is stirred at $-20°C$ and n-butyllithium (1.5 M solution in hexane, 103 mL) is added dropwise. After the addition is complete the solution is warm to 0°C and stirred for 0.5 h.

4. The LDA solution prepared in step 3 is added dropwise to the solution obtained in step 2 and the reaction mixture stirred for 1 h at -70 to $-60°C$.

5. To the reaction mixture is added a solution of freshly distilled trimethyl phosphite (5.3 mL) in anhydrous tetrahydrofuran (5 mL) and absolute oxygen (dried over H_2SO_4 and P_4O_{10}) is passed through the solution at a rate of 50–100 mL/min for 2–3 h.[a]

6. Pour the reaction mixture into dilute HCl (0.5 M, 600 mL) while cooling with ice and acidify the mixture to pH 3–4.

7. Transfer the mixture into a separating funnel, separate the phases and extract the aqueous phase with ethyl acetate (4 × 500 mL). Combine the organic extracts, wash them with water (3 × 300 mL), dry the organic phase over anhydrous magnesium sulfate. Filter to remove magnesium sulfate and remove the solvents on a rotary evaporator to give a residue.[b]

8. Take up the residue with diethyl ether (50–100 mL) and stir the mixture until crystallization takes place. Add pentane slowly with stirring until no further cloudiness is observed and leave the mixture in a refrigerator overnight. Filter with suction to give crude product (17 g, containing 35–40% starting material) and recrystallize twice from 2-propanol to give 7.65 g (41%) of clausenamide

2: Oxidation of the C=C bond

(about 95% pure by ^{1}H NMR): m.p. 236–237.5 °C. The following spectra are diagnostic: IR (KBr) 3402, 3321, 1689 cm^{-1}; ^{1}H NMR (300 MHz, DMSO-d_6) δ 3.01 (s, 3H), 3.50 (dd, J = 8 Hz, J = 10.5 Hz, 1H), 3.82 (dd, J = 10 Hz, J = 7 Hz, 1H), 4.30 (dd, J = 8 Hz, J = 2 Hz, 1H), 4.65 (dd, J = 2 Hz, J = 3 Hz, 1H), 5.39 (d, J = 7 Hz, 1H), 5.45 (d, J = 3 Hz, 1H), 6.61–6.64 (m, 2H), 7.03–7.28 (m, 8H).

[a] At which time the product to starting material ratio no longer changes (checked by TLC, SiO$_2$; 2:1 ethyl acetate–methanol, product R_f 0.3, starting material R_f 0.37, molybdatophosphoric acid spray reagent for visualization).

[b] Alternatively, the crude product can be chromatographed on aluminium oxide (neutral): the crude product is absorbed onto silica gel (dissolving in warm MeOH, addition of 5 parts by weight of silica gel, concentration on a rotary evaporator, and evaporation several times with ethyl acetate until a MeOH-free product results). The absorbate is introduced onto a column containing Al$_2$O$_3$ (neutral, 50 parts by weight). Starting material (5 g) is eluted first with ethyl acetate (flash chromatography, checked by HPLC). Elution with ethyl acetate–methanol mixtures (40:1, 20:1, and then 10:1) gives 8.6 g (46.1%) of clausenamide, 98% pure according to ^{1}H NMR.

A reaction related to the autoxidation of enolates is the oxidation of ketones with potassium superoxide.[5] The first step of this reaction involves enolate formation by potassium superoxide with simultaneous release of one molecule of oxygen. The resulting oxygen thus affects the hydroxylation to afford α-hydroxy ketone (Scheme 2.8.3).

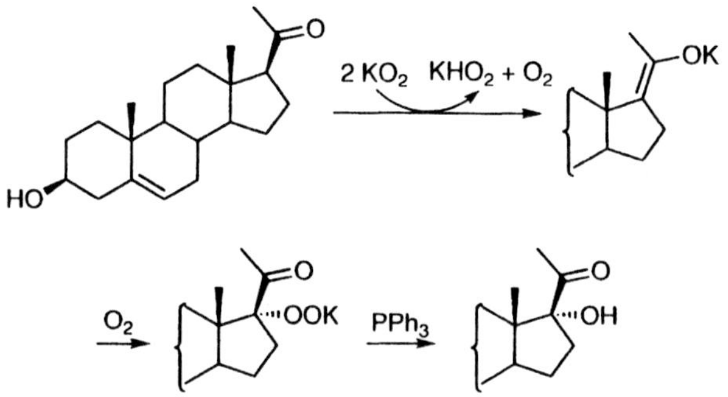

Scheme 2.8.3

Another widely used reagent in the α-hydroxylation of enolates is oxo-diperoxymolybdenum(pyridine)hexamethylphosphoric triamide (MoO$_5$·Py·HMPA, MoOPH). The MoOPH reagent oxidizes a variety of enolates to give the corresponding α-hydroxy carbonyl products.[1] For example, oxidation of the lithium enolate of epoxy ketone **3** with MoOPH gives the α-hydroxy ketone **4** in 86% yield (Scheme 2.8.4).[6] Overoxidation is an occasional problem with this reagent.[7,8] In general, MoOPH fails to oxidize enolates derived from β-dicarbonyl compounds probably because of the formation of the 1,3-dicarbonyl complex of Mo^{VI} which resists oxidation.[7,8]

Protocol 2.
Synthesis of (2S,4S,5E,9R,10R)-9,10-epoxy-2-hydroxy-4-isopropyl-7-methylene-5-cyclodecen-1-one (Structure 4, Scheme 2.8.4)

Caution! Although MoOPH has not been reported to be explosive, precautions should be taken, as with other molybdenum complexes, when crystalline MoOPH is in contact with oxidizable materials. Also the HMPA ligand is a potential carcinogen. Carry out all procedures in a well-ventilated hood, and wear disposable vinyl or latex gloves and chemical-resistant safety goggles.

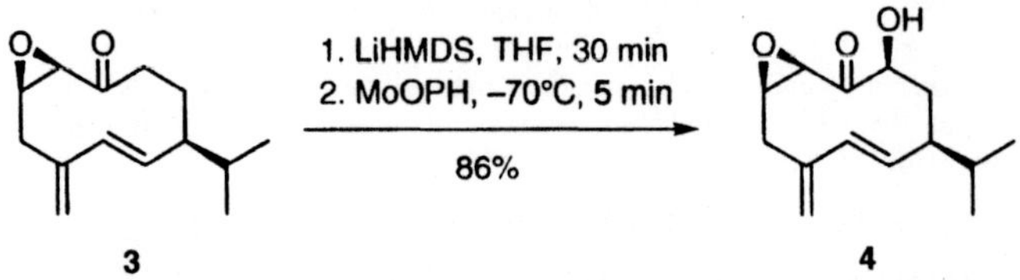

Scheme 2.8.4

This is an example of diastereoselective α-hydroxylation of a chiral ketone enolate with MoOPH reagent.

Equipment
- Three-necked, round-bottomed flask (100 mL)
- Separating funnel (250 mL)
- Magnetic stirrer
- Sintered glass filter funnel (30 mL)
- Cooling bath (ice–water)
- Source of dry argon
- Cooling bath (dry-ice/acetone)
- Teflon-coated magnetic stirrer bar
- Pressure-equalizing addition funnel (25 mL)
- Syringes (2 mL and 5 mL)
- Rubber septum

Materials
- Dry tetrahydrofuran (31 mL) — **flammable, irritant**
- Hexamethyldisilazane (HMDS), 1.2 mL, 5.69 mmol — **flammable, corrosive**
- n-Butyllithium, 1.57 M in n-hexane, 3.5 mL, 5.50 mmol — **pyrophoric**
- (4S,5E,9R,10R)-9,10-Epoxy-4-isopropyl-7-methylene-5-cyclodecen-1-one, 0.965 g, 4.34 mmol — **unknown toxicity, treat as toxic**
- Acetone, as needed — **flammable, irritant**
- Dry ice, as needed
- MoOPH, 3.10 g, 7.14 mmol — **unknown toxicity, treat as toxic**
- 10% Sodium sulfite, 50 mL — **irritant**
- Diethyl ether, 3 × 50 mL — **flammable, toxic**
- Brine, 50 mL
- Magnesium sulfate, anhydrous, as needed
- Silica gel, 50 g — **hazardous dust**
- n-Hexane, as needed — **flammable, irritant**
- Isopropyl ether, as needed — **flammable, irritant**

Method
1. Equip a 100-mL three-necked round-bottomed flask with an argon bubbler, a 25-mL additional funnel, a rubber septum and a magnetic stirring bar. The apparatus is maintained under a positive pressure of argon gas.

2: Oxidation of the C=C bond

2. Add to the reaction vessel HMDS (1.2 mL, 5.69 mmol) and dry THF (22 mL) via syringes. Stir the solution and cool it to 0–5 °C using an ice–water bath.

3. Add dropwise n-BuLi (1.57 M in n-hexane, 3.5 mL, 5.50 mmol) via syringe, stir the resulted solution at 0–5 °C for 15–30 min and then cool to −78 °C using a dry-ice/acetone bath.

4. Dissolve (4S,5E,9R,10R)-9,10-epoxy-4-isopropyl-7-methylene-5-cyclodecen-1-one (0.965 g, 4.34 mmol) in dry THF (9 mL) and add this solution dropwise via the addition funnel to the LiHMDS solution prepared in step 3. After addition is complete, continue stirring the reaction mixture for 1 h at −78 °C and then warm the reaction mixture to −20 °C.

5. Add solid MoOPH (3.10g, 7.14 mmol) in a single portion and stir the reaction mixture for 25 min. Quench the reaction by addition of 10% aqueous sodium sulfite (50 mL).

6. Transfer the reaction mixture into a separating funnel, add brine and extract with diethyl ether (3 × 50 mL).

7. Combine the organic extracts and wash them with brine (50 mL) and dry the organic phase over anhydrous magnesium sulfate.

8. Filter to remove magnesium sulfate and concentrate the solution on a rotary evaporator. Purify the product using silica gel chromatography (50 g silica gel, n-hexane–diethyl ether as eluent) to give (2S,4S,5E,9R,10R)-9,10-epoxy-2-hydroxy-4-isopropyl-7-methylene-5-cyclodecen-1-one as crystals; 0.638 g (86%). A portion of the product was recrystallized from hexane/isopropyl ether to give pure product as needles: m.p. 116–119 °C (hexane/isopropyl ether); $[\alpha]_D^{20} = -422°$ ($c = 0.915$, diethyl ether). The following spectra are diagnostic: IR 3360, 3090, 3030, 2970, 2880, 1715, 1610, 1445, 1255, 1040, 1010, 980, 970, 905, 805 cm^{-1}; ^{1}H NMR (100 MHz) δ 0.82 (d, $J = 6.4$ Hz, 3H), 0.94 (d, $J = 6.4$ Hz, 3H), 1.35–2.55 (m, 6H), 2.86 (dd, $J = 13.2, 3.3$ Hz, 1H), 3.21 (ddd, $J = 10.2, 4.7, 3.3$ Hz, 1H), 3.84 (d, $J = 4.7$ Hz, 1H), 40.9 (br d, $J = 9.9$ Hz, 1H), 4.98 (s, 1H), 5.00 (dd, $J = 16.3, 10.5$ Hz, 1H), 5.14 (s, 1H), 5.97 (d, $J = 16.3$ Hz, 1H).

A safer alternative oxidant to MoOPH is oxodiperoxymolybdenum(pyridine)-1,3-dimethyl-3,4,5,6-tetrahydro-2(1H)-pyrimidinone (MoO$_5$•Py•DMPU, MoOPD) in which the carcinogenic ligand HMPA is replaced by 1,3-dimethyl-3,4,5,6-tetrahydro-2(1H)-pyrimidinone.[9] This stable crystalline compound developed more recently for the α-hydroxylation of carbonyl compounds is less reactive than the MoOPH reagent due to its poorer solubility and/or change in ligand.[9] However, the MoOPD reagent successfully oxidizes those enolates generated from ketones, esters, and lactones to give α-hydroxy carbonyl products.[2,9] The reaction is generally cleaner and, in most cases, the yields are comparable to MoOPH.

An indirect α-hydroxylation of enolates involves the use of hypervalent iodine reagents such as iodobenzene diacetate. Treatment of tetralone **5** with

KOH in methanol followed by addition of iodobenzene diacetate affords α-hydroxy tetralone **6** in 67% yield (Scheme 2.8.5).[10] In this reaction the enolate, formed from the ketone and potassium hydroxide, attacks iodobenzene diacetate to give an α-iodo(III) carbonyl intermediate which upon reaction with methanol and subsequent rearrangement affords an α-hydroxy dimethyl acetal. Hydrolysis of α-hydroxy dimethyl ketal then gives the α-hydroxy ketone. Noteworthy is the overall stereochemical outcome of this reaction in that the α-hydroxy group is installed at the more hindered face of the enolate.

Protocol 3.
Synthesis of *cis*-2-hydroxy-6-methoxy-4,7-dimethyltetral-1-one (Structure 6, Scheme 2.8.5)

Caution! Carry out all procedures in a well-ventilated hood, and wear disposable vinyl or latex gloves and chemical-resistant safety goggles.

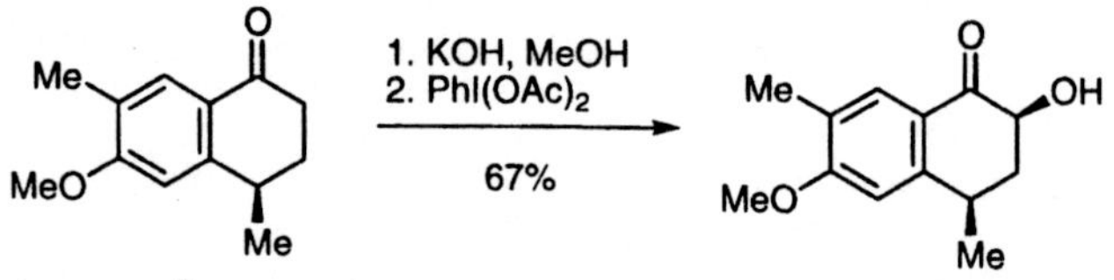

Scheme 2.8.5

This is an example of diastereoselective α-hydroxylation of a chiral ketone enolate with hypervalent iodine reagents.

Equipment
- Round-bottomed flask (100 mL)
- Separating funnel (250 mL)
- Magnetic stirrer
- Teflon-coated magnetic stirrer bar
- Cooling bath (ice/water, salt)
- Sintered glass filter funnel (30 mL)

Materials
- Methanol, 20 mL — **flammable, toxic**
- Potassium hydroxide, 2.1 g, 36.8 mmol — **corrosive, toxic**
- 6-Methoxy-4,7-dimethyltetral-1-one, 0.5g, 2.45 mmol — **unknown toxicity, treat as toxic**
- Iodobenzene diacetate, 0.947g, 2.94 mmol — **unknown toxicity, treat as toxic**
- Diethyl ether, 70 mL — **flammable, toxic**
- Sodium bicarbonate, 3%, 25 mL
- Magnesium sulfate, anhydrous, as needed

Method
1. Equip a 100-mL round-bottomed flask with an argon bubbler and a magnetic stirring bar.

2: Oxidation of the C=C bond

2. Add 6-methoxy-4,7-dimethyltetral-1-one (0.5g, 2.45 mmol), MeOH (20 mL) and KOH (2.1 g, 36.8 mmol) to the reaction vessel, cool with an ice–salt bath and stir the solution for 10 min.

3. Add iodobenzene diacetate (0.947 g, 2.94 mmol) as a solid in one portion. Stir the reaction mixture at 0°C for 1 h and then at room temperature overnight.

4. Remove the solvent on a rotary evaporator and dissolve the residue in diethyl ether (70 mL).

5. The diethyl ether solution is washed with 3% sodium bicarbonate (25 mL) and water (25 mL).

6. Dry the organic phase over anhydrous magnesium sulfate.

7. Filter to remove magnesium sulfate.

8. Remove solvent to dryness to give *cis*-2-hydroxy-6-methoxy-4,7-dimethyltetral-1-one as colourless prisms (0.361 g, 67%): m.p. 103°C (from *n*-hexane–diethyl ether). The following spectra are diagnostic: IR ($CHCl_3$) 3450, 1670 cm^{-1}; ^{1}H NMR (90 MHz, $CDCl_3$) δ 1.36 (d, J = 6.8 Hz, 3H), 1.74 (m, 1H), 2.21 (s, 3H), 2.49 (ddd, J = 12.6, 5.4, 4.2 Hz, 1H), 3.10 (m, 1H), 3.91 (s, 1H), 3.91 (s, 3H), 4.33 (ddd, J = 12.6, 5.4, 1.3 Hz, 1H), 6.77 (s, 1H), 7.83 (s, 1H).

Other hypervalent iodine reagents capable of oxidizing enolates to α-hydroxy ketones include iodosylbenzene and *o*-iodosylbenzoic acid.[11] Although enolate oxidations using hypervalent iodine reagents are tolerant of many functional groups including alkenes, amines, sulfides, epoxides, aziridines, and lactams, the reaction is usually limited to ketone substrates. The exception is iodobenzene diacetate which in some cases also reacts with esters to give α-hydroxy acids.[12] Certain sterically hindered ketones fail to give the desired products. The presence of internal hydroxy groups (phenols or alcohols) often results in intramolecular alkoxylation when stereoelectronically allowed.[12] Hypervalent iodine compounds are usually not suitable for the α-oxygenation of β-dicarbonyl substrates due to the formation of iodonium ylids.[14]

N-Sulfonyloxaziridines are a special class of oxaziridines having a sulfonyl group at the ring-nitrogen atom. Due to the presence of the bulky and strongly electron-attracting sulfonyl group these reagents are highly chemoselective, electrophilic, aprotic oxidizing reagents.[1,2,15] In addition to the α-hydroxylation of enolates, *N*-sulfonyloxaziridines also oxidize carbanions to alcohols.[1,2] For example, treatment of imide **7** with NaHMDS followed by reacting the resulted enolate with racemic *trans*-2-(phenylsulfonyl)-3-phenyloxaziridine affords the corresponding α-hydroxy product **8** in 83% yield (Scheme 2.8.6).[16]

Protocol 4.
Synthesis of 3-[5-[3-(2,5-dimethyl-1H-pyrrol-1-yl)-2,5-dimethoxyphenyl]-5-methoxy-4-methyl-2-hydroxy-1-oxopentyl]-4-(1-methylethyl)-2-oxazolidinone (Structure 8, Scheme 2.8.6)

Caution! Carry out all procedures in a well-ventilated hood, and wear disposable vinyl or latex gloves and chemical-resistant safety goggles.

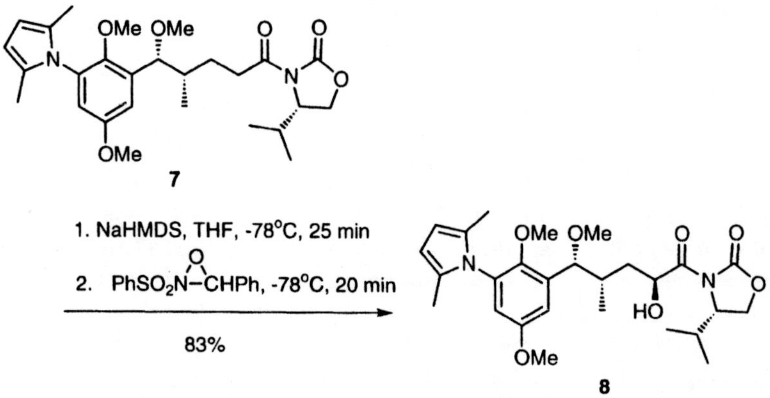

Scheme 2.8.6

Reagents: 1. NaHMDS, THF, -78°C, 25 min; 2. PhSO$_2$N−CHPh (oxaziridine), -78°C, 20 min. Yield 83%.

This is an example of a chiral auxiliary controlled diastereoselective α-hydroxylation of enolate with an N-sulfonyloxaziridine.

Equipment
- Three-necked, round-bottomed flask (1 L)
- Separating funnel (2 L)
- Additional funnel (250 mL)
- Source of dry argon
- Magnetic stirrer
- Sintered glass filter funnel (100 mL)
- Round-bottomed flask (100 mL)
- Teflon-coated magnetic stirrer bar
- Syringes (20 mL, 50 mL, and 200 mL)
- Cooling baths (dry ice–acetone)
- Rubber septums

Materials

Material	Hazard
Dry tetrahydrofuran, 330 mL	flammable, irritant
NaHMDS, 1 M in THF, 36.4 mL, 36.4 mmol	corrosive, flammable
3-[5-[3-(2,5-Dimethyl-1H-pyrrol-1-yl)−2,5-dimethoxyphenyl]-5-methoxy-4-methyl-1-oxopentyl]-4-(1-methylethyl)-2-oxazolidinone, 13,59 g, 27.96 mmol	unknown toxicity, treat as toxic
2-(Phenylsulfonyl)-3-phenyloxaziridine, 13.14 g, 50.33 mmol	unknown toxicity, treat as toxic
Triethylamine, 11.7 mL, 83.9 mmol	flammable, corrosive
Acetic acid, 16 mL, 280 mmol	corrosive
Water, 100 mL	
Diethyl ether, 250 mL	flammable, toxic
Sodium bicarbonate, 5%, 120 mL	
Brine, 70 mL	
Magnesium sulfate, as needed	
Dichloromethane, as needed	toxic, irritant
Hexane, as needed	flammable, irritant

2: Oxidation of the C=C bond

Method

1. Equip a 1-L three-necked round-bottomed flask with a nitrogen bubbler, a 250-mL additional funnel, a rubber septum and a magnetic stirring bar. The reaction is maintained under a positive pressure of dry argon.
2. Add to the reaction vessel anhydrous tetrahydrofuran (200 mL) and a solution of NaHMDS (1 M in THF, 36.4 mL) via syringes. Cool the solution to −78°C with a dry-ice–acetone bath.
3. Dissolve 3-[5-[3-(2,5-dimethyl-1H-pyrrol-1-yl)-2,5-dimethoxyphenyl]-5-methoxy-4-methyl-1-oxopentyl]-4-(1-methylethyl)-2-oxazolidinone (13.59 g, 27.96 mmol) in anhydrous THF (80 mL) in the additional funnel.
4. Add the solution prepared in step 3 to the cooled solution obtained in step 2 over 20 min. Stir the reaction mixture for 25 min after the addition is complete.
5. Dissolve 2-(phenylsulfonyl)-3-phenyloxaziridine (13.14 g, 50.33 mmol) in anhydrous THF (50 mL) in a 100-mL round-bottomed flask under argon. Cool the resulting solution to −78°C with a dry-ice–acetone bath.
6. Add the precooled solution obtained in step 5 via cannula over 7 min to the enolate solution obtained in step 4. After the addition is complete, continue stirring for an additional 20 min.
7. Add dry triethylamine (11.7 mL, 83.9 mmol) to the solution obtained in step 6 via syringe and stir the reaction mixture at −78°C for 20 min. Add acetic acid (16 mL, 280 mmol) via syringe to quench the reaction, remove the cooling bath and allow the solution to warm to room temperature.
8. Add water (100 mL) to the reaction mixture obtained in step 7, stir the resulted biphasic mixture vigorously for 15 min and then add diethyl ether (250 mL).
9. Transfer the mixture into a separating funnel and separate the phases. Wash the organic phase with water (100 mL), 5% aqueous sodium bicarbonate (2 × 60 mL) and brine (70 mL). Dry the organic phase over magnesium sulfate.
10. Filter to remove magnesium sulfate and concentrate on a rotary evaporator to give a residue. Purify the resulting residue via flash chromatography (silica gel, 1 × CH_2Cl_2–5% Et_2O and 1 × hexane–55% Et_2O) to give 11.6 g (83%) of 3-[5-[3-(2,5-dimethyl-1H-pyrrol-1-yl)-2,5-dimethoxyphenyl]-5-methoxy-4-methyl-2-hydroxy-1-oxopentyl]-4-(1-methylethyl)-2-oxazolidinone as a white solid which crystallized from Et_2O–hexane, m.p. 131.5–132°C, $[\alpha]_D = +81°$ ($c = 0.40$, CH_2Cl_2). The following spectroscopic data are diagnostic: IR (CH_2Cl_2) 3600–3400, 1970, 1940, 1785, 1695, 1610, 1480, 1450, 1430, 1390, 1210, 1100, 1050, 1020, 1010 cm^{-1}; ^{1}H NMR δ 6.93 (d, $J = 3.1$ Hz, 1H), 6.63 (d, $J = 3.1$ Hz, 1H), 5.90 (s, 2H), 5.07 (br t, 1H), 4.73 (d, $J = 5.8$ Hz, 1H), 4.40–4.26 (m, 3H), 3.79 (s, 3H), 3.30 (d, $J = 7.5$ Hz, 1H), 3.26 (s, 3H), 3.17 (s, 3H), 2.43

Protocol 4. Continued

(m,1H), 2.24–2.10 (m, 1H), 2.07 (s, 3H), 1.99 (s, 3H), 1.69–1.48 (m, 2H), 1.03 (d, J = 6.8 Hz, 3H), 0.93 (d, J = 7.1 Hz, 3H), 0.88 (d, J = 6.9 Hz, 3H); ^{13}C NMR δ 175.10, 155.19, 153.69, 148.43, 135.70, 131.31, 129.26, 128.42, 114.13, 112.34, 106.02, 105.83, 81.71, 69.04, 64.07, 59.65, 58.94, 57.31, 55.66, 37.59, 35.74, 28.23, 17.94, 14.53, 13.96, 12.80, 12.57; MS m/z 502.2700 (M$^+$, 92%), 375 (19), 274 (100), 244 (22).

One of the important features of N-sulfonyloxaziridines is that they are available in enantiopure form making it possible to prepare optically active α-hydroxy carbonyl compounds from prochiral enolates. Furthermore, the absolute configuration of the product is determined by the oxaziridine so that either enantiomer is available and predictable based on a steric control model.[1,2] For example, reaction of the sodium enolate of deoxybenzoin (**9**) with (+)-(camphorylsulfonyl)oxaziridine (**10**) gives (*S*)-benzoin (**11**) in >97.7% ee (Scheme 2.8.7).[17] (*R*)-Benzoin can be obtained similarly by using the antipodal (−)-(camphorylsulfonyl)oxaziridine reagent.

Protocol 5.
Synthesis of (+)-(*S*)-benzoin (Structure 11, Scheme 2.8.7)

Caution! Carry out all procedures in a well-ventilated hood, and wear disposable vinyl or latex gloves and chemical-resistant safety goggles.

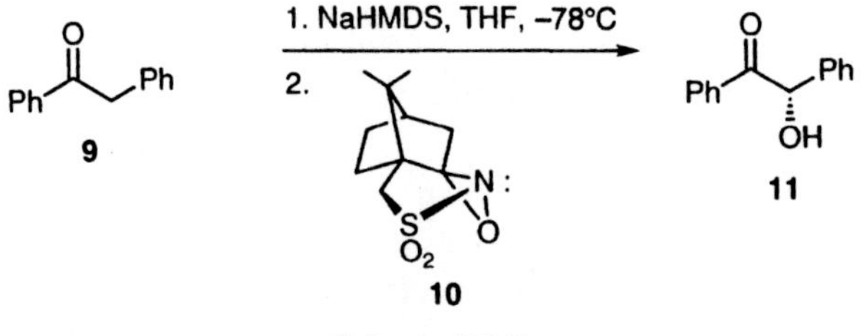

Scheme 2.8.7

This is an example of the enantioselective α-hydroxylation of a prochiral ketone enolate with an enantiomerically pure (camphorylsulfonyl)oxaziridines

Equipment

- Three-necked, round-bottomed flask (1 L)
- Source of dry argon
- Thermostated hotplate stirrer
- Reflux condenser
- Cooling bath (dry-ice–acetone)
- Separating funnel
- Pressure-equalizing addition funnels (200 mL)
- Sintered glass filter funnel
- Rubber septum
- Erlenmeyer flask
- Syringe (100 mL)
- Teflon-coated magnetic stirrer bar
- Round-bottomed flask (200 mL)

2: Oxidation of the C=C bond

Materials

- Dry tetrahydrofuran (550 mL) — flammable, irritant
- NaHMDS, 0.75 M solution in THF, 88 mL, 66 mmol — corrosive, flammable
- Deoxybenzoin, 10.0 g, 51 mmol — unknown toxicity, treat as toxic
- (+)-(2R,8aS)-10-(Camphorylsulfonyl)oxaziridine, 17.5 g, 77 mmol — unknown toxicity, treat as toxic
- Trifluoroacetic anhydride, 25 g, 119 mmol — corrosive
- Saturated ammonium iodide, 25 mL — irritant
- Diethyl ether, 650 mL — flammable, toxic
- Sodium thiosulfate, 0.1 M, 150 mL — irritant
- Saturated sodium bicarbonate, 375 mL
- Sulfuric acid, 1%, 175 mL — oxidizer, corrosive
- Brine, 120 mL
- Magnesium sulfate, anhydrous
- n-Pentane, 700 mL — flammable, irritant
- Dichloromethane, 100 mL — toxic, irritant
- Sodium hydroxide, 5%, 125 mL — corrosive, toxic
- Ethanol, as needed — highly toxic, flammable
- Acetone, as needed — flammable, irritant
- Dry ice, as needed

Method

1. Equip a 1-L three-necked round-bottomed flask with an argon gas bubbler, a 200-mL addition funnel, a rubber septum and a magnetic stirring bar. Maintain the apparatus under a positive pressure of dry argon gas.

2. Add freshly distilled THF (250 mL) followed by sodium bis(trimethylsilyl)-amide (0.75 M in THF, 88 mL, 66 mmol) via syringe to the reaction vessel. Cool the solution to −78 °C with a dry-ice–acetone bath.

3. Prepare a solution of deoxybenzoin (10.0 g, 51 mmol) in dry THF (100 mL) and transfer it to the pressure-equalizing addition funnel. Add this mixture dropwise to the solution of sodium bis(trimethylsilyl)amide obtained in step 2. After the addition is complete, continue stirring for 30 min at −78 °C.

4. Prepare a solution of (+)-(2R,8aS)-10-(camphorylsulfonyl)oxaziridine (17.5 g, 77 mmol) in dry THF (150 mL) and add it dropwise to the reaction mixture prepared in step 3 over 20 min. After the addition is complete, continue stirring the dark violet solution for 15 min at −78 °C.

5. Add trifluoroacetic anhydride (25 g, 119 mmol) via syringe to the reaction mixture.[a] Remove the cooling bath and allow the solution to warm to room temperature.

6. Replace the addition funnel with a reflux condenser and heat the reaction mixture to reflux. After stirring at reflux for 2 h, cool the reaction mixture to room temperature and quench it by addition of saturated ammonium iodide (25 mL).

7. Reduce the volume of the reaction solution to about 200 mL using a rotary evaporator and dilute with diethyl ether (400 mL).

8. Transfer the reaction mixture into a separating funnel (1.5 L) and wash successively with sodium thiosulfate solution (0.1 M, 2 × 75 mL), saturated

Protocol 5. Continued

sodium bicarbonate solution (5 × 75 mL), cold sulfuric acid (1%, 75 mL) and saturated brine (60 mL). Dry the organic phase over anhydrous magnesium sulfate (10 g).

9. Remove the solvent on a rotary evaporator and take up the product in *n*-pentane (10 × 50 mL). Combine the *n*-pentane extracts and dilute with *n*-pentane (200 mL). After filtration, remove the solvent on a rotary evaporator to give the crude trifluoroacetate ester (15 g). The residue consisted of 11.8 g (72%) of (camphorsulfonyl)imine.
10. Dissolve the crude trifluoroacetate ester with dichloromethane (100 mL), cool the solution to 5°C and add a 5% sodium hydroxide solution (125 mL) slowly. After addition is complete, stir the reaction mixture for 45 min.
11. Transfer the reaction mixture into a separating funnel (1 L), add THF (50 mL) and diethyl ether (250 mL). Wash the mixture successively with water (2 × 50 mL), 1% cold sulfuric acid (2 × 50 mL) and brine (60 mL). Dry the organic phase over anhydrous magnesium sulfate (10 g) for 30 min. After filtration, remove the solvent on a rotary evaporator and crystallize the residue from *n*-pentane–ethanol to give 7.3 g (67.5% from deoxybenzoin) in two crops; m.p. 135°C, $[\alpha]_D = +114.9°$ ($c = 1.5$, acetone), >97.7% ee.[b]

[a] In this protocol, the resulted α-hydroxy product was trapped with trifluoroacetic anhydride for easier product isolation on a large scale. This is not necessary if the product is isolated by chromatography.
[b] The optical purity is determined by HPLC using a Daicel Chiral Pak OT(+) HPLC column, 25 cm × 0.46 cm; solvent MeOH; flow rate 0.5 mL/min. First to be eluted is (+)-(*S*)-benzoin.

By choice of appropriate reaction conditions and (camphorylsulfonyl)-oxaziridine reagent, high enantiomeric excess can be obtained not only with acyclic enolates but also with cyclic substrates.[1-3] One example is the enantioselective oxidation of the lithium enolate of 2-methyl-1-tetralone (**12**) with (+)-[(8,8-dichlorocamphoryl)sulfonyl]oxaziridine (**13**) where (*R*)-2-hydroxy-2-methyl-1-tetralone (**14**) is obtained in 95% ee and 66% yield.[19] Another important feature of the *N*-sulfonyloxaziridines oxidizing reagents is that they are reactive towards stabilized enolates such as those derived from β-dicarbonyl compounds.[1,2] For example, reaction of the potassium enolate of β-keto ester **15** with [(8,8-dimethoxycamphoryl)sulfonyl]oxaziridine (**16**) gives **17** in >95% ee.[19]

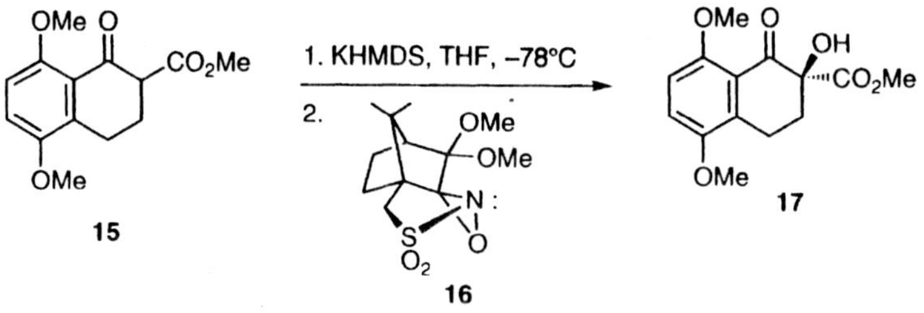

3. Asymmetric α-hydroxylation of enol derivatives

Complimentary to the direct α-hydroxylation of enolates is the oxidation of enol derivatives. Although an extra step is necessary to prepare the enol derivative, this approach does offer some advantages. For example, with enol derivatives protic oxidizing reagents such as peracids can be used and the enol has a defined geometry.

The most commonly employed enol derivatives for the α-hydroxylation of carbonyl compounds are silyl enol ethers which can be obtained by a number of different methods. *m*-Chloroperoxybenzoic acid (*m*-CPBA) is generally used for the α-hydroxylation of enol derivatives. An example is illustrated in Scheme 2.8.8. Reaction of silyl enol ether **18** with *m*-CPBA in dichloromethane affords α-hydroxy ketone **19** in 57% yield.[20]

Protocol 6.
Synthesis of methyl (3aα,6β,7β,7aα)-2,3,3a,6,7,7a-hexahydro-3a-hydroxy-6-methyl-3-oxo-7-[[2-(trimethylsilyl)ethoxy]methoxy]-4-benzofurancarboxylate (Structure 19, Scheme 2.8.8)

Caution! Carry out all procedures in a well-ventilated hood, and wear disposable vinyl or latex gloves and chemical-resistant safety goggles.

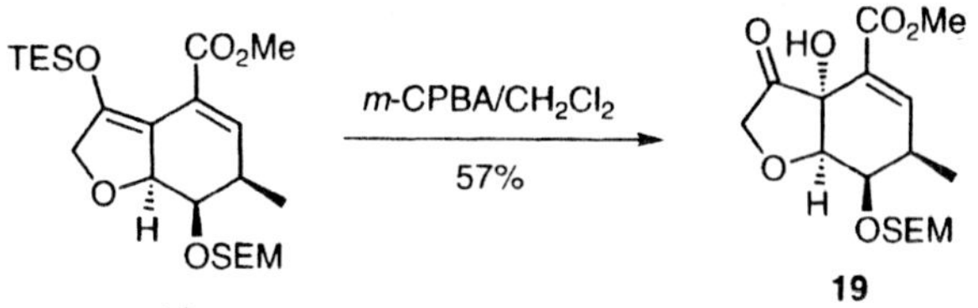

Scheme 2.8.8

This is an example of diastereoselective α-hydroxylation of a chiral cyclic ketone silyl enol ether with *m*-chloroperbenzoic acid.

Equipment
- Double-necked, round-bottomed flask (10 mL)
- Round-bottomed flask (5 mL)
- Teflon-coated magnetic stirrer bar
- Rubber septum

Protocol 6. Continued

- Separating funnel
- Thermostated hotplate stirrer
- Source of dry argon
- Cooling bath (ice–water)
- Sintered glass filter funnel
- Syringe (1 mL)
- Erlenmeyer flask

Materials

- Dry dichloromethane (1 mL) — **toxic, irritant**
- Methyl (6β,7β,7aα)-2,3,3a,6,7,7a-tetrahydro-3-[(triethylsilyl)oxy]-6-methyl-7-[[2-(trimethylsilyl)ethoxy]methoxy]-4-benzofurancarboxylate, 11.5 mg, 0.024 mmol — **unknown toxicity, treat as toxic**
- m-Chloroperoxybenzoic acid, 5.5 mg, 0.027 mmol — **oxidizer, irritant**
- Sodium sulfite, 10%, 5 mL — **irritant**
- Saturated sodium bicarbonate, 5 mL
- Brine, 5 mL
- Magnesium sulfate, anhydrous, as needed
- Hexane, as needed — **flammable, irritant**
- Ethyl acetate, as needed — **flammable, irritant**
- Silica gel, as needed

Method

1. Equip a 10-mL double-necked round-bottomed flask with an argon gas bubbler, a rubber septum and a magnetic stirring bar. Maintain the apparatus under a positive pressure of dry argon gas.
2. Add methyl (6β,7β,7aα)-2,3,3a,6,7,7a-tetrahydro-3-[(triethylsilyl)oxy]-6-methyl-7-[[2-(trimethylsilyl)ethoxy]methoxy]-4-benzofurancarboxylate (11.5 mg, 0.024 mmol) to the reaction vessel.
3. Add dry dichloromethane (0.5 mL) using a syringe.
4. Stir the resulted solution and cool to 0 °C with an ice–water bath.
5. In a 5-mL round-bottomed flask add m-chloroperoxybenzoic acid (5.5 mg, 0.027 mmol) and dry dichloromethane (0.5 mL).
6. Transfer the solution prepared in step 5 using a 1-mL syringe to the reaction mixture obtained in step 4.
7. Stir the reaction mixture at 0 °C for 30 min.
8. Add dichloromethane (10 mL) to the reaction mixture and transfer the solution mixture into a 25-mL separating funnel.
9. Wash the reaction mixture with 10% aqueous sodium sulfite (5 mL), saturated sodium bicarbonate (5 mL) and brine (5 mL).
10. Dry the organic phase over anhydrous magnesium sulfate.
11. Filter to remove the magnesium sulfate.
12. Remove solvent on a rotary evaporator to give a residue.
13. Purify the product by silica gel column using 35% ethyl acetate in hexane to afford methyl (3aα,6β,7β,7aα)-2,3,3a,6,7,7a-hexahydro-3a-hydroxy-6-methyl-3-oxo-7-[[2-(trimethylsilyl)ethoxy]methoxy]-4-benzofurancarboxyte, 5.2 mg

2: Oxidation of the C=C bond

(57%): $[\alpha]_D = +66.0°$ ($c = 0.52$, CHCl$_3$). The following spectroscopic data are diagnostic: IR (neat) 3500, 2954, 1773, 1701, 1264, 1249, 1021, 861, 837 cm^{-1}; ^{1}H NMR (CDCl$_3$) δ 0.04 (9H, s), 0.93 (2H, m), 1.27 (3H, d, J = 7 Hz), 2.75 (1H, m), 3.55 (1H, m), 3.74 (1H, m), 3.83 (3H, s), 4.05 (1H, dd, J = 4 and 3 Hz), 4.13 (1H, d, J = 17 Hz), 4.31 (1H, d, J = 17 Hz), 4.35 (1H, J = 3 Hz), 4.70 (1H, d, J = 7 Hz), 4.83 (1H, d, J = 7 Hz), 5.27 (1H, s), 6.95 (1H, d, J = 3Hz); ^{13}C NMR (CDCl$_3$) δ −1.5, 15.3, 18.0, 34.2, 52.3, 66.1, 68.9, 74.4, 76.0, 83.9, 95.4, 124.4, 148.0, 166.9, 208.7; MS m/z 300, 282, 268, 242, 224, 199, 168, 152, 136, 109, 73.

Another class of oxidizing reagents capable of hydroxylating enolates to α-hydroxy carbonyl compounds comprises peroxides such as dimethyl dioxirane (DMDO).[3,21] Reaction of silyl enol ether **20** with DMDO gives α-hydroxy ketone **21** in 51% yield as a single isomer.[21]

An indirect α-hydroxylation of silyl enol ethers involves asymmetric dihydroxylation of enol ethers.[22] Using the commercially available Sharpless' AD reagents, AD-mix-α or AD-mix-β, high enantiomeric excess of α-hydroxy ketones can be obtained from silyl enol ethers or methyl vinyl ethers (see Section 2.5 for asymmetric dihydroxylation of olefins).[22] For example, reaction of silyl enol ether **22** with AD-mix-β affords α-hydroxy ketone **23** in 99% ee.

Osmium tetroxide-catalysed dihydroxylation of silyl enol ethers gives α-hydroxy carbonyl compounds. When chiral non-racemic silyl enol ethers are used, a diastereomerically enriched α-hydroxy product can be obtained (Scheme 2.8.9).[23]

Protocol 7.
Synthesis of 6-hydroxy-3-ethoxy-7-methyl-2-oxabicyclo[3.3.0]octan-6-carboxaldehyde (Structure 25, Scheme 2.8.9)

Caution! Osmium tetroxide is highly toxic. Carry out all procedures in a well-ventilated hood, and wear disposable vinyl or latex gloves and chemical-resistant safety goggles.

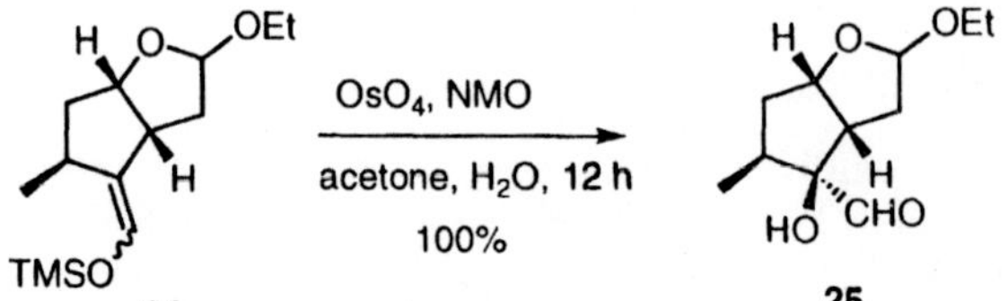

Scheme 2.8.9

This is an example of diastereoselective α-hydroxylation of a chiral aldehyde silyl enol ether with osmium tetroxide/N-methylmorpholine N-oxide (NMO).

Equipment

- Round-bottomed flask (50 mL)
- Separating funnel
- Thermostated hotplate stirrer
- Sintered glass filter funnel
- Teflon-coated magnetic stirrer bar
- Erlenmeyer flask
- Rubber septum
- Source of nitrogen

Materials

Osmium tetroxide, 0.017 g, 0.067 mmol	highly toxic, oxidizer
N-Methylmorpholine N-oxide, 0.51 g, 3.76 mmol	irritant
Acetone, 16 mL	flammable, irritant
Water, 8 mL	
6-Trimethylsilyloxymethylene-3-ethoxy-7-methyl-2-oxabicyclo[3.3.0]octane, 0.93 g, 3.42 mmol	unknown toxicity, treat as toxic
Sodium hydrosulfite, 0.70 g	flammable solid, corrosive
Florisil, 2.70 g	irritant
Sodium sulfate, as needed	irritant
Diethyl ether, 4 × 20 mL	flammable, toxic
Magnesium sulfate, anhydrous, as needed	

Method

1. Equip a 50-mL double-necked round-bottomed flask with a nitrogen gas bubbler and a magnetic stirring bar. Maintain the apparatus under a positive pressure of nitrogen gas.
2. Add osmium tetroxide (0.017 g, 0.067 mmol), N-methylmorpholine N-oxide (0.51 g, 3.76 mmol), acetone (10 mL) and water (8 mL) to the reaction vessel.
3. Stir the reaction mixture at room temperature for 10 min.

2: Oxidation of the C=C bond

4. Dissolve 6-trimethylsilyloxymethylene-3-ethoxy-7-methyl-2-oxabicyclo[3.3.0]octane (0.93 g, 3.42 mmol) in acetone (6 mL) and add this solution to the reaction mixture obtained in step 3.
5. Stir the reaction mixture at room temperature for 12 h.
6. Add sodium hydrosulfite (0.70 g) and Florisil (2.70 g) to the reaction solution and stir for 30 min.
7. Filter the reaction mixture and remove the acetone filtrate on a rotary evaporator. Add sodium sulfate to the aqueous solution until it is saturated.
8. Extract the aqueous portion with diethyl ether (4 × 20 mL).
9. Dry the combined etheral extracts over anhydrous magnesium sulfate and filter.
10. Remove solvent to dryness to give 6-hydroxy-3-ethoxy-7-methyl-2-oxabicyclo[3.3.0]octan-6-carboxaldehyde, 0.73g (100%).

References

1. Davis, F. A.; Chen, B.-C. *Chem. Rev.* **1992**, *92*, 919.
2. Davis, F. A.; Chen, B.-C. *Methoden der Organischen Chemie*, vol. E21e, Houben-Wegl: **1995**; pp. 4497–4518.
3. Bailey, E, J.; Barton, D. H. R.; Elks, J.; Templeton, J. F. *J. Chem. Soc.* **1962**, 1578.
4. Hartwig, W.; Born, L. *J. Org. Chem.* **1987**, *52*, 4352.
5. Betancor, C.; Francisco, C. G.; Freire, R.; Suarez, E. *J. Chem. Soc., Chem. Commun.* **1988**, 947.
6. Kuwahara, S.; Mori, K. *Tetrahedron* **1990**, *46*, 8075.
7. Vedejs, E.; Engler, D. A.; Telschow, J. E. *J. Org. Chem.* **1978**, *43*, 188.
8. Vedejs, E. *J. Am. Chem. Soc.* **1974**, *96*, 5944.
9. Anderson, J. C.; Smith, S. C. *Synlett* **1990**, 107.
10. Irie, H.; Matsumoto, R.; Nishimura, M.; Zhang, Y. *Chem. Pharm. Bull.* **1990**, *38*, 1852.
11. Moriarty, R. M.; Prakash, O. *Acc. Chem. Res.* **1986**, *19*, 244.
12. Moriarty, R. M.; Hu, H. *Tetrahedron Lett.* **1981**, *22*, 2747.
13. Turuta, A. M.; Kamernitzky, A. V.; Fadeeva, T. M.; Zhulin, A. V. *Synthesis* **1985**, 1129.
14. Neilands, O.; Karele, B. *Zh. Org. Khim.* **1971**, *7*, 1611.
15. Davis, F. A.; Sheppard, A. C. *Tetrahedron* **1989**, *45*, 5703.
16. Baker, R.; Castro, J. L. *J. Chem. Soc., Perkin Trans. 1* **1990**, 47.
17. Davis, F. A.; Haque, M. S.; Przeslawski, R. M. *J. Org. Chem.* **1989**, *54*, 2021.
18. Davis, F. A.; Weismiller, M. C. *J. Org. Chem.* **1990**, *55*, 3715.
19. Davis, F. A.; Clark, C.; Kumar, A.; Chen, B.-C. *J. Org. Chem.* **1994**, *59*, 1184.
20. White, J. D.; Bolton, G. L.; Dantanarayana, A. P.; Fox, C. M. J.; Hiner, R. N.; Jackson, R. W.; Sakuma, K.; Warrier, U. S. *J. Am. Chem. Soc.* **1995**, *117*, 1908.
21. Underiner, T. L.; Paquette, L. A. *J. Org. Chem.* **1992**, *57*, 5438.
22. Hashiyama, T.; Morikawa, K.; Sharpless, K. B. *J. Org. Chem.* **1992**, *57*, 5067.
23. Kraus, G. A.; Thurston, J. *J. Am. Chem. Soc.* **1989**, *111*, 9203.

3

Oxidation of carbonyl compounds

3.1 Asymmetric Baeyer–Villiger oxidation

C. BOLM, T. K. K. LUONG, and O. BECKMANN

1. Introduction

In 1899, Baeyer and Villiger investigated reactions of ketones with Caro's reagent, peroxomonosulfuric acid, and they observed a previously unknown transformation: an oxygen atom was inserted into the C—C bond α to the carbonyl group, affording esters and lactones.[1] Later, for this new type of oxidation the term 'Baeyer–Villiger reaction' was coined.[2]

Various peroxy compounds such as peracids, hydrogen peroxide, and alkyl peroxides can be used as oxidants yielding esters and lactones from acyclic and cyclic ketones, respectively. Acids, bases, enzymes (see Chapter 5, Section 5.3 for biological Baeyer–Villiger reaction), and metal-containing reagents catalyse Baeyer–Villiger reactions. As to the mechanism of the Baeyer–Villiger reaction of ketones with organic peracids, a two-step process originally proposed by Criegee[3] is generally accepted: the first step is a reversible proton-catalysed addition of the peroxy compound to the ketone carbonyl affording an intermediate tetrahedral peroxyhemiketal. The following, in most cases, rate-determining step involves the migration of one of the two substituents to the next oxygen atom, simultaneous O—O bond cleavage and release of the acid moiety.

The high regioselectivity of the second step, the migration, is an essential feature, for it makes the Baeyer–Villiger reaction a valuable tool in organic synthesis. In general, the migratory aptitude is related to the stabilization of the positive charge developing at the migrating group in the transition state. Thus, for alkyl groups in both cyclic and straight-chain ketones, the migratory propensity decreases in the order of tertiary > secondary > primary > methyl. Furthermore, the Baeyer–Villiger oxidation occurs with retention of configuration at the migrating substituent as in other [1,2]-rearrangements.

Metals can either catalyse the addition of peroxy species to the carbonyl group or promote the rearrangement of the resulting peroxyhemiketal. The use of simple metal-containing reagents as Lewis acids, e.g. $BF_3 \cdot Et_2O$ or $SnCl_4$, elucidates this type of activation.[4]

2. Asymmetric metal-catalysed Baeyer–Villiger oxidations

In 1991, Yamada, Mukaiyama, and co-workers described a catalytic system which employed nickel(II) complexes as catalysts and a combination of an aliphatic aldehyde and dioxygen as coreductant and oxidant, respectively.[5] In related studies, Murahashi et al. used Fe_2O_3 as catalyst and applied the aldehyde/dioxygen method to the synthesis of esters from ketones via Baeyer–Villiger oxidation.[6] An asymmetric version of the metal-catalysed Baeyer–Villiger reaction is the enantioselective oxidation of racemic 2-phenylcyclohexanone (**2**) to the corresponding lactone **3** in the presence of 1 mol% of chiral copper complex (*S,S*)-**1** and pivaldehyde under an oxygen atmosphere at room temperature, which was first reported in 1994 (Scheme 3.1.1).[7] The optically active lactone **3** was formed in 41% yield with up to 69% ee (depending on reaction time and the amount of pivaldehyde). The best results were obtained with benzene as solvent and pivaldehyde as coreductant under an oxygen atmosphere. As the ligand structure was optimized, variations in the ligand's substitution pattern revealed that in the catalysis both the nitro and the *t*-butyl group at the phenyl ring are essential to achieve effective asymmetric induction. Substituents at the oxazoline ring other than *t*-butyl result in lower yields and decreased enantioselectivity.

Protocol 1.
Asymmetric copper-catalysed synthesis of 3-phenyl-2-oxepanone (Structure 3, Scheme 3.1.1)

Caution! Benzene is toxic and a potent carcinogen. This protocol must be undertaken in an efficient hood and protective gloves and chemical-resistant safety goggles must be worn at all times.

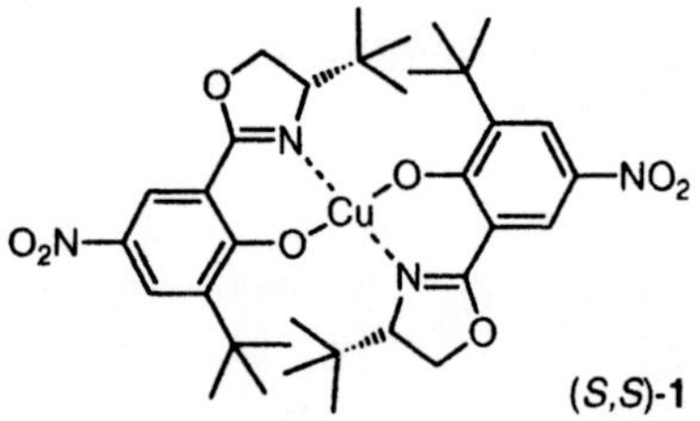

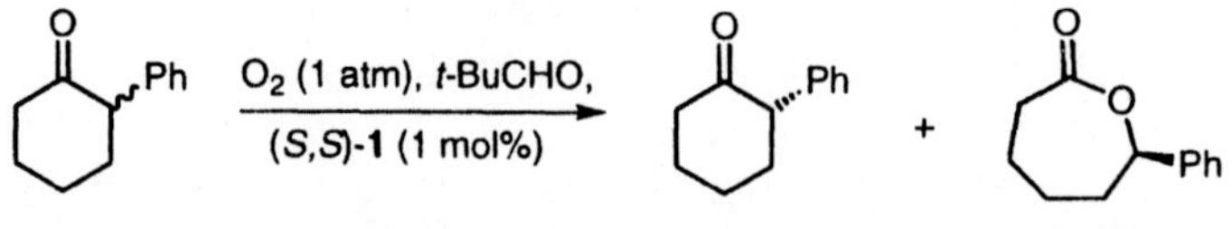

Scheme 3.1.1

3: Oxidation of carbonyl compounds

This reaction is representative of the enantioselective synthesis of lactones from cyclic ketones catalyzed by the copper(II) complex **1**.

Equipment

- Round-bottomed Schlenk-type flask (25 mL)
- Teflon-coated magnetic stirrer bar
- Rubber septum
- Source of oxygen
- Balloon
- Column for flash chromatography (20 cm × 2 cm)
- Syringe (500 μL)
- Clamp for the balloon
- High-vacuum line
- Separating funnel (250 mL)
- Filter papers
- Filter funnel
- Syringe (5 mL)

Materials

- 2-Phenylcyclohexanone, 174 mg, 1 mmol
- Copper-catalyst **1**, 7 mg, 10 μmol
- Pivaldehyde (2,2-dimethylpropanal), 55 μL, 0.5 mmo   **flammable**
- Benzene, 2.5 mL   **toxic, carcinogenic, flammable**
- Diethyl ether   **flammable**
- Sodium bicarbonate
- Anhydrous sodium sulfate
- Silica gel for flash chromatography
- Petrol ether
- Cotton wool

Method

1. To a round-bottomed Schlenk-type flask (25 mL) equipped with a magnetic stirrer bar add copper-catalyst **1** (7 mg, 10 μmol) and 2-phenylcyclohexanone (**2**) (174 mg, 1 mmol). Stopper the flask with a rubber septum (see Fig. 3.1.1).

2. Evacuate the flask, affix a balloon filled with oxygen and pinched with a clamp to the tap of the flask and let the oxygen into it.

3. Add via syringe successively benzene (2.5 mL) and pivaldehyde (55 μL, 0.5 mmol) through the septum into the flask. Stir the green reaction mixture vigorously at room temperature for 16 h.

4. Prior to the work-up, carefully evaporate the benzene under reduced pressure into a vacuum-trap. Then dissolve the residue in diethyl ether (150 mL), filter through wool, transfer the green solution into a separating funnel (250 mL), wash with saturated aqueous sodium bicarbonate (50 mL) and dry the organic layer with anhydrous sodium sulfate (*ca.* 30 g, 1 h).

5. Filter the 'dried' solution and remove the solvent on a rotary evaporator. Prepare a silica gel chromatography column using a mixture of petrol ether and diethyl ether (1:1, v/v), and load the column by dissolving the residue in the minimum volume of diethyl ether. The lactone elutes with an R_f of 0.25[a] (silica, petrol ether–diethyl ether 1:1, v/v). Evaporate the eluates

Protocol 1. *Continued*

containing this material to yield 39 mg (41%)[b] of the solid lactone **3** with 65 % ee.[c]

[a] Visualization was accomplished by an ethanol solution of phosphomolybdic acid.
[b] With respect to the amount of pivaldehyde used.
[c] The enantiomeric excess can be determined by HPLC (see Ref. 7) or gas chromatography (Lipodex E, 1.1 bar N_2, 140–170°C, rate: 0.4°C/min, 77.1 min [(+)-(R), major], 78.6 min [(−)-(S), minor]).

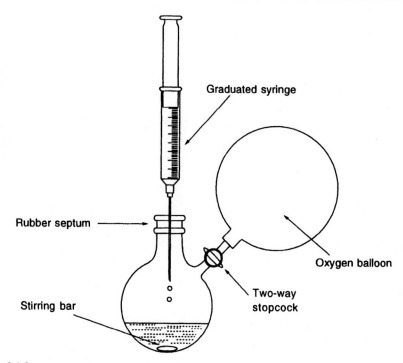

Fig. 3.1.1

With respect to the cyclohexanones, the scope of the reaction remains limited since only 2-aryl substituted cyclohexanones are reactive enough to give the corresponding optically active lactones. Cyclobutanone derivatives, however, are readily oxidized by (S,S)-**1** under the same reaction conditions with up to 95% ee.[8] Cyclobutanones have also been oxidized by chiral titanium reagents using *t*-butyl hydroperoxide or 3-chloroperoxybenzoic acid as oxidants.[9]

Yet another asymmetric method for the metal-catalysed Baeyer–Villiger reaction was reported in the same year as the copper catalysis outlined above: Strukul and co-workers used hydrogen peroxide in combination with a chirally modified platinum complex for the oxidation of cyclic ketones (Scheme

3: Oxidation of carbonyl compounds

3.1.2).[10] The best result (58% ee) was achieved in the oxidation of 2-pentylcyclopentanone (5) using cationic complex 4 bearing an optically active diphosphine ligand.

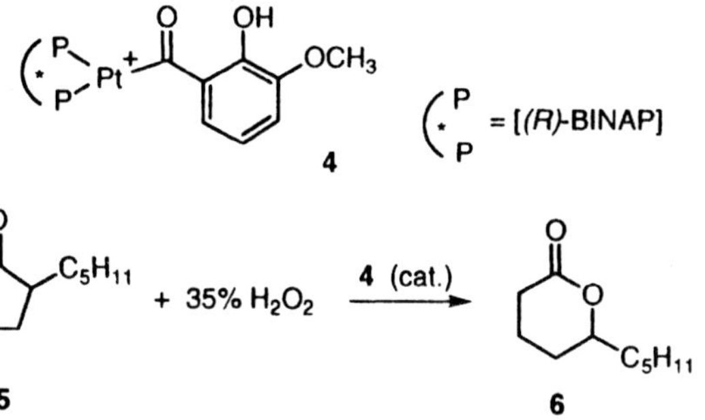

Scheme 3.1.2

Detailed studies revealed that the transformation involved coordination of the ketone to a vacant coordination site of the platinum complex followed by nucleophilic attack of free hydrogen peroxide on the carbonyl. Lactone 6 was formed regioselectively and resulted from a kinetic resolution of racemic ketone 5.

References

1. Baeyer, A.; Villiger, V. *Berichte* **1899**, *32*, 3625.
2. (a) Hassal, C. H. *Org. React.* **1957**, *9*, 73. (b) Krow, G. R. *Org. React.* **1993**, *43*, 251.
3. Criegee, R. *Justus Liebigs Ann. Chem.* **1948**, *560*, 127.
4. (a) McClure, J. D.; Williams, P. H. *J. Org. Chem.* **1962**, *27*, 24. (b) Matsubara, S.; Takai, K.; Nozaki, H. *Bull. Chem. Soc. Jpn.* **1983**, *56*, 2029. (c) Suzuki, M.; Takada, H.; Noyori, R. *J. Org. Chem.* **1982**, *47*, 902. (d) Göttlich, R.; Yamakoshi, K.; Sasai, H.; Shibasaki, M. *Synlett* **1997**, 971.
5. Yamada, T.; Takahashi, K.; Kato, K.; Takai, T.; Inoki, S.; Mukaiyama, T. *Chem. Lett.* **1991**, 641.
6. Murahashi, S.-I.; Oda, Y.; Naota, T. *Tetrahedron Lett.* **1992**, *33*, 7557.
7. Bolm, C.; Schlingloff, G.; Weickhardt, K. *Angew. Chem.* **1994**, *106*, 1944; *Angew. Chem., Int. Ed. Engl.* **1994**, *33*, 1848.
8. (a) Bolm, C.; Schlingloff, G. *J. Chem. Soc., Chem. Commun.* **1995**, 1247. (b) Bolm, C.; Schlingloff, G.; Bienewald, F. *J. Mol. Catal.* **1997**, *117*, 347.
9. Lopp, M.; Paju, A.; Kanger, T.; Pehk, T. *Tetrahedron Lett.* **1996**, *37*, 7583.
10. (a) Gusso, A.; Baccin, C.; Pinna, F.; Strukul, G. *Organometallics* **1994**, *13*, 3442. (b) Strukul, G.; Varagnolo, A.; Pinna, F. *J. Mol. Catal.* **1997**, *117*, 413.

4

Oxidation of hetero atoms

4.1 Asymmetric oxidation of sulfides and selenides

H. B. KAGAN

1. Introduction

Chiral sulfoxides are compounds of great important in asymmetric synthesis since the sulfinyl moiety can play the role of a chiral auxiliary and be removed later.[1-3] Moreover, biologically active compounds are being prepared increasingly with a sulfoxide functionality.[4] For these reasons it is useful to possess a set of methods of preparative value to obtain various kinds of chiral sulfoxides. The most general and flexible approach is the asymmetric oxidation of sulfides (asymmetric sulfoxidation) as depicted in Scheme 4.1.1. A wide structural variation can be achieved by modification of the R^1 and R^2 groups. Recently, catalytic *sulfimidation* (**1→3**) has been described with ee values up to 70%.[5]

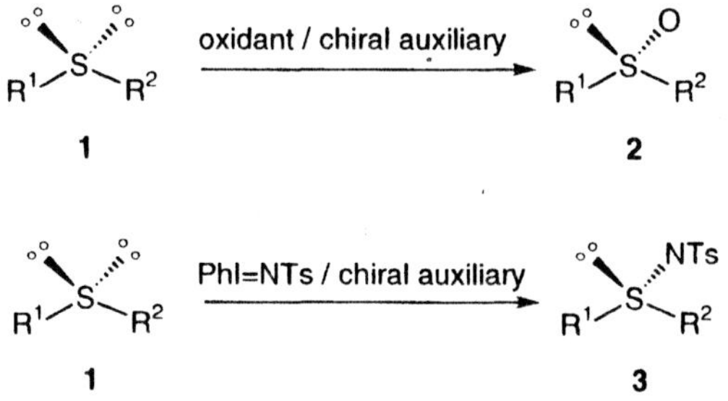

Scheme 4.1.1

Chiral selenoxides have less synthetic applications than sulfoxides but are of academic interest.[6] Asymmetric oxidation of selenides (asymmetric selenoxidation) is a straightforward route to chiral selenoxides (Scheme 4.1.2) and takes advantage of the knowledge concerning asymmetric sulfoxidation. Some examples of enantioselective oxidation of tellurides to *chiral telluroxides* have been reported recently but will not be presented here.

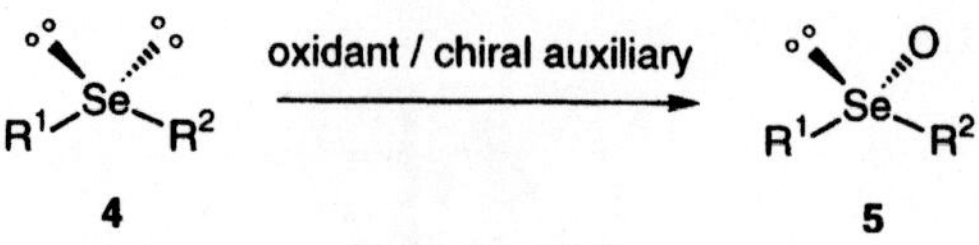

Scheme 4.1.2

$R^1\text{-Se-}R^2$ (4) → $R^1\text{-Se(O)-}R^2$ (5)
oxidant / chiral auxiliary

Only the topic of enantioselective oxidation is covered here; the case of diastereoselective oxidation is out of the scope of this chapter and has been reviewed recently.[7]

2. Reactions

2.1 Asymmetric sulfoxidation (stoichiometric)

The reactions of synthetic interest are mostly asymmetric synthesis where the chiral auxiliary is used in stoichiometric amounts. However, catalytic asymmetric sulfoxidation is of increasing efficiency (for a review see Ref. 8) and ultimately could be the most convenient approach to chiral sulfoxides. Some of the most efficient reactions using stoichiometric or catalytic amounts of a chiral auxiliary will be presented here briefly.

2.1.1 Chiral oxaziridines

This class of chiral oxidants has been devised by Davies and co-workers.[9,10] Two categories of chiral oxaziridines (**6** and **7**) have been investigated (Scheme 4.1.3), both deriving from terpene derivatives (for the preparation of **6**, see Section 4.2). Some results are listed in Table 4.1.1. The reactions were usually performed in dichloromethane or carbon tetrachloride at 20 °C. The sulfoxide is recovered after separation of the chiral imine **8** or **9**, which in principle may be reoxidized to the initial chiral oxaziridine **6** or **7**.

Table 4.1.1 Asymmetric oxidation of sulfides by oxaziridines **6** or **7** or by aq. H_2O_2/DBU and imine **8**

Entry	Oxidant[a,b]	Sulfoxide	ee (%) sulfoxide	Yield[a] (%)	Ref.
1	6a	9-Anthryl-S(O)-Me	80 (S)	85–90	9
2	6a	p-Tol-S(O)-Me	8 (S)	85–90	9
3	6b or 6c	p-Tol-S(O)-Me	67 (S)	85–90	9
4	7a	9-Anthryl-S(O)-Me	64 (S)	70	10
5	7a	p-Tol-S(O)-Me	28 (S)	80	10
6	7b	9-Anthryl-S(O)-Me	95 (S)	90	10
7	7b	p-Tol-S(O)-Me	>95 (S)	80	10
8	7b	p-Tol-S(O)-CH_2Ph	94 (S)	88	10
9	7b	Me-S(O)-t-Bu	94 (S)	84	10
10	8d/H_2O_2	Me-S(O)-t-Bu	86 (S)	>98	12
11	8d/H_2O_2	p-Tol-S(O)-Me	60 (S)	96	12
12	8d/H_2O_2	2-Phenyl-1,3-dithiane monoxide	≥98 (S)	>98	12

[a] Entries 1–9: oxidation at 20 °C in CCl_4 with **6a–6c**, oxidation at −20 °C in CCl_4 for **7a** and **7b** (except entries 4–6 where solvent is CH_2Cl_2).
[b] Entries 10–12; 1 eq. **8d**, 4 eq. H_2O_2/H_2O, 4 eq. DBU, in CH_2Cl_2 at −20 °C.

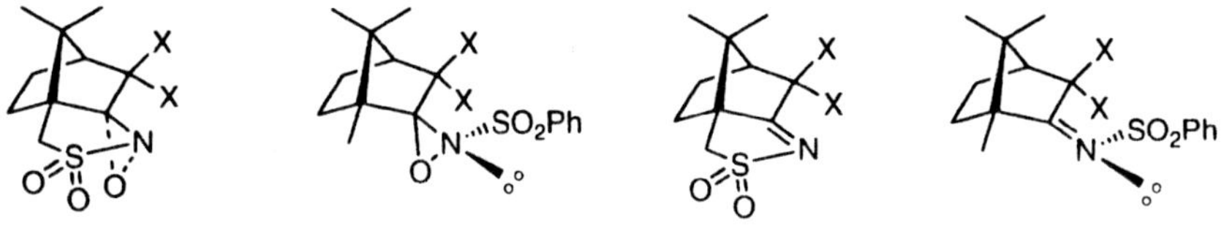

4: Oxidation of hetero atoms

6
a: X = H
b: X = Cl
c: X = Br
d: X = OMe

7
a: X = H
b: X = Cl
c: X = Br
d: X = OMe

8
a: X = H
b: X = Cl
c: X = Br
d: X = OMe

9
a: X = H
b: X = Cl
c: X = Br
d: X = OMe

Scheme 4.1.3

Oxaziridines **7** are of particular interest and are able to give sulfoxides with ee values close or higher than 90% (entries 6–9). It is possible to get a fine-tuning of the enantioselectivity by the proper choice of the reagent. The two enantiomers of oxaziridine **6a** and **7b** are now commercially available.

Page and co-workers prepared several imines **8**.[11,12] These imines were used in basic conditions in the presence of aqueous hydrogen peroxide to generate presumably an α-hydroperoxyamine which oxidizes directly a sulfide into a sulfoxide instead of closing into an oxaziridine **6**. This mechanism recalls the Payne oxidation using nitriles and H_2O_2. The above procedure was very successful (see Table 4.1.1 for some examples). The reaction, in principle, may be catalytic in imine **8**, but the experimental procedure used it in stoichiometric amounts.

Protocol 1.
Sulfoxidation by chiral oxaziridines: synthesis of (S)-methyl p-tolyl sulfoxide (Structure 2: R^1 = Me, R^2 = p-Tol, Scheme 4.1.1)

Caution! Cary out all procedures in a well-ventilated hood and wear disposable gloves and chemical-resistant safety glasses.

The oxaziridine **6b** and the enantiomers of **7a** are commercially available (Fluka and Aldrich). Other oxaziridines may be prepared according to Refs 9 and 10. The usual solvents to run the oxidation are toluene, dichloromethane, tetrachloromethane, or acetonitrile. The reactions are performed at room temperature or at −20 °C. A typical example is the oxidation of methyl p-tolyl sulfide into (S)-methyl p-tolyl sulfoxide as described below according to Ref. 10.

Equipment
- Double-necked, round-bottomed flask (50 mL)
- Source of dry argon
- Teflon-coated magnetic stirrer bar
- Plates for preparative silica gel thin-layer chromatography

Protocol 1. Continued

Materials

- Oxaziridine **7b**, 85 mg, 0.25 mmol
- Tetrachloromethane — harmful by inhalation
- Diethyl ether — flammable
- Methyl p-tolyl sulfide, 37.9 mg, 275 mmol — stench
- Eu(hfc)$_3$ — keep in anhydrous conditions
- Ethyl acetate — HPLC grade, flammable
- Hexane — HPLC grade, flammable

Method

1. Place 85.0 mg (0.25 mmol) of oxaziridine (−)-**7b** in 10 mL of CCl$_4$ into a 50-mL round-bottomed flask with a magnetic stirring bar and an argon inlet and subsequently add methyl p-tolyl sulfide (37.9 mg, 0.275 mmol) in 5 mL of CCl$_4$.
2. Keep the reaction mixture at 20 °C for 4 h and concentrate.
3. Prepare a silica gel chromatography plate, and apply the reaction mixture dissolved in the minimum of diethyl ether. Run the plate by diethyl ether. After diethyl ether evaporation methyl p-tolyl sulfoxide is obtained (37 mg, 95% yield). Caution: the sulfoxide is hygroscopic.
4. Measure the specific rotation of the recovered (*S*)-methyl p-tolyl sulfoxide by dissolving in acetone (*c* = 1.6). $[\alpha]_D = -139.1°$. Measure the ee (≥95%) by ^{1}H NMR in CDCl$_3$ in presence of Eu(hfc)$_3$ (signal for the methyl group adjacent to the sulfoxide moiety).

Protocol 2.
Sulfoxidation in the presence of a chiral imine: synthesis of (S_S)-*trans*-2-phenyl-1,3-dithiane monoxide[12]

Caution! Carry out all procedures in a well-ventilated hood and wear disposable gloves and chemical-resistant safety glasses.

Equipment

- Double-necked, round-bottomed flask (50 mL)
- Teflon-coated magnetic stirrer bar
- Silica gel column for chromatography (Merck 9385)

Materials

- Imine **8d**
- Dichloromethane — harmful by inhalation
- Aqueous solution (30%) of H$_2$O$_2$, 10 mmol — caution as with all peroxides
- 1,8-Diazabicyclo[5.4.0]undex-7-ene (DBU) — corrosive
- 2-Phenyl-1,3-dithiane, 490 mg, 2.5 mmol — stench
- 2,2,2-Trifluoro-1-(9-anthryl)ethanol, 5–10 eq. — keep in anhydrous conditions, irritant
- Ethyl acetate — HPLC grade, flammable

4: Oxidation of hetero atoms

- Hexane — HPLC grade, flammable
- Silica gel for flash chromatography — harmful by inhalation

Method

1. Place DBU in dichloromethane into a 5-mL round-bottomed flask with a magnetic stirring bar.
2. Keep the reaction mixture at $-20\,°C$ for 4 h.
3. Add a commercial aqueous solution (30%) of hydrogen peroxide (10 mmol) to the stirred solution of DBU in CH_2Cl_2 maintaining at $-20\,°C$.
4. Add imine **8d** (842 mg, 2.5 mmol) and subsequently the substrate (490 mg, 2.5 mmol). Stir the mixture for 24 h.
5. Add a solution of aqueous sodium sulfite (destruction of residual hydrogen peroxide). Extract with CH_2Cl_2 (2 × 20 mL). Solvent evaporation gives a mixture of the desired product and imine **8d**.
6. Prepare a short silica gel chromatography column. Apply the reaction mixture dissolved in the minimum of dichloromethane. Elute the column with dichloromethane. Imine **8d** is eluted first (95% recovery), followed by the sulfoxide. After solvent evaporation (S_S)-*trans*-2-phenyl-1,3-dithiane monoxide is obtained in quantitative yield.
7. Enantiomeric excess (≥98%) is measured by ^1H NMR spectroscopy in $CDCl_3$ at 400 MHz in the presence of 5–10 molar equivalents of 2,2,2-trifluoro-1-(9-anthryl)ethanol.

2.1.2 Hydroperoxides in the presence of chiral titanium complexes

The efficient asymmetric epoxidation of allylic alcohols was described by Sharpless in 1980, involving *t*-butyl hydroperoxide (TBHP) as oxidant and the combination of $Ti(OiPr)_4$ and diethyl tartrate (DET) as chiral mediator.[13] The structure of the Sharpless reagent formed *in situ* has been established as the bimetallic species $[DET\text{-}Ti(OiPr)_2]_2$ with the two titanium atoms bridged by oxygen (alkoxide) of DET.[14] The Sharpless reagent may be used in catalytic amounts in the presence of molecular sieves which trap traces of water (for epoxidation using the Sharpless reagent, see Chapter 2, Section 2.3). In 1984 two modifications of the Sharpless reagent were described as useful chiral mediators in the oxidation of alkyl aryl sulfides by TBHP, while the Sharpless reagent itself did not provide any significant enantioselectivity.

In Orsay Kagan and co-workers found that the introduction of one equivalent of water (with respect to titanium) significantly increased the enantioselectivity of various sulfoxidations by TBHP.[15] The optimized combination was $Ti(OiPr)_4/(R,R)\text{-}DET/H_2O = 1:2:1$ and will be called here the Orsay water-modified reagent since it was used in stoichiometric amounts with respect to the sulfide. Some results are listed in Table 4.1.2.

Table 4.1.2 Asymmetric oxidation of sulfides mediated by chiral titanium complexes

Entry	Complex (1 eq.)	Sulfoxide	ee (%) sulfoxide	Yield (%)	Ref.
1	Orsay reagent[a,b]	p-Tol-S(O)-Me	88 (R)	89	15
2	Orsay reagent[a,c]	p-Tol-S(O)-Me	99.5 (R)	75	24
3	Orsay reagent[a]	Me-S(O)-n-Octyl	80 (R)	71	17
4	Orsay reagent[a,c]	Ferrocenyl-S(O)-p-tol	80 (R)	71	17
5	Orsay reagent[a,c]	Ferrocenyl-S(O)-t-Bu	95 (R)	81	22
6	Orsay reagent[a]	$(CO)_3CrC_6H_5$-S(O)-Me	83 (S)	65	23
7	Padova reagent[b,d,e]	p-Tol-S(O)-Me	88 (R)	60	16
8	Padova reagent[b,d,e] (1 eq.)	trans-2-Methyl-2-phenyl 1,3-dithiane monoxide	83	66	19
9	Padova reagent[d] (2 eq.)	trans-2-Ethoxycarbonyl-1,3-dithiane 1,3-dioxide	97	80	21

[a] Ti(Oi-Pr)$_4$/(R,R)-DET/H$_2$O = 1:2:1, cumene hydroperoxide (CHP) as oxidant unless stated otherwise. Reaction run at −20 °C in dichloromethane.
[b] t-BuOOH as oxidant.
[c] Modified procedure of preparation of Orsay reagent.
[d] Ti(Oi-Pr)$_4$/(R,R)-DET/H$_2$O = 1:4. Reaction run at −20 °C in dichloromethane. CHP as oxidant unless stated otherwise.
[e] Solvent: 1,2-dichloroethane.

In Padova Modena and co-workers discovered that a large excess of diethyl tartrate similarly enhanced the enantioselectivity of sulfoxidations by TBHP.[16] These authors used the combination Ti(OiPr)$_4$/(R,R)-DET = 1:4, called here the Padova reagent; some results are collected in Table 4.1.2.

The above two reagents behave similarly, although in some special cases it is better to use one or the other. The ee values are the highest for aryl methyl sulfides, with a strong decrease in enantioselectivity in the oxidation of the series: Ar–S–Me > Ar–S–Et > Ar–S–nPr > Ar–S–nBu. An empirical rule of asymmetric induction has been proposed to predict the absolute configuration of the sulfoxides and is of wide generality (Scheme 4.1.4).[17] It also applies to the Padova reagent.

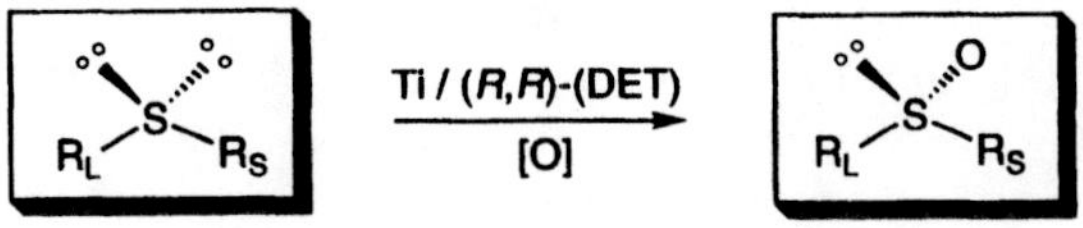

R_L: large group or π-system (Ar, C=C, or C≡C)

R_S: small group (alkyl)

Scheme 4.1.4

Cumene hydroperoxide (CHP) gives significantly higher ee values than TBHP in many cases.[17] It is now used widely with both the Orsay and Padova reagents.

4: Oxidation of hetero atoms

Orsay and Padova reagents have found some applications for the preparation of sulfoxides with quite high ee values; for some selected results see Refs 18–25.

Careful control of the conditions of preparation of the Orsay reagent may give up to 99% ee[24] (Table 4.1.2, entries 4 and 5). It is interesting to note that the asymmetric oxidation of sulfide **10** on a kilogram scale has been performed by Rhône-Poulenc Co. using a half equivalent of the combination Ti(OiPr)$_4$/(S,S)-DET = 1:2 and cumene hydroperoxide (Scheme 4.1.5).[4,25]

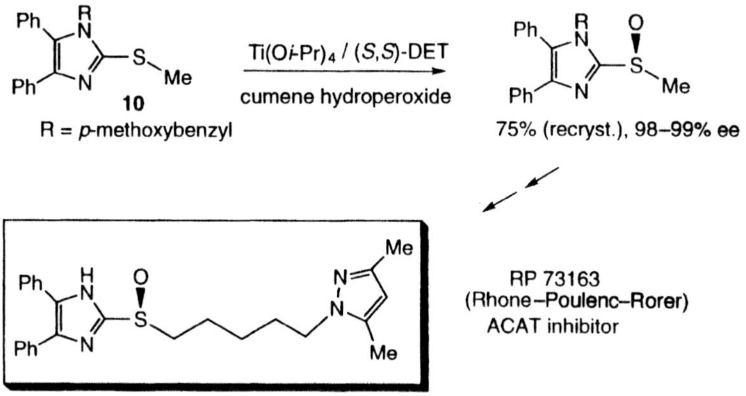

Scheme 4.1.5

Protocol 3.
Sulfoxidation mediated by the Orsay reagent: synthesis of (*S*)-methyl *p*-tolyl sulfoxide (Structure 2: R^1 = Me, R^2 = *p*-Tol, Scheme 4.1.1)[42]

Caution! Carry out all procedures in a well-ventilated hood and wear disposable gloves and chemical-resistant safety glasses.

Equipment
- Double-necked, round-bottomed flask (1 L) with septum caps
- Source of dry argon
- Syringe (10 mL)
- Teflon-coated magnetic stirrer bar
- Sintered-glass funnel, 9-cm diameter
- Flash column chromatography, 75-mm diameter

Materials
- (*S,S*)-Diethyl tartrate, 6.19 g, 30 mmol — freshly distilled
- Titanium tetraisopropoxide, 4.48 mL, 15 mmol — freshly distilled, hygroscopic
- Dichloromethane, 130 mL — harmful by inhalation
- Methyl *p*-tolyl sulfide, 4.20 g, 4.09 mL, 30 mmol — stench
- Cumene hydroperoxide 80%, 5.79 g, 30 mmol — commercial product
- Ethyl acetate — HPLC grade, flammable
- Cyclohexane — HPLC grade, flammable
- Hexane, 60 mL — harmful by inhalation

Protocol 3. *Continued*

- Sodium hydroxide, 2 M, 80 mL   **corrosive**
- Magnesium sulfate
- Silica gel for flash chromatography   **harmful by inhalation**

Method

1. In a 1-L round-bottomed flask with a magnetic stirring bar and argon inlet filled with 125 mL of dichloromethane at 20°C, inject through septum with a 10 mL syringe (*S,S*)-diethyl tartrate (6.19 g, 30 mmol).

2. After an argon purge introduce Ti(O*i*Pr)$_4$ (4.48 mL, 15 mmol) through the septum cap. The stirred solution turns yellow. After a few minutes add distilled water (0.27 mL, 15 mmol) dropwise with a syringe. After 25 min of strong stirring, a pale yellow solution is obtained.

3. Add methyl *p*-tolyl sulfide (4.20 g, 4.09 mL, 30 mmol) dissolved in 5 mL of CH$_2$Cl$_2$ with a syringe into the solution. After cooling to −30°C, introduce 80% cumene hydroperoxide (5.70 g, 5.54 mL, 30 mmol) dropwise with stirring, during 5 min. Place the flask in a freezer (−23°C) overnight (15 h).

4. Add 5.05 mL of water followed by a strong stirring at room temperature for 1.5 h. Filter the resulting suspension on Celite on a 9-cm-diameter sintered-glass funnel (porosity 2). Treat the filtrate (300 mL) with 80 mL of 2 N NaOH and 40 mL of saturated aqueous NaCl with vigorous stirring. Decant the organic phase (negative test for peroxides), dry it on MgSO$_4$ and concentrate to give 10 g oil (sulfoxide, some sulfide and 2-phenyl-2-propanol).

5. Prepare a flash column chromatography [75-mm diameter, filled with 120 g silica gel (Merck 230-240 mesh), and EtOAc–cyclohexane 9:1 as solvent]. Apply at the top of the column the crude product dissolved in the minimum amount of solvent. Elute with 200 mL of EtOAc–cyclohexane 9:1 followed by 200 mL of EtOAc. Methyl *p*-tolyl sulfide mixed with 2-phenyl-2-propanol is first eluted (150 mL), followed by methyl *p*-tolyl sulfoxide (300 mL). Evaporation gives 4.0 g of (−)-(*S*)-methyl *p*-tolyl sulfoxide, 89% ee, in 85% yield. Caution: the sulfoxide is hygroscopic. The above material crystallized from 60 mL of hot hexane affords 3.14 g (68%) of (*S*)-methyl *p*-tolyl sulfoxide, 99.5% ee. $[\alpha]_D = -142°$ ($c = 1$, acetone); m.p. = 73–76°C.

6. Measure the ee (89%) by HPLC with a Daicel Chiralcel OD-H phase (elution with AcOEt–hexane 1:9). To avoid errors in the ee determination it is important to mix the various fractions of sulfoxide and then to measure the ee of an aliquot.[43]

An optimized procedure (but more difficult to handle) has been set up for the asymmetric oxidation of some sulfides of up to 99% ee.[22,24] This procedure involves careful control of the conditions of the preparation of the titanium reagent (temperature, rate of addition of water, ageing-time, precooling of

4: Oxidation of hetero atoms

the reactants before addition to the titanium complex). In this way, methyl *p*-tolyl sulfoxide with >99.5% ee has been prepared on a 3-mmol scale.[24]

Protocol 4.
Sulfoxidation mediated by the Padova reagent: synthesis of *trans*-2-ethoxycarbonyl-1,3-dithiane 1,3-dioxide

Caution! Carry out all procedures in a well-ventilated hood and wear disposable gloves and chemical-resistant safety glasses.

The usefulness of this reagent will be exemplified in the formation of the *trans*-disulfoxide derived from ethyl 1,3-dithiane-2-carboxylate.[21]

Equipment
- Double-necked, round-bottomed flask (50 mL) with septum caps
- Source of dry nitrogen
- Syringe (1 mL)
- Teflon-coated magnetic stirrer bar
- Sintered-glass funnel
- Column for flash chromatography

Materials
- (*R,R*)-Diethyl tartrate, 0.85 mL, 5 mmol — **freshly distilled**
- Titanium tetraisopropoxide, 0.37 mL, 1.25 mmol — **freshly distilled, hygroscopic**
- Dichloromethane, 15 mL — **harmful by inhalation**
- Ethyl 1,3-dithiane-2-carboxylate, 0.20 mL, 1.25 mmol — **stench**
- Cumene hydroperoxide 80%, 1 mL, 5 mmol — **commercial product**
- Ethyl acetate — **HPLC grade, flammable**
- Acetone — **HPLC grade, flammable**
- Sodium sulfate
- Silica gel for flash chromatography — **harmful by inhalation**

Method

1. In a 50-mL round-bottomed flask equipped with a magnetic stirring bar and nitrogen inlet place 15 mL of dichloromethane at 20 °C.

2. Inject through the septum with a 1-mL syringe (*R,R*)-diethyl tartrate (0.85 mL, 5 mmol).

3. After a nitrogen purge introduce Ti(O*i*Pr)$_4$ (0.37 mL, 1.25 mmol) through the septum cap. The stirred solution turns yellow. Stir for 20 min.

4. Introduce ethyl 1,3-dithiane-2-carboxylate (0.24 g, 0.20 mL, 1.25 mmol). Cool the mixture to −35 °C and keep it at this temperature for 1 h.

5. Add 80% cumene hydroperoxide (1.0 mL, 5.0 mmol). Stir at −35 °C under nitrogen for 48 h.

6. Add 0.5 mL of water. Warm the reaction mixture to room temperature and stir for 1 h. Filter off the TiO$_2$ gel through a pad of Celite. Dry the filtrate over Na$_2$SO$_4$ and concentrate.

7. Prepare a flash column chromatography (filled with silica gel, with ethyl

Protocol 4. Continued

acetate as solvent). Apply to the top of the column the crude product dissolved in the minimum amount of solvent. Elute with EtOAc–acetone (gradient 1:0→0:1). Recover the monoxide (32 mg, 12% as a *cis/trans* mixture) and then the *trans*-disulfoxide [223 mg, 80%, $[\alpha]_D = +240°$ ($c = 3$, $CHCl_3$)].

8. Measure the enantiomeric excess (>97%) by ^1H NMR using 0.2 eq. of Eu(hfc)$_3$.

2.2 Asymmetric sulfoxidation (catalytic)

In this section most of the best non-enzymatic oxidations currently known are reported (see Table 4.1.3). Enzymatic sulfoxidations will not be considered here but are described in Chapter 5, Section 5.4.

2.2.1 Hydroperoxides in the presence of chiral titanium complexes

Some of the above sulfoxidations mediated by chiral titanium complexes could be transformed into catalytic processes by addition of molecular sieves. Thus the combination Ti(O*i*Pr)$_4$/(*R,R*)-DET/*i*-PrOH 1:4:4, which derives from the Padova reagent by addition of 2-propanol, acts as a catalyst (10% mol eq.) in the presence of MS 4-Å affording aryl methyl sulfoxides around 90% ee.[26] Some results are listed in Table 4.1.3. BINOL **11** has been used by

Table 4.1.3 Asymmetric oxidation of sulfides catalysed by chiral transition metal complexes

Entry	Catalyst (0.1 eq. unless stated)	Sulfoxide	ee (%) sulfoxide	Yield (%)	Ref.
1	Ti/(*R,R*)-DET[a]	Ph-S(O)-Me	91 (*R*)	81	26
2	Ti/(*R,R*)-DET[a]	*o*-Anisyl-S(O)-Me	89 (*R*)	72	26
3	Ti/(*R,R*)-DET[a]	Benzyl-S(O)-Me	90 (*R*)	72	26
4	Ti/Binol **11**[b]	*p*-Tol-S(O)-Me	53→96	80→6	27
5	Ti/octahydrobinol[c]	*p*-Tol-S(O)-Me	86	52	29
6	Ti/aminotriol **12**[d]	Benzyl phenyl sulfoxide	84	94	30
7	Ti/diol **13**[d]	*p*-Tol-S(O)-Me	95(*S*)	42	31
8	Mn salen **14**[e]	Methyl 2-nitrophenyl sulfoxide	94 (−)	94	32
9	V Schiff base (0.01 eq.)[f]	2-Phenyl-1,3-dithiane monoxide	85	84	33
10	V Schiff base (0.01 eq.)[f]	*t*-Bu-S(O)-S-*t*-Bu	91(*S*)	98	34

[a] Ti(O*i*-Pr)$_4$/(*R,R*)-DET/*i*-PrOH = 1:4:4. Reaction in CH_2Cl_2 at −20°C in presence of 4-Å molecular sieves. Oxidant: cumene hydroperoxide (2 eq.).
[b] Ti(O*i*-Pr)$_4$/(*R*)-binol **11**/H_2O = 1:1:4. Reaction in CH_2Cl_2 at −20°C. Oxidant: cumene hydroperoxide (2 eq.). Some kinetic resolution amplified the initial ee.
[c] Ti(O*i*-Pr)$_4$/(*R*)-octahydrobinol.
[d] Some kinetic resolution amplified the initial ee. Oxidant: cumene hydroperoxide.
[e] Oxidant: PhIO. Reaction in acetonitrile.
[f] VO(acac)$_2$ + 1 eq. **15**. Oxidant: 30% aq. H_2O_2.

4: Oxidation of hetero atoms

Uemura and co-workers instead of diethyl tartrate in a water-modified reagent.[27,28] This catalyst is not very enantioselective giving the sulfoxide of 53% ee in the oxidation of methyl *p*-tolyl sulfide by CHP. However, partial overoxidation to the sulfone is easily attained, with some kinetic resolution which amplifies the ee. Thus 96% ee is obtained for 66% conversion. Similar observations have been made by Reetz *et al.* for titanium complexes having some (*R*)-octahydrobinol derivatives as ligands.[29]

Trialkanolamines **12** have been found to be good ligands of titanium. The resulting complexes catalyse the oxidation of sulfides by TBHP with initial moderate ee values which are subsequently amplified by kinetic resolution.[30]

Diol **13** has been used by Yamanoi and Imamoto in a catalyst combination Ti(O*i*Pr)$_4$/**13** = 1/2, with CHP as oxidant.[31] The observed ee values are also the result of the combined enantioselective oxidation and kinetic resolution.

Protocol 5.
Sulfoxidations catalysed by a titanium reagent: synthesis of (*R*)-methyl phenyl sulfoxide (Structure 2: R^1 = phenyl, R^2 = Me, Scheme 4.1.1)[26]

Caution! Carry out all procedures in a well-ventilated hood and wear disposable gloves and chemical-resistant safety glasses.

Equipment

- Schlenk tube (25 mL) with septum cap
- Vacuum line
- Source of dry argon
- Syringe (1 mL)
- Teflon-coated magnetic stirrer bar
- Column for flash chromatography

Materials

• (*R,R*)-Diethyl tartrate, 205 μL, 1.2 mmol	freshly distilled
• Titanium tetraisopropoxide, 90 μL, 0.3 mmol	freshly distilled, hygroscopic
• Dichloromethane (anhydrous), 15 mL	harmful by inhalation
• Molecular sieves 4-Å	activated at 200°C under vacuum for 16 h, stored under argon
• Methyl phenyl sulfide, 3 mmol	stench
• Cumene hydroperoxide 80%, 1.2 mL, 6 mmol	commercial product
• Hexane	HPLC grade, flammable
• 2-Propanol	HPLC grade, flammable
• Diethyl ether, 145 mL	flammable
• Magnesium sulfate	
• 1,4-Dioxane, 15 mL	flammable
• Sodium hydroxide, 2M, 50 mL	corrosive
• Silica gel for flash chromatography	harmful by inhalation

Method

1. Place 10 mL of dichloromethane and 1 weight equivalent (with respect to the sulfide) of molecular sieves 4-Å at 20°C in a 25-mL Schlenk tube with a magnetic stirring bar and argon inlet.

2. Inject through the septum cap with a 1-mL syringe (*R,R*)-diethyl tartrate (205 μL, 1.2 mmol). Stir the mixture for 2.5 min.

Protocol 5. Continued

3. Introduce slowly Ti(OiPr)$_4$ (90 μL, 0.3 mmol) through the septum cap. The stirred solution turns yellow. Stir for 10 min at 16°C.

4. Introduce slowly 2-propanol (92 μL, 1.2 mmol). Maintain stirring for 10 min followed by cooling in a freezer (−22°C) without stirring for an additional 20 min.

5. Start the reaction by the rapid addition of methyl phenyl sulfide (3 mmol) and then the precooled (−22°C) cumene hydroperoxide (1.2 mL, 6 mmol). Store the flask in the refrigerator (−22°C) without stirring for 16 h.

6. Pour the content of the flask into a solution of 3 g ferrous sulfate heptahydrate (10.8 mmol) and 1 g citric acid (4.8 mmol) in water (30 mL), 1,4-dioxane (15 mL), and diethyl ether (25 mL). Stir the mixture for 15 min. Extract the aqueous phase with diethyl ether (3 × 20 mL). Stir vigorously the combined organic phases with 2 M aqueous sodium hydroxide (50 mL) for 1 h. Extract the aqueous solution with diethyl ether (3 × 20 mL). The combined organic solutions are washed with brine (25 mL), dried over MgSO$_4$, filtered and evaporated under reduced pressure.

7. Prepare a column for flash chromatography (filled with silica gel, with ethyl acetate as solvent). Apply to the top of the column the crude product dissolved in the minimum amount of solvent. Elute with ethyl acetate. Recover first the unreacted sulfide, the overoxidized sulfone and 2-phenyl-2-propanol (cumylic alcohol) and then pure sulfoxide (374 mg, 81%). [α]$_D$ = 124.1° (c = 1, acetone).

8. Measure the enantiomeric excess (91.2%) by chiral HPLC analysis on Daicel Chiralcel OD-H: λ = 254 nm, 0.5 mL min^{-1} (eluent: hexane–i-PrOH 9:1). Mix all the fractions including the sulfoxide before ee measurement.

2.2.2 Hydroperoxides in the presence of chiral Schiff base complexes

Many metallic complexes with Schiff bases as ligands (mostly chiral salen) have been found to have good catalytic properties in the oxidation of sulfides by hydroperoxides or hydrogen peroxide. However, the enantioselectivities often remain very low (for reviews, see Refs 7 and 8). Here, results will be only considered where ee values are higher than 80% (Scheme 4.1.6).

A (salen)manganese complex **14** prepared by Katsuki and co-workers is very efficient (1% mol eq.) and enantioselective in the oxidation of various alkyl aryl sulfides (up to 94% ee).[32] Unfortunately the oxidant is iodosylbenzene, which means that this method is not practical. A promising catalyst was set up by Bolm and Bienewald based on a vanadium complex (1% mol eq.) prepared by mixing VO(acac)$_2$ and Schiff base **15**.[33] The oxidations are conveniently performed in dichloromethane by aqueous H$_2$O$_2$ (30%). Methyl p-tolyl sulfide gives the corresponding sulfoxide in 70% ee while 2-phenyl-1,3-

4: Oxidation of hetero atoms

dithiane affords its monoxide in 85% ee. The Bolm catalyst has been used recently to prepare t-BuS-S(O)-t-Bu from (t-BuS)$_2$ with 91% ee.[34]

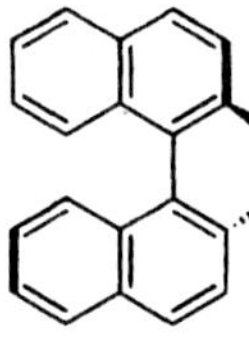

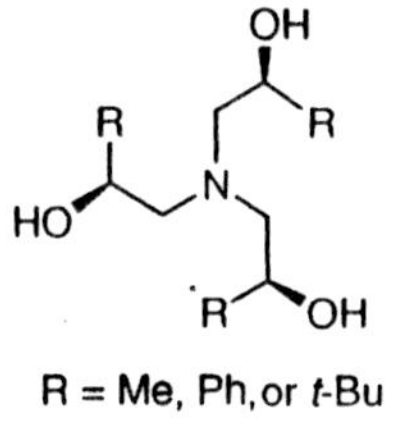

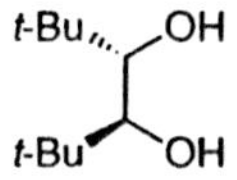

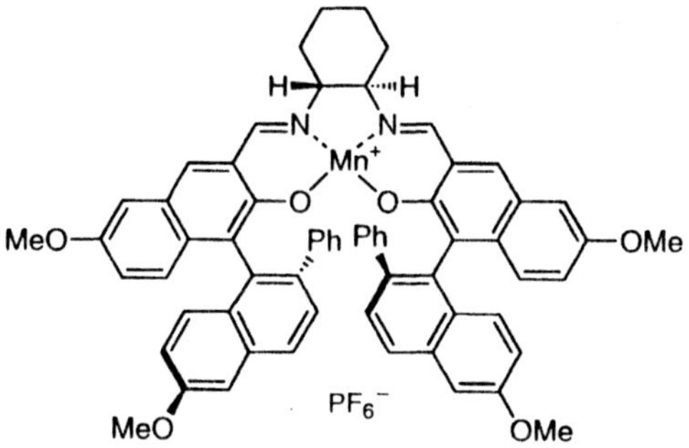

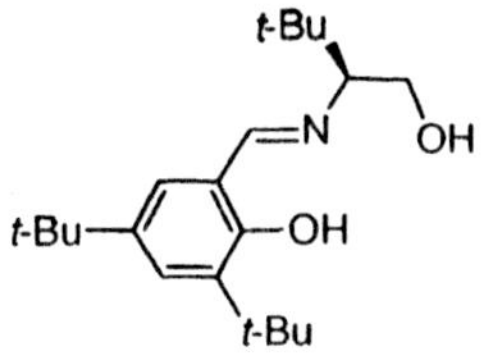

Scheme 4.1.6

Protocol 6.
Sulfoxidations catalysed by a chiral vanadium complex:[a] synthesis of 2-phenyl-1,3-dithiane monoxide[33]

Caution! Carry out all procedures in a well-ventilated hood and wear disposable gloves and chemical-resistant safety glasses

Equipment
- One test-tube (5 mL)
- Flash column chromatography

Materials
- (S)-**15**, 5.0 mg, 15 μmol
- Vanadyl bis(acetylacetonate), 1.6 mg, 0.01 mmol
- 2-Phenyl-1,3-dithiane, 19.6 mg, 1 mmol — stench
- Dichloromethane, 3 mL — harmful by inhalation
- Aqueous solution (30%) of H$_2$O$_2$, 1.1 mmol — caution as with all peroxides
- Hexane — HPLC grade, flammable
- 2-Propanol — HPLC grade, flammable
- Silica gel for flash chromatography — harmful by inhalation

Protocol 6. *Continued*

Method

1. Prepare the chiral catalyst from VO(acac)$_2$ (1.6 mg, 0.01 mmol) and Schiff base **15** (15 μmol, 5.0 mg) which are dissolved in a test-tube in CH$_2$Cl$_2$ (2 mL) at room temperature and stirred.
2. Add 2-phenyl-1,3-dithiane (19.6 mg, 1 mmol), followed by dropwise addition of 30% H$_2$O$_2$ (1.1 mmol). Stir for 16 h at room temperature.
3. Remove the aqueous phase. Recover the organic phase and evaporate the solvent.
4. Isolate the *trans*-2-phenyl-1,3-dithiane monoxide (17.8 mg, 84%) by chromatography on silica gel.
5. Measure the enantiomeric excess (85%) by HPLC on Chiralcel OD-H (UV detector 254 nm; hexane–2-propanol 7:3, flow rate 0.5 mL min^{-1}, retention times 15.0 and 34.5 min).

[a] This procedure also applies to the transformation of a disulfide into a thiosulfinate.[34]

2.3 Asymmetric selenoxidation

Enantiomerically enriched simple selenoxides were prepared for the first time by Davis *et al.* in 1983.[35] These compounds are prone to racemization in the presence of water; they are more stable when bulky groups are connected to selenium.

Chiral oxaziridines are able to oxidize various selenides into selenoxides with up to 96% ee (Scheme 4.1.7).[36,37]

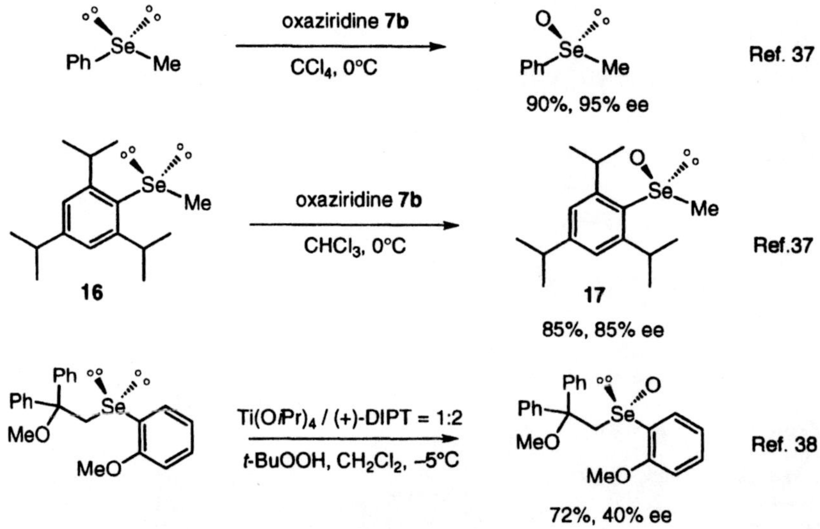

Ph–Se–Me →(oxaziridine **7b**, CCl$_4$, 0°C)→ Ph–Se(O)–Me   Ref. 37
90%, 95% ee

16 (2,4,6-triisopropylphenyl-Se-Me) →(oxaziridine **7b**, CHCl$_3$, 0°C)→ **17** Ref. 37
85%, 85% ee

Ph$_2$(MeO)C–CH$_2$–Se–(2-MeO-C$_6$H$_4$) →(Ti(O*i*Pr)$_4$ / (+)-DIPT = 1:2, *t*-BuOOH, CH$_2$Cl$_2$, –5°C)→ selenoxide   Ref. 38
72%, 40% ee

Scheme 4.1.7

4: Oxidation of hetero atoms

A chiral titanium reagent (Ti(O*i*Pr)$_4$/(*R,R*)-DIPT 1:2) has also been used with less success (≤ 40% ee) for asymmetric selenoxidation by *t*-BuOOH (Scheme 4.1.7).[38,39]

However, up to 92% ee has been obtained in asymmetric oxidation of some allyl aryl selenides, in which the initial selenoxide undergoes a facile and highly stereoselective [2,3]-sigmatropic rearrangement to an allylic alcohol (Scheme 4.1.8).[40]

Asymmetric synthesis of various cyclohexylidenemethyl ketones is also possible by the tandem asymmetric selenoxidation (with oxaziridines or titanium reagents) and selenoxide elimination.[41] One example is indicated in Scheme 4.1.8.

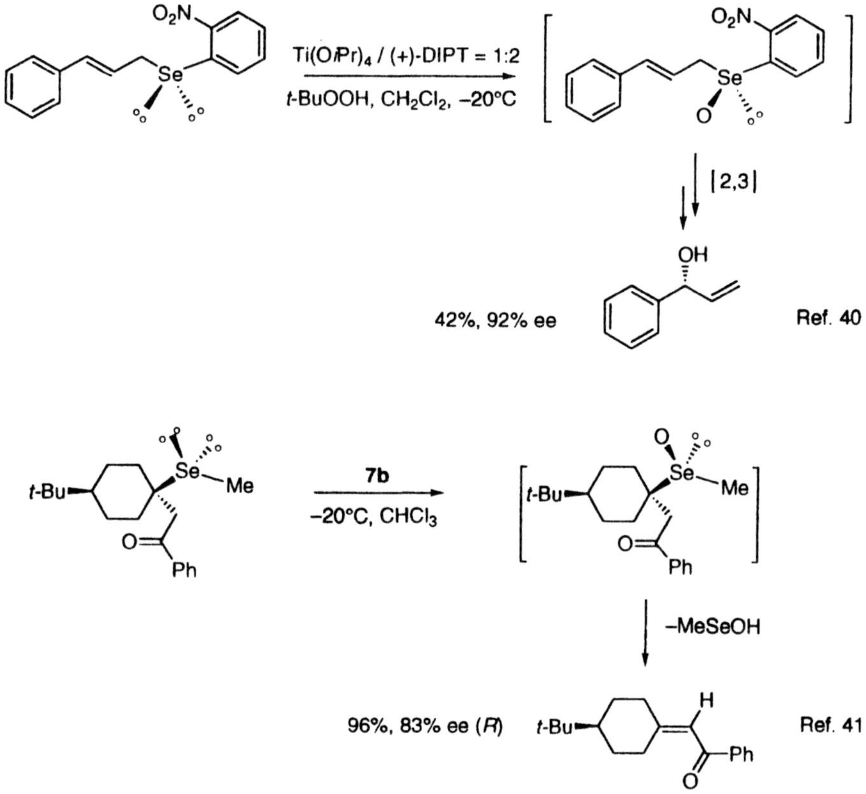

Scheme 4.1.8

No catalytic enantioselective selenoxidation is known yet.

Protocol 7.
Asymmetric selenoxidation: synthesis of methyl 2,4,6-triisopropylphenyl selenoxide (Structure 17, Scheme 4.1.7)[37]

Caution! Carry out all procedures in a well-ventilated hood and wear disposable gloves and chemical-resistant safety glasses.

The following procedure is typical of the conversion of hindered alkyl aryl selenides into the corresponding selenoxides.[37] Great care must be taken to avoid moisture which favours racemization of selenoxides.

Equipment
- Two dry double-necked, round-bottomed flasks (25 mL), each with a three-way stopcock
- Source of dry argon
- Teflon-coated magnetic stirrer bar
- Syringe (20 mL)
- Column for flash chromatography

Materials
- Oxaziridine **7b**, 204 mg, 0.6 mmol
- $CDCl_3$, 10 mL   dry
- Diethyl ether   flammable
- Methyl 2,4,6-triisopropylphenyl selenide, 146.5 mg, 0.5 mmol   irritant
- (S)-(+)-2,2,2-Trifluoro-1-(9-anthryl)ethanol
- Aluminium oxide, 60 g   basic, activated 120 °C, 3 h
- n-Pentane, 15 mL   dry solvent
- 3-Å Molecular sieves, 2 g

Method
1. Place 2.0 g of 3-Å molecular sieves in a dry 25-mL two-necked round-bottomed flask equipped with a three-way stopcock attached to an argon balloon, a magnetic stirrer bar and a septum.
2. Flush with argon, add 3 mL of dried $CDCl_3$ and cool to 0 °C.
3. Add methyl 2,4,6-triisopropylphenyl selenide (146.5 mg, 0.5 mmol) dissolved in 3 mL of dried $CDCl_3$.
4. Add dropwise via syringe oxaziridine (−)-**7b** (204 mg, 0.6 mmol) in 4 mL of $CDCl_3$.
5. After 1.5 h (check by TLC that the reaction is complete) evaporate the solvent to dryness under 10 mmHg at 0 °C. Flush the flask with dry argon, add via syringe 15 mL of dry n-pentane and stir.
6. Transfer via syringe the clear n-pentane solution to a second dry similar round-bottomed flask. Remove the solvent at 0 °C to give solid methyl 2,4,6-triisopropylphenyl selenoxide **17**. Collect a small portion and measure its enantiomeric excess (>95%) by ^1H NMR with (S)-(+)-2,2,2-trifluoro-1-(9-anthryl)ethanol.

7. Isolate (S)-17 of 85% ee by flash chromatography on 60 g of basic aluminium oxide, activated at 120°C for 3 h, under argon using a 98:2 dry CH_2Cl_2-MeOH mixture. Isolated yield 85%, (S)-17, $[\alpha]_D = -57.0°$ ($c = 1.1$, $CHCl_3$).

References

1. Solladié, G. *Synthesis* **1981**, 185–196.
2. Posner, G. H. In *The Chemistry of Sulphones and Sulphoxides*; Patai, S.; Rappoport, Z.; Sterling, C. J. M., eds, John Wiley and Sons: Chichester, UK, **1998**, Chapter 16.
3. Carreno, M. C. *Chem. Rev.* **1995**, *95*, 1717–1760.
4. (a) Pitchen, P.; France, C. J.; McFarlane, I. M.; Newton, C. G.; Thompson, D. M. *Tetrahedron Lett.* **1994**, *35*, 485–488. (b) Morita, S.; Matsubara, J.; Otsubo, K.; Kitano, K.; Ohtani, T.; Kawano, Y.; Uchida, M. *Tetrahedron: Asymmetry* **1997**, *8*, 3707–3710.
5. Takada, H.; Nishibayashi, Y.; Ohe, K.; Uemura, S.; Baird, P.; Sparey, T. J.; Taylor, P. C. *J. Org. Chem.* **1997**, *62*, 6512–6518.
6. Komatsu, N.; Uemura, S. In *Advances in Detailed Reaction Mechanisms*, vol. 4; JAI Press Inc.: Tokyo, **1995**, pp. 73–92.
7. Diter, P.; Kagan, H. B. In *Organosulfur Chemistry*, Bulman Page, P. C., ed., Academic Press, vol. 2, **1997**, pp. 1–37.
8. Kagan, H. B. In *Catalytic Asymmetric Synthesis*, Ojima, I., ed., Wiley-VCH: New York, **2000**, Chapter 6C.
9. David, F. A.; Weismiller, M. C.; Murphy, C. K.; Thimma Reddy, R.; Chen, B. C. *J. Org. Chem.* **1992**, *57*, 7274–7285.
10. Davis, F. A.; Thimma Reddy, R.; Han, W.; Carroll, P. J. *J. Am. Chem. Soc.* **1992**, *114*, 1428–1437.
11. Bulman Page, P. C.; McKenzie, M. J.; Buckle, D. R. *J. Chem. Soc., Perkin Trans I* **1995**, 2673–2676.
12. Bulman Page, P. C.; Heer, J. P.; Bethell, D.; Collington, E. W.; Andrews, D. M. *Synlett* **1995**, 773–775.
13. Katsuki, T.; Sharpless, K. B. *J. Am. Chem Soc.* **1980**, *102*, 5974–5976.
14. Finn, M. G.; Sharpless, K. B. *J. Am. Chem. Soc.* **1992**, *113*, 113–126.
15. Pitchen, P.; Deshmukh, M.; Dunach, E.; Kagan, H. B. *J. Am. Chem. Soc.* **1984**, *106*, 8188–8193.
16. Di Furia, F.; Modena, G.; Seraglia, R. *Synthesis* **1984**, 325–326.
17. Zhao, S.; Samuel, O.; Kagan, H. B. *Tetrahedron* **1987**, *43*, 5135–5144.
18. Samuel, O.; Ronan, B.; Kagan, H. B. *J. Organomet. Chem.* **1989**, *370*, 43–50.
19. Bortolini, O.; Di Furia, F.; Licini, G.; Modena, G.; Rossi, M. *Tetrahedron Lett.* **1986**, *27*, 6257–6260.
20. Marino, J. P.; Bogdan, S.; Kimura, K. *J. Am. Chem. Soc.* **1992**, *144*, 5566–5572.
21. Aggarwal, V. K.; Evans, G.; Moya, E.; Dowden, J. *J. Org. Chem.* **1992**, *57*, 6390–6391.
22. Diter, P.; Samuel, O.; Taudien, S.; Kagan, H. B. *Tetrahedron: Asymmetry* **1994**, *5*, 549–552.
23. Griffiths, S. L.; Perrio, S.; Thomas, S. E. *Tetrahedron: Asymmetry* **1994**, *5*, 545–548.

24. Brunel, J. M.; Dieter, P.; Deutsch, M.; Kagan, H. B. *J. Org. Chem.* **1995**, *60*, 8086–8088.
25. Pitchen, P. *Chem. Ind.* **1994**, 636–639.
26. (a) Brunel, J. M.; Kagan, H. B. *Synlett* **1996**, 404–406. (b) Brunel, J. M.; Kagan, H. B. *Bull. Soc. Chim. Fr.* **1996**, *133*, 1109–1115.
27. Komatsu, N.; Hashizume, M.; Sugita, T.; Uemura, S. *J. Org. Chem.* **1993**, *58*, 4529–4533.
28. Komatsu, N.; Hashizume, M.; Sugita, T.; Uemura, S. *J. Org. Chem.* **1993**, *58*, 7624–7626.
29. Reetz, M. T.; Merk, C.; Naberfeld, G.; Rudolph, J.; Griebenow, N.; Goddard, R. *Tetrahedron Lett.* **1997**, *38*, 5273–5276.
30. Di Furia, F.; Licini, G.; Modena, G.; Motterle, R.; Nugent, W. A. *J. Org. Chem.* **1996**, *61*, 5175–5177.
31. Yamanoi, Y.; Imamoto, T. *J. Org. Chem.* **1997**, *62*, 8560–8564.
32. (a) Noda, K.; Hosoya, N.; Yanai, K.; Irie, R.; Katsuki, T. *Tetrahedron Lett.* **1994**, *35*, 1887–1890. (b) Kokubo, C.; Katsuki, T. *Tetrahedron* **1996**, *44*, 13895–13900.
33. Bolm, C.; Bienewald, F. *Angew. Chem., Int. Ed. Engl.* **1996**, *34*, 2640–2642.
34. Liu, G.; Cogan, D. A.; Ellman, J. A. *J. Am. Chem. Soc.* **1997**, *119*, 9913–9914.
35. Davis, F. A.; Billmers, J. M.; Stringer, O. O. *Tetrahedron Lett.* **1983**, *24*, 3191–3194.
36. Davis, F. A.; Reddy, R. T.; Weismiller, M. C. *J. Am. Chem. Soc.* **1989**, *111*, 5964–5965.
37. Davis, F. A.; Reddy, R. T. *J. Org. Chem.* **1992**, *57*, 2599–2606.
38. Tiecco, M.; Tingoli, M.; Testaferri, L.; Bartoli, D. *Tetrahedron Lett.* **1987**, *29*, 3849–3852.
39. Shimizu, T.; Kobayashi, M.; Kamigata, N. *Bull. Chem. Soc. Jpn.* **1989**, *62*, 2099.
40. Komatsu, N.; Nishibayashi, Y.; Uemura, S. *Tetrahedron Lett.* **1993**, *34*, 2339–2342.
41. Komatsu, N.; Matsunaga, S.; Sugita, T.; Uemura, S. *J. Am. Chem. Soc.* **1993**, *115*, 5847–5848.
42. Zhao, S.; Samuel, O.; Kagan, H. B. *Org. Synth.* **1989**, *68*, 49–55.
43. Diter, P.; Samuel, O.; Taudien, S.; Kagan, H. B. *J. Org. Chem.* **1994**, *59*, 370–373.

4: Oxidation of hetero atoms

4.2 Asymmetric oxidation of other heteroatoms

S. MIYANO and T. HATTORI

1. Introduction

The heteroatoms which should be picked up in this section include all the elements of groups 14 and 15 except carbon. However, an attempt to cover the literature up to early 1997 has revealed that asymmetric oxidations of this class of heteroatoms are virtually limited to only nitrogen from the viewpoint of synthetic usefulness or at least having the potential for synthetic use in the laboratory. Thus, this section will discuss asymmetric oxidations of nitrogen-containing compounds, with particular emphasis on asymmetric or, more specifically, stereospecific oxidation of chiral imines to oxaziridines[1,2] as well as kinetic resolution of C-chiral β-hydroxy tertiary amines by *N*-oxide formation.[3]

It is known that the three substituents of an amine are arranged pyramidally around nitrogen in a geometry similar to that of tetravalent carbon.[4] Consequently, if the three substituents are different, the nitrogen atom constitutes a chiral centre and, in principle, such amines are capable of existing as enantiomers. However, the energy barrier to pyramidal inversion of the enantiomers is usually so low for simple amines that they are interconverted rapidly to prevent them from being isolated as stable enantiomers. In contrast to this, the barrier to pyramidal inversion of amine *N*-oxides is high enough to permit the isolation of each enantiomer. Therefore, direct oxidation of a tertiary amine having three different substituents with a chiral oxidizing reagent is, in principle, a route to optically active amine oxides, albeit in low optical yields (Scheme 4.2.1).[5] For example, treatment of *N-trans*-but-2-enyl-*N*-ethyl-*p*-toluidine with (−)-*O,O*-dibenzoyl-D-pertartaric acid gave the corresponding *N*-oxide **1** of $[\alpha]_D^{25}$ = +5.5° (*c* = 0.95, MeOH), although the extent of asymmetric induction was not determined.[6]

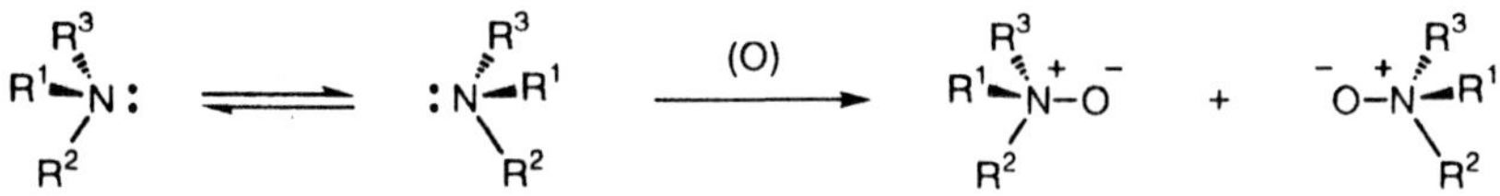

Scheme 4.2.1

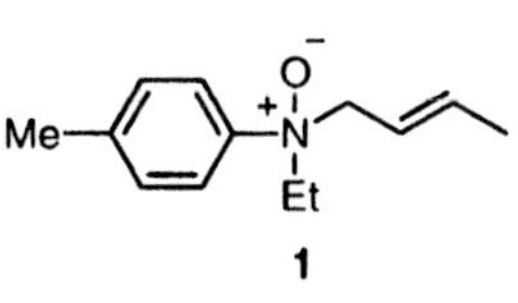

1

Trivalent organophosphorus compounds having three different substituents also contain a P-chiral centre, which, differing from nitrogen, is conformationally rather stable, allowing the isolation of enantiomers. Here again, however, attempts to kinetically resolve the phosphines through partial oxidation with chiral peracids or amine oxides met with very little success.[7] Ugi *et al.* claimed that chiral phosphites could be obtained by stereoselective oxidation of the racemates by enantiomerically pure (3-oxocamphorsulfonyl)oxadirizine **2a**, but without giving details.[8]

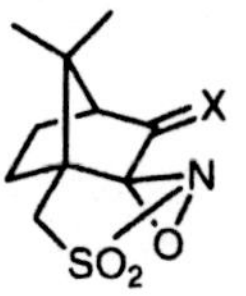

	X
2a	O
2b	H_2
2c	Cl_2
2d	$(OMe)_2$

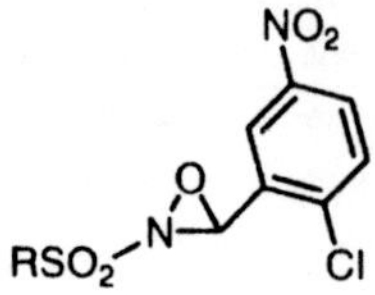

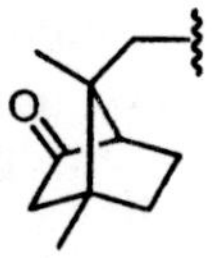

2. Asymmetric oxidation of imines

The barrier to pyramidal inversion in trivalent nitrogen can be increased substantially to permit the isolation of N-chiral nitrogen compounds in an optically active form by incorporating the nitrogen into a three-membered ring system such as oxaziridine comprising carbon, nitrogen, and oxygen atoms.[4,5] Oxaziridines can be constructed by peracid oxidation of imines, and chiral peracids were utilized for the first asymmetric synthesis of oxaziridines in 1968 by the groups of Boyd[9] and Montanari.[10] Oxidation of aldimines with (+)-monoperoxycamphoric acid proceeded with the asymmetric induction as high as 60% ee for 2-*t*-butyl-3-(*p*-bromophenyl)oxaziridine (Scheme 4.2.2).[11] Optically active oxaziridines are also obtainable by oxidation of chiral imines with achiral peracids,[12] or by oxidation of imines in chiral media.[13]

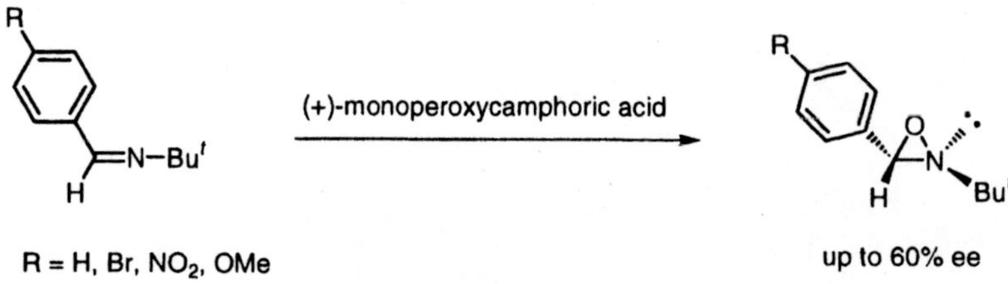

R = H, Br, NO_2, OMe

up to 60% ee

Scheme 4.2.2

4: Oxidation of hetero atoms

In 1978, Davis et al.[14] found that the ring oxygen atom of 2-sulfonyloxaziridines adjacent to chiral carbon and nitrogen atoms is highly electrophilic, resulting in a variety of oxygen-transfer reactions with better enantioselectivity than the peracids.[15] Moreover, the sulfonyloxaziridines are aprotic and neutral so that they can be applied to a wider range of oxidations. For example, chiral oxaziridines such as **2** and **3** generally result in better asymmetric induction than chiral peracids in the oxidation of sulfides to sulfoxides, selenides to selenoxides (see Section 4.1), enolates to α-hydroxycarbonyl compounds (see Chapter 2, Section 2.8), and in the asymmetric epoxidation of alkenes.[1,2,16]

The biphasic basic oxidation of acyclic chiral imines with peracids affords mixtures of oxaziridine diastereoisomers **3** requiring separation by crystallization which limits their production on a large scale. By contrast, the diastereomerically pure (camphorsulfonyl)oxaziridine derivatives **2** are readily available in substantial quantities because the *exo* face of the C—N double bond in sulfonyl imine, **4**, is blocked, affording on oxidation a single oxaziridine isomer (Scheme 4.2.3).

Protocol 1.[17]
Synthesis of (+)-(2R,8aS)-10-(camphorsulfonyl)oxaziridine (Structure 2b, Scheme 4.2.3)

Caution! All procedures should be carried out in a well-ventilated hood. Note that toxic and irritant sulfur dioxide and hydrogen chloride evolve in step 2. Wear disposable gloves and safety goggles.

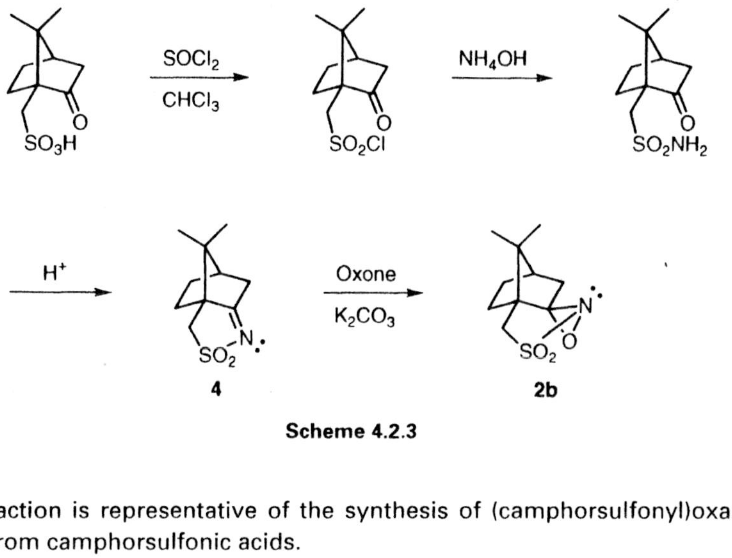

Scheme 4.2.3

This reaction is representative of the synthesis of (camphorsulfonyl)oxaziridines **2** from camphorsulfonic acids.

Protocol 1. Continued

Equipment

- Two-necked round-bottomed flask (2 L)
- Dropping funnels (2 × 250 mL)
- Magnetic stirring bar
- Reflux condenser
- Two-necked round-bottomed flask (5 L)
- Mechanical stirrer
- Ice bath
- Rotary evaporator
- Round-bottomed flask (1 L)
- Egg-shaped magnetic stirring bar (2-inch)
- Dean–Stark water separator
- Double-walled condenser
- Inert gas source
- Sintered glass funnels (coarse porosity) (2 × 150 mL)
- Three-necked round-bottomed Morton flask (5 L)
- Efficient mechanical stirrer
- 125-mm Teflon stirring blade
- Safe Lab stirring bearing[a]
- Addition funnel (500 mL)
- Separatory funnel (3 L)

Materials

- Camphorsulfonic acid, 116 g, 0.5 mol — **corrosive**
- Reagent-grade chloroform, 750 mL — **harmful by inhalation**
- Thionyl chloride, 71.4 g, 0.6 mol — **toxic, irritant**
- Ammonium hydroxide solution, 1.6 L — **corrosive**
- Dichloromethane, 1525 mL — **harmful by inhalation**
- Brine, 250 mL
- Anhydrous magnesium sulfate
- Amberlyst 15 ion-exchange resin, 5 g
- Toluene, 1 L — **flammable**
- Absolute ethanol, 750 mL — **flammable**
- Potassium carbonate, 543 g, 3.93 mol — **irritant**
- Water, 2 L
- Oxone[b] ($2KHSO_5 \cdot KHSO_4 \cdot K_2SO_4$), 345 g, 0.56 mol — **irritant, oxidizer**
- Saturated sodium sulfite solution, 100 mL — **irritant**
- 2-Propanol, ca. 500 mL — **flammable, irritant**

1. Equip a two-necked round-bottomed flask (2 L) with a dropping funnel (250 mL), a magnetic stirring bar, and a reflux condenser fitted with an outlet connected to a disposable pipette dipped in 2 mL of chloroform in a test-tube for monitoring gas evolution. Add to the flask camphorsulfonic acid (116 g, 0.5 mol) and reagent-grade chloroform (750 mL).

2. Heat the suspension of camphorsulfonic acid to reflux and add thionyl chloride (71.4 g, 43.8 mL, 0.6 mol, 1.2 eq.) dropwise over 1 h. Continue the heating until gas evolution (sulfur dioxide and hydrogen chloride) ceases (ca. 9–10 h) to give a solution of camphorsulfonyl chloride in chloroform.

3. In a two-necked round-bottomed flask (5 L) fitted with a dropping funnel (250 mL) and a mechanical stirrer, place a reagent-grade ammonium hydroxide solution (1.6 L) and then cool the flask to 0 °C in an ice bath.

4. Add dropwise the solution of the camphorsulfonyl chloride prepared in step 2 at 0–10 °C over a period of 1 h. Allow the reaction mixture to warm to room temperature, stir for 4 h, separate the organic layer, and extract the aqueous layer with dichloromethane (3 × 250 mL). Wash the combined organic layers with brine (250 mL) and dry over anhydrous magnesium sulfate. Remove the solvent on a rotary evaporator to give the crude camphorsulfonamide (104 g, 90%).

4: Oxidation of hetero atoms

5. Equip a round-bottomed flask (1 L) with a 2-inch egg-shaped magnetic stirring bar, a Dean–Stark water separator, and a double-walled condenser containing a mineral oil bubbler connected to an inert gas source. Into the flask, place Amberlyst 15 ion-exchange resin (5 g) and a toluene solution (500 mL) of the crude (+)-(1S)-camphorsulfonamide (41.5 g) prepared in step 4.

6. Heat the reaction mixture at reflux for 4 h. After the reaction, cool the flask, but while it is still warm (40–50 °C), slowly add dichloromethane (200 mL) to dissolve any (camphorsulfonyl)imine that crystallizes. Filter the solution through a 150-mL sintered glass funnel, and wash the reaction flask and filter funnel with additional dichloromethane (75 mL). Remove the solvent on a rotary evaporator to isolate the (−)-(camphorsulfonyl)imine. Recrystallize the resulting solid from absolute ethanol (750 mL) to give white crystals (34.5–36.4 g, 90–95%), m.p. 225–228 °C, $[\alpha]_D = -32.7°$ ($c = 1.9$, $CHCl_3$).

7. Equip a three-necked round-bottomed Morton flask (5 L) with an efficient mechanical stirrer, a 125-mm Teflon stirring blade, a Safe Lab stirring bearing, and an addition funnel (500 mL). Into the flask, place the toluene solution (500 mL) of (−)-(camphorsulfonyl)imine (39.9 g, 0.187 mol) prepared in step 6 and a room-temperature solution of anhydrous potassium carbonate (543 g, 3.93 mol, 7 eq. based on Oxone) dissolved in water (750 mL).

8. Vigorously stir the reaction mixture and add dropwise in three portions over 45 min a solution of Oxone (345 g, 0.56 mol, 6 eq. of $KHSO_5$) dissolved in water (1250 mL). Confirm completion of the oxidation by TLC.[c]

9. Filter the reaction mixture through a 150-mL sintered-glass funnel to remove solids. Transfer the filtrate to a 3-L separatory funnel to separate the toluene phase, and wash the aqueous phase with dichloromethane (3 × 100 mL). Wash the filtered solids and any solids remaining in the Morton flask with additional dichloromethane (200 mL). Wash the combined organic extracts with saturated sodium sulfite (100 mL), dry over anhydrous magnesium sulfate for 15–20 min, filter, and concentrate on a rotary evaporator. Crystallize the resulting white solid from hot 2-propanol (ca. 500 mL) to afford, after drying under vacuum in a desiccator, white needles (35.9 g, 84%), m.p. 165–167 °C, $[\alpha]_D = +44.6°$ ($c = 2.2$, $CHCl_3$).

[a] Safe Lab stirring bearing can be purchased from Aldrich Chemical Company, Inc.
[b] Oxone is available from Aldrich Chemical Company, Inc.
[c] Occasionally batches of Oxone exhibit reduced reactivity in this oxidation. If this occurs, Oxone is added until oxidation is complete as determined by TLC.

3. Kinetic resolution of β-hydroxy amines by *N*-oxide formation

Chiral β-hydroxy amines are characteristic structural features of many natural products and drugs, and play an important role as chiral building blocks, auxiliaries, and ligands in metal-catalysed reactions.[18] Therefore, the availability of efficient methods for the preparation of enantiopure structural arrays of this class of amino alcohols is of considerable current interest.[19,20]

A practical and fairly general procedure for the kinetic resolution of C-chiral β-hydroxy tertiary amines[21,22] is provided by the selective oxidation of one enantiomer to the *N*-oxide by using *t*-butyl hydroperoxide (TBHP) and the titanium tetraisopropoxide–diisopropyl tartrate (DIPT) catalyst system.[3] However, two important changes are made in this reaction compared with the original catalyst system for asymmetric epoxidation and kinetic resolution of allylic alcohols (for the original system, see Chapter 2, Section 2.3).[23] The first is the use of a 2:1 (Ti/ligand) catalyst for the *N*-oxide formation rather than the 2:2 catalyst used in the epoxidation reaction.[24] Although optimum values of the titanium/tartrate ratio differ slightly from substrate to substrate between

Table 4.2.1 Enantiomeric excess of recovered β-hydroxy amines after *ca.* 60% conversion[21,22]

Structure	ee (%)	Structure	ee (%)
n-C₈H₁₇–CH(OH)–CH₂–NMe₂	91	Ph–CH(OH)–CH₂–NMe₂	95
n-C₈H₁₇–CH(OH)–CH₂–N(pyrrolidine)	94	BnO–CH(OH)–CH₂–NMe₂	91
Cyclohexyl–CH(OH)–CH₂–N(pyrrolidine)	92	1-Naphthyl-O–CH₂–CH(OH)–CH₂–NMe₂	92
trans-2-NMe₂-cyclohexanol	92	Ph–CH(OH)–CH₂–N(pyrrolidine)	95
cis-2-NMe₂-cyclohexanol	95	Ph–CH(OH)–CH₂–N(piperidine)	97
Ph(Me)C(OH)–CH(NMe₂)	95	Buᵗ–CH(OH)–CH₂–NMe₂	93
Ph(Me)C(OH)–CH(NMe₂) diastereomer	93	Buᵗ–CH(OH)–CH₂–N(piperidine)	96

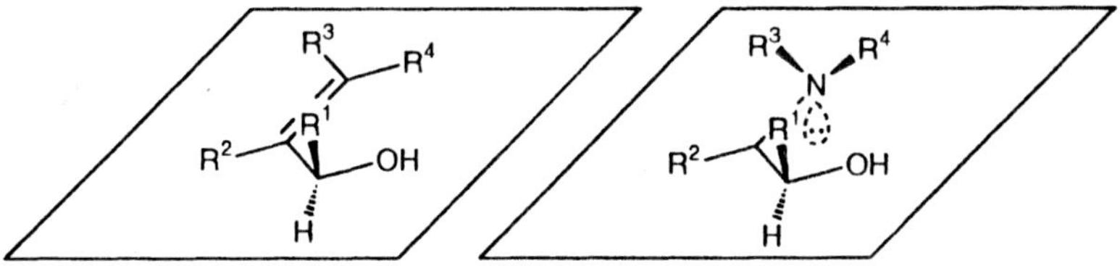

4: Oxidation of hetero atoms

Fig. 4.2.1 Fast-reacting allylic alcohol and β-hydroxy amine with (+)-DIPT.

the range from 2:1.2 to 2:1.5, the 2:1.2 ratio should be tried first as the majority of cases give the best results with this ratio. Second, an '*ageing period*' is indispensable to gain high stereoselectivity; an amino alcohol, DIPT, and titanium alkoxide are all allowed to mix in dichloromethane at room temperature for at least 30 min before the solution is cooled and the oxidant added. The oxidations are generally run to 60% conversion which requires less than 2 h at $-20\,°C$, and the enantiomeric excesses of the slow-reacting (i.e. recovered) enantiomers of the amino alcohols often exceed 90% (Table 4.2.1). When using (+)-DIPT the absolute configuration at the carbinol centre in the fast-reacting enantiomers is always the same as predicted by considering the stereochemical features of the kinetic resolution of secondary allylic alcohols (Fig. 4.2.1).

Protocol 2.
Kinetic resolution of *N,N*-dimethyl(2-hydroxydecyl)amine (Structure 5, Scheme 4.2.4)[21]

Caution! All the procedures should be carried out in a well-ventilated hood. Wear disposable gloves and chemical-resistant safety goggles.

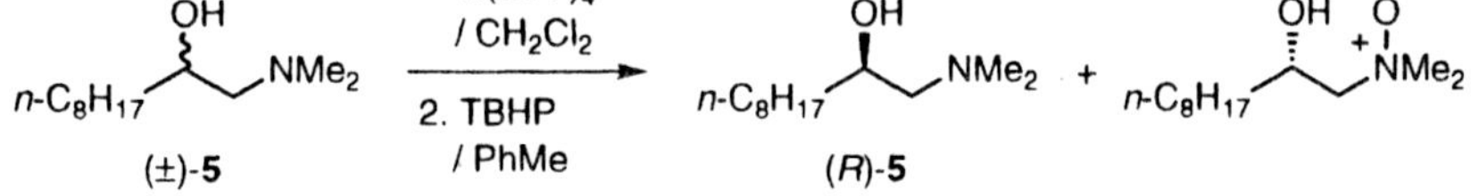

Scheme 4.2.4

This reaction is representative of the kinetic resolution of racemic β-hydroxy amines.

Equipment

- One-necked, round-bottomed flask (50 mL)
- Silicone-rubber septum
- Teflon-coated magnetic stirring bar
- Syringes (0.5 mL, 2 mL, 20 mL)
- Dry ice/carbon tetrachloride slush bath
- Source of dry nitrogen
- Two sintered glass filters (porosity 3)
- Rotary evaporator

Protocol 2. *Continued*

Materials

• N,N-Dimethyl(2-hydroxydecyl)amine,[a] 404 mg, 2.01 mmol	irritant
• (+)-Diisopropyl tartrate [(+)-DIPT],[b] 570 mg, 2.43 mmol	
• Dry dichloromethane,[c] 20 mL	harmful by inhalation
• Titanium tetraisopropoxide,[d] 1.20 mL, 4.09 mmol	flammable, irritant
• *t*-Butyl hydroperoxide (TBHP),[e] 3.29 M solution in toluene, 365 μL, 1.20 mmol	flammable, corrosive, oxidizer
• Diethyl ether for extraction, 25 mL	flammable, irritant
• Water, 1.2 mL	
• 40% Sodium hydroxide solution, 0.8 mL	corrosive
• Celite	
• Chloroform for extraction, 5 mL	harmful by inhalation
• Hexane for extraction, 40 mL	flammable, irritant
• Anhydrous sodium sulfate	

1. Dry a one-necked round-bottomed flask (50 mL) equipped with a Teflon-coated magnetic stirring bar in an electric oven (110°C, <3 h) and allow to cool to room temperature while flushing with dry nitrogen.

2. To the flask, add (±)-*N,N*-dimethyl(2-hydroxydecyl)amine (404 mg, 2.01 mmol) and (+)-DIPT (570 mg, 2.43 mmol, 1.21 eq.), and flush briefly with dry nitrogen. Equip the neck of the flask with a septum and maintain the mixture under a positive pressure of dry nitrogen.

3. Charge the flask with dichloromethane (20 mL) and then titanium tetraisopropoxide (1.20 mL, 4.09 mmol, 2.04 eq.) via syringes.

4. Stir the mixture for 30 min at room temperature. After completion of the ageing period, cool the flask, while stirring, in a dry ice/carbon tetrachloride slush bath (*ca.* −20°C). To this solution, add 0.6 eq. of TBHP (365 μL, 1.20 mmol, 3.29 M solution in toluene).

5. After stirring for 2 h at −20°C, quench the reaction by adding diethyl ether (20 mL), water (0.8 mL), and a 40% sodium hydroxide solution (0.8 mL).

6. Vigorously stir the mixture for 4–5 h to yield a gelatinous precipitate, and decant the supernatant liquid through a short pad of Celite (1 cm on top of a sintered glass filter funnel). Vigorously stir the precipitate in refluxing chloroform (*ca.* 5 mL) for 5 min before filtering it through the Celite pad.

7. Concentrate the combined filtrates on a rotary evaporator to leave a pale yellow, viscous oil, and dry it under high vacuum. Triturate the oil in hexane (20 mL), filter the clear supernatant solution through a sintered glass filter funnel, and wash the solid, which is the optically active *N*-oxide of *N,N*-dimethyl(β-hydroxydecyl)amine (233 mg, 53.5%), with hexane (20 mL).

8. Dilute the hexane extracts with diethyl ether (5 mL), wash with water (2 × *ca.* 200 μL) to remove any trace of the remaining *N*-oxide,[f] and dry over

4: Oxidation of hetero atoms

anhydrous Na_2SO_4. Evaporate the solvent to afford $(-)$-$(\beta$-hydroxydecyl)-amine as an oil (144 mg, 35.7%): $[\alpha]_D^{20} = -3.58°$ ($c = 1.65$, EtOH) (91% ee).[g]

[a] $(\pm)$-N,N-Dimethyl(2-hydroxydecyl)amine is prepared by heating at reflux a mixture of 1,2-epoxydecane and dimethylamine (40% aqueous solution) in THF.
[b] (+)-DIPT is distilled under high vacuum by using a wiped film molecular still, and stored under a nitrogen atmosphere.
[c] Dichloromethane is freshly distilled from calcium hydride.
[d] Titanium tetraisopropoxide is distilled under high vacuum and stored under a nitrogen atmosphere.
[e] TBPH in toluene is prepared and stored as described before.[25]
[f] A syringe with a needle rather than a separating funnel may be conveniently used to take up the organic phase.
[g] Enantiomeric excess is determined by acetylating a small sample in the usual way followed by NMR shift study using Eu(hfbc)$_3$ in C_6D_6.

References

1. Davis. F. A.; Sheppard, A. C. *Tetrahedron* **1989**, *45*, 5703.
2. Davis. F. A.; Chen, B.-C. *Chem. Rev.* **1992**, *92*, 919.
3. Katsuki, T.; Martín, V. S. *Org. React.* **1996**, *48*, Chapter 1.
4. Davis, F. A.; Jenkins, Jr., R. H., In *Asymmetric Synthesis*; Morrison, J. D.; Scott, J. W., eds.; Academic Press: Orland, **1984**; *4*, Chapter 4.
5. Morrison, J. D.; Mosher, H. S. *Asymmetric Organic Reactions*; Prentice-Hall: London, **1971**; Chapter 8.
6. Moriwaki, M.; Sawada, S.; Inouye, Y. *Chem. Commun.* **1970**, 419.
7. Pietrusiewicz, K. M.; Zablocka, M. *Chem. Rev.* **1994**, *94*, 1375.
8. Ugi, I.; Jacob, P.; Landgraf, B.; Rupp, C.; Lemmen, P.; Verfürth, U. *Nucleosides Nucleotides* **1988**, *7*, 605.
9. Boyd, D. R. *Tetrahedron Lett.* **1968**, 4561.
10. Montanari, F.; Moretti, I.; Torre, G. *Chem. Commun.* **1968**, 1694.
11. Pirkle, W. H.; Rinaldi, P. L. *J. Org. Chem.* **1978**, *43*, 4475.
12. (a) Bucciarelli, M.; Forni, A.; Moretti, I.; Torre, G. *J. Chem. Soc., Perkin Trans. 2* **1977**, 1339. (b) Aggarwal, V. K.; Wang, M. F. *J. Chem. Soc., Chem. Commun.* **1996**, 191.
13. Bucciarelli, M.; Forni, A.; Moretti, I.; Torre, G. *J. Chem. Soc., Perkin Trans. 1* **1980**, 2152.
14. Davis, F. A.; Jenkins, Jr., R. H.; Awad, S. B.; Stinger, O. D.; Watson, W. H.; Galloy, J. *J. Am. Chem. Soc.* **1982**, *104*, 5412.
15. Vishwakarma, L. C.; Stringer, O. D.; Davis, F. A. *Org. Synth. Col. Vol. VIII*, **1993**, 546.
16. Chen, B.-C.; Murphy, C. K.; Kumar, A.; Reddy, R. T.; Clark, C.; Zhou, P.; Lewis, B. M.; Gala, D.; Mergelsberg, I.; Scherer, D.; Buckley, J.; DiBenedetto, D.; Davis, F. A. *Org. Synth.* **1996**, *73*, 159.
17. Towson, J. C.; Weismiller, M. C.; Lal, G. S.; Sheppard, A. C.; Kumar, A.; Davis, F. A. *Org. Synth. Col. Vol. VIII*, **1993**, 104.
18. Ager, D. J. *Chem. Rev.* **1996**, *96*, 835.
19. Enders, D.; Haertwig, A.; Raabe, G.; Runsink, J. *Angew. Chem., Int. Ed. Engl.* **1996**, *35*, 2388.
20. Li, G.; Angert, H. H.; Sharpless, K. B. *Angew. Chem., Int. Ed. Engl.* **1996**, *35*, 2813.

21. Miyano, S.; Lu, L. D.-L.; Viti, S. M.; Sharpless, K. B. *J. Org. Chem.* **1985**, *50*, 4350.
22. Hayashi, M.; Okamura, F.; Toba, T.; Oguni, N.; Sharpless, K. B. *Chem. Lett.* **1990**, 547.
23. Katsuki, T.; Sharpless, K. B. *J. Am. Chem. Soc.* **1980**, *102*, 5974.
24. Berrisford, D. J.; Bolm, C.; Sharpless, K. B. *Angew. Chem., Int. Ed. Engl.* **1995**, *34*, 1059.
25. Hill. J. G.; Rossiter, B. E.; Sharpless, K. B. *J. Org. Chem.*, **1983,** *48*, 3607.

5

Oxidation using a biocatalyst

5.1 Hydroxylation at a saturated carbon atom

R. AZERAD

1. Introduction

Since the first demonstration by Murray and Peterson[1-3] that a filamentous fungus, *Rhizopus arrhizus*, was able to convert progesterone to 11α-hydroxyprogesterone efficiently (Scheme 5.1.1), many substances, either of natural or synthetic origin, have been examined using a variety of microorganisms (bacteria or fungi). Only a few examples will be given here. For a more exhaustive coverage, the reader is referred to books [4-6] or reviews dedicated to the microbial hydroxylation of the most common substrate groups such as steroids,[7-13] alkaloids,[8,14-16] terpenes,[17,18] and synthetic materials.[19,20]

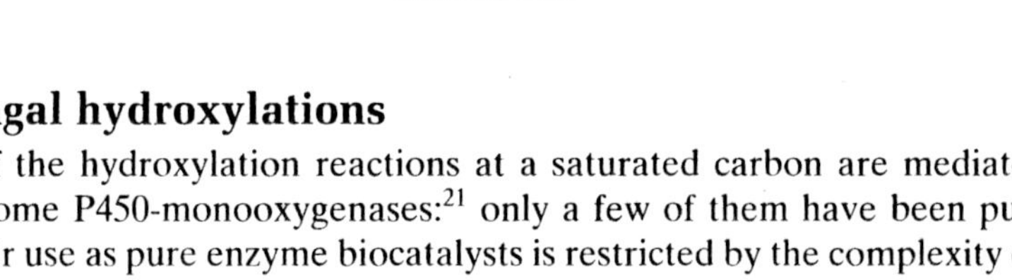

Scheme 5.1.1

2. Fungal hydroxylations

Most of the hydroxylation reactions at a saturated carbon are mediated by cytochrome P450-monooxygenases:[21] only a few of them have been purified and their use as pure enzyme biocatalysts is restricted by the complexity of the conditions needed for full activity (coenzymes, associated oxidoreductases, lipid factors, etc.), which derives from their natural involvement in membrane-bound multienzyme systems. The bulk of preparative monooxygenase-mediated hydroxylation reactions has thus been essentially performed with whole microorganisms, with some of the drawbacks potentially associated with such techniques: secondary enzymic activities, difficult isolation and separation

Table 5.1.1 Some National Fungal Culture Collections

ATCC	American Type Culture Collection, 12301 Parklawn Drive, Rockville, MD 208522, USA
CBS	Centralbureau voor Schimmelcultures, Oosterstraat 1, PO Box 273, 3740 AG, Baarn, The Netherlands
DMS	Deutsche Sammlung von Mikroorganismen, Mascheroder Weg 1b, D-38124, Branschweig, Germany
IAM	Institute of Applied Microbiology, University of Tokyo, Tokyo, Japan
IFO	Institute for Fermentstion, 17–85, Juso Honmachi 2-chome, Yodogawa-ku, Osaka, Japan
IMI	International Mycological Institute, Bakeham Lane, Egham, Surrey TW20 9TY, UK
JCM	Japan Collection of Microorganisms, The Institute of Physical and Chemical Research (RIKEN), Wako, Saitama 351-01, Japan
LCP	Laboratoire de Cryptogamie, Museum National d'Histoire Naturelle, 12 rue Buffon, 75005—Paris, France
MUCL	Mycothèque de l'Université Catholique de Louvain-La-Neuve, Place Croix du Sud 3, B-1348 Louvain-La-Neuve, Belgium
NRRL	Northern Regional Research Laboratories, Agricultural Research Service, US Department of Agriculture, 1815 N. University Street, Peoria, IL 61604, USA

of desired products in the presence of other cell products, permeability problems, and so on.

On the other hand, baker's yeast, which is easily and commercially available as grown cells (pressed or lyophilized) and which has been currently used in other bioconversion reactions, is a poor hydroxylating reagent. It is thus necessary to grow other microorganisms, mainly moulds (filamentous fungi) and sometimes bacteria, to obtain an active biomass susceptible to performing hydroxylation reactions. Fungal biomass is generally preferred, because it can be easily separated from the growth (or incubation) medium by simple filtration, whereas the collection of yeast or bacterial cells generally requires centrifugation or complex filtration techniques. Moreover, contrary to bacteria that frequently use the hydroxylation products as a carbon (and/or nitrogen) source, fungi generally perform the hydroxylation reaction only as a detoxification process for eliminating hydrophobic substances, and accumulate the more hydrophilic hydroxylation products in the incubation medium, without further metabolism. Many strains of such microorganisms are available and can be acquired through easily accessible international collections* (Table 5.1.1).

Representative fungal strains frequently used in the literature for hydroxylation reactions and available in unrestricted conditions from the corresponding collections are given in Table 5.1.2. These strains may constitute the basic

*The World Data Centre for Microorganisms (WDCM), freely accessed via the Internet (http://wdcm.nig.ac.jp), provides a comprehensive directory of world culture collections and data bases of microbes.

5: Oxidation using a biocatalyst

Table 5.1.2 Representative fungal microorganisms frequently used for preparative hydroxylation reactions

Microorganisms	Typical strain numbers
Absidia cylindrospora	LCP 57.1569
Aspergillus niger	ATCC 9142
Aspergillus ochraceus	ATCC 1008
Beauveria bassiana	ATCC 7159
Cephalosporium aphidicola	IMI 68689
Cunninghamella blakesleeana	ATCC 8688a
Cunninghamella echinulata	ATCC 9244 or NRRL 3655
Cunninghamella elegans	ATCC 9245 or ATCC 36112
Curvularia lunata	NRRL 2380
Mortierella isabellina	NRRL 1757
Mucor plumbeus	CBS 110-16 or ATCC 4740
Rhizopus arrhizus	ATCC 11145

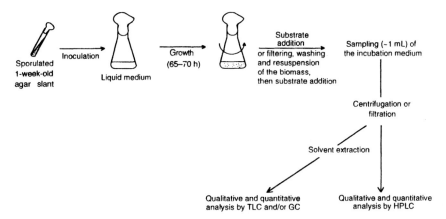

Fig. 5.1.1 General method for screening and analysing the production of hydroxylated metabolites by fungal microorganisms.

choice in the necessary screening studies for hydroxylation of a new substrate. A schematic summary of the general methodology for screening or production of hydroxylated metabolites is illustrated in Fig. 5.1.1.

3. General methods for biohydroxylation

The reader should be aware that the basic experimental methods used for manipulating microorganisms are not difficult, but require a high standard of cleanliness to avoid contamination and growth of undesired microbes (spores of contaminating microorganisms are present in air and on all surfaces in the laboratory). Unless indicated, all transfer operations related to the preparation and manipulation of pure cultures have to be performed in a sterile cabinet, with sterilized glassware and sterilized instruments. Microorganisms from the collections (see Table 5.1.1) are generally available as living cultures on solid medium or as freeze-dried specimens. The former can be transferred directly to fresh nutrient slants, while the latter have to be rehydrated using a sterile liquid medium before transfer onto a new growth medium. Most collections provide a detailed procedure for reviving the freeze-dried microorganisms supplied. The way to maintain pure cultures is described in Protocol 1.

Protocol 1.
Maintaining pure cultures of fungal strains

Caution! Carry out all procedures in a well-ventilated hood and wear disposable gloves and safety goggles.

Equipment
- Laminar air-flow cabinet
- Inoculation knife (scalpel) or needle
- Autoclave, or domestic pressure cooker
- Bunsen burner or portable gas burner
- Aluminium foil
- Cotton wool
- 18 cm × 1.8 cm glass tubes

Materials
- Yeast extract (Difco), 5 g
- Malt extract (Difco), 5 g
- Glucose, 20 g
- Agar (technical, Difco), 20 g
- Bacto-peptone (Difco), 5 g

Method
1. Prepare solid growth media by adding yeast extract (5 g), malt extract (5 g), Bacto-peptone (5 g), glucose (20 g), and agar (20 g), to deionized (or distilled) water (1 L). Heat the mixture with shaking in a boiling water-bath until clear and homogeneous, or heat it in a microwave oven for 6 min. The hot mixture is evenly distributed in tubes (about 7 mL per tube) which are stoppered with cotton-wool plugs and covered with aluminium foil. Autoclave at 120 °C for 20 min.
2. Remove the tubes from the autoclave while still hot (>50 °C) and cool them at room temperature in a slanting position (for example by using a 1-cm-diameter glass rod slipped under the open ends of the tubes), in order to obtain, after cooling, a solid surface extending to about two-thirds of the length of the tube.
3. Sterilize the scalpel blade (or an inoculation loop) by heating in the flame of the Bunsen burner. Take a sample of the original culture from the surface of the initial slant with the scalpel (or an inoculation loop from a grown suspension in a liquid medium) and after carefully opening a cooled tube, with the lid being lifted to the rear of the sterile cabinet, introduce the inoculum sample onto the surface of the solid medium.[a]

5: Oxidation using a biocatalyst

4. Keep the tube at room temperature (between 20 and 28 °C) or, better, keep it in an incubator adjusted to 25–26 °C. Maximal growth, corresponding to the covering of the whole surface by the mycelium, followed by sporulation, occurs in 5–8 days.

5. The agar slopes obtained (stock cultures) are stored at 4–6 °C and can be used for inoculation of liquid medium for 2–4 weeks. They then have to be subcultured as described in steps 3–4, to get the maximal survival and optimal activity of the inoculum.[b]

[a] The inoculation procedure is described in more detail in all microbiology textbooks (see, for example, Ref. 22).
[b] Various techniques[23] are available for extended storage times (up to several years), the most commonly used being storage in the cold of agar slopes under mineral oil, or storage of spore suspension in a glycerol–water mixture at −80 °C.

The liquid growth medium given in Protocol 2 is suitable as a universal growth medium for the culture of fungal strains. It can be used either in shaken Erlenmeyer flask cultures, as described in Protocol 2, or in stirred fermenter cultures, as illustrated in Protocols 4–6.

Protocol 2.
Growing the strains for use in biohydroxylations

Caution! Carry out all procedures in a well-ventilated hood and wear disposable gloves and safety goggles.

Equipment
- Erlenmeyer flasks (250 mL)
- Autoclave, or domestic pressure cooker
- Sterile pipettes
- Laminar air-flow cabinet
- Thermostated rotary shaker
- Bunsen burner or portable gas burner

Materials
- Corn steep liquor (Solulys 48 L, Roquette, France), 10 g
- Glucose, 30 g
- Potassium dihydrogenphosphate (KH_2PO_4), 1 g
- Potassium hydrogenphosphate (K_2HPO_4), 2 g
- Sodium nitrate, 2 g   **oxidizer, toxic**
- Potassium chloride, 0.5 g
- Magnesium sulfate heptahydrate, 0.5 g
- Iron(II) sulfate heptahydrate, 0.02 g.   **irritant**
- Cotton wool
- Aluminium foil

Method
1. Prepare a solution of corn steep liquor (10 g), sodium nitrate (2 g), potassium chloride (0.5 g), magnesium sulfate heptahydrate (0.5 g), and iron(II) sulfate heptahydrate (0.02 g) in 900 mL of deionized water and distribute it

Protocol 2 *Continued*

   by 90-mL volumes into 250-mL Erlenmeyer flasks which are stoppered with cotton-wool plugs and covered with aluminium foil. Autoclave at 120°C for 20 min.
2. Dissolve glucose (50 g) in water to make a final volume of 100 mL. Prepare a solution of KH_2PO_4 (1 g) and K_2HPO_4 (2 g) in water (40 mL). Both solutions are sterilized separately in Erlenmeyer flasks stoppered with cotton-wool plugs and covered with aluminium foil by autoclaving at 120°C for 20 min.
3. In sterile conditions, add 4 mL of the phosphate buffer solution and 6 mL of the glucose solution to each basic medium flask (prepared in step 1) to constitute the final liquid medium (pH about 6.5).[a]
4. In sterile conditions, add 0.5 mL of a sterile glycerol–water (1:3) mixture to a freshly grown and sporulated agar slope of the strain to cultivate and, with the pipette tip, scrape the surface of the slant to produce a concentrated spore suspension.
5. Inoculate with 1-2 drops of the spore suspension several 250-mL Erlenmeyer flasks containing 100 mL of liquid medium and put the flasks into the thermostated rotary shaker, adjusted at 27°C and 200 r.p.m. After 60–70 h, maximal growth is generally obtained[b] and the biomass present in the culture flasks may be used for biohydroxylation in the following incubation step.

[a] The sterilization of glucose solutions in the presence of salts, and particularly phosphate salts, produces a pronounced browning of the medium and may generate potential inhibitory substances for the microorganism growth.
[b] In such conditions, the biomass is present either as a number of 'pellets' of variable size (from less than 1 mm to several mm in diameter), or as a swollen, large filamentous pad. Moreover, sporulation has sometimes occurred. Nevertheless, such inhomogeneous fungal biomass is generally fully suitable for biohydroxylation reactions.

The solubility of the substrate in the aqueous bioconversion medium is generally low, but is not a critical factor. It can be added as a concentrated solution in a small volume of a water-miscible non-toxic organic solvent such as ethanol, acetone, dimethyl sulfoxide, or N,N-dimethylformamide, corresponding to 1–2% of the final aqueous volume. Addition of a small amount of a neutral emulsifying agent, such as Tween 80, is sometimes advantageous (see Protocol 3). A precipitate of the solute generally occurs upon addition of the organic solution to the aqueous incubation medium, but the finely divided suspension of substrate is generally rapidly adsorbed into the lipids of the cell membrane and becomes available for metabolization, unless crystallization takes place. Concentrations of substrate are generally in the 0.2–1 g/L range; higher levels of substrate being toxic for the microorganism may preclude any hydroxylation reaction.

Although addition of the substrate to the culture medium, at the beginning of the growth period, is the simplest mode of performing the biotransforma-

5: Oxidation using a biocatalyst

tion, it is generally preferred to grow the microorganism first to obtain maximum biomass, avoiding any inhibition effect of the substrate, then to use the grown biomass for the bioconversion, either directly in the growth medium, which minimizes biomass manipulation, or as a washed and resuspended biomass ('resting cells') in water or buffered solution. The latter method (see Protocol 5) may be useful when the incubation conditions have to be closely controlled and optimized (pH, increase of biomass concentration, etc.). Moreover, the non-growing 'resting cells' conditions generally afford cleaner final products, uncontaminated with medium components and excreted substances from the microorganism. Incubation times may extend to 7–10 days, in the absence of any energy source added, because such microorganisms have accumulated during growth sufficient internal nutrients to maintain primary metabolic functions. Sterile conditions are generally not required, since microbial growth medium is absent.

Privileged saturated hydroxylation sites are generally found at (i) allylic or benzylic positions, and (ii) tertiary carbon atoms. In addition, heteroatom dealkylations occur as a consequence of the oxidative removal of alkyl groups, proceeding via α-oxidation to an unstable intermediate. However, in spite of some knowledge of the general factors which can influence the site of hydroxylation of a given substrate, the prediction of the regio- or stereochemical outcome of microbial hydroxylation reactions is difficult.

4. Examples of biohydroxylations at saturated carbon atoms
4.1 Hydroxylation of terpenes

The cyclic terpenes of the labdane [24–33] and drimane families [34–36] have been examined particularly as substrates for biohydroxylation in a synthetic approach to various bioactive compounds such as warburganal or forskolin. In this series, as for other 4,4-dimethyl-substituted terpenoid compounds, 3β-hydroxylated products are produced easily, sometimes in very high yields. Such functionalized derivatives may be of important value by themselves, or can be further converted by simple chemical methods to other hydroxylated derivatives, such as, for example, 1α-hydroxy compounds (Scheme 5.1.2).[37,38] An example

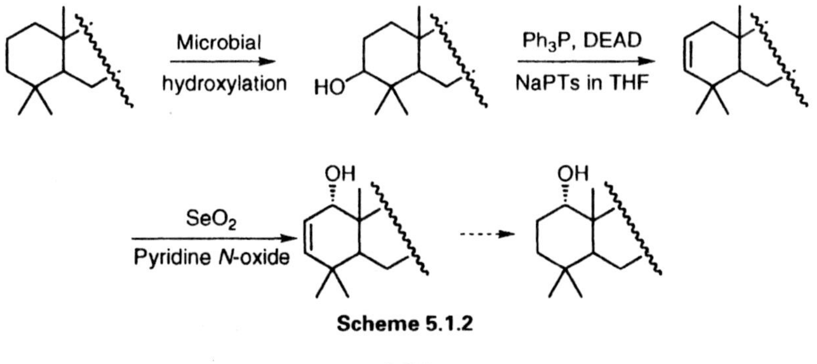

Scheme 5.1.2

of the 3β-hydroxylation of a common natural labdane diterpene, sclareol **1**, is described in Protocol 3.[28,29]

Protocol 3.
Biohydroxylation of sclareol (1) by *Mucor plumbeus* ATCC 4740: 3β-hydroxysclareol (Structure 2, Scheme 5.1.3)

Caution! Carry out all procedures in a well-ventilated hood and wear disposable gloves and safety goggles.

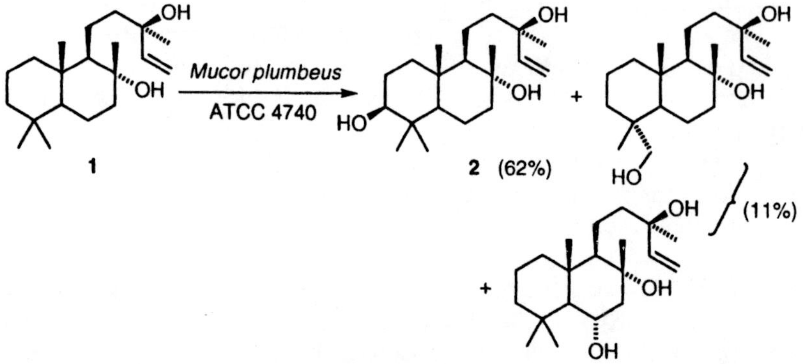

Scheme 5.1.3

Equipment

- 10 Erlenmeyer flasks (250 mL) containing 100 mL of sterile liquid medium
- Thermostated rotary shaker
- Sterile pipettes
- Laminar air-flow cabinet
- Sintered glass filter funnel (250 mL, porosity 3)
- Separatory funnel (1 L)
- Liquid–liquid continuous extractor (500 mL)
- Flash-chromatography column (4 × 30 cm)

Materials

- Sclareol (95%),[a] 500 mg
- Tween 80, 1 mL
- Ethanol (99%), 5 mL — flammable
- Freshly grown and sporulated agar slope of *M. plumbeus* ATCC 4740
- 10% Sodium hydroxide solution — corrosive
- Saturated sodium chloride solution
- Diethyl ether — flammable
- Anhydrous sodium sulfate
- Silica gel 60 (Merck, 230–400 mesh), 30 g
- Ethyl acetate — flammable, irritant
- Pentane — flammable, irritant

Method

1. Inoculate with a spore suspension of *M. plumbeus* ATCC 4740 ten 250-mL Erlenmeyer flasks of liquid medium and put the flasks into the thermostated rotary shaker, adjusted at 27 °C and 200 r.p.m. After 60 h, maximal growth is

obtained and the culture flasks may be used in the following incubation step.

2. Add to each flask, in sterile conditions, 50 mg of sclareol dissolved in ethanol (0.5 mL) and Tween 80 (0.1 mL) and carry on the incubation in the rotary shaker, as in step 1. It is possible to monitor the product(s) formation by periodically taking a 0.5-mL sample, extracting it with ethyl acetate and following the disappearance of substrate and appearance of product(s) by TLC on a silica gel plate (diethyl ether–ethyl acetate 9:1).

3. After 3 days, filter the incubation mixtures on a Buchner funnel and extract the filtrate thoroughly, by portions, with diethyl ether in a continuous liquid–liquid extractor (2 days, about 500 mL).[b] In a separatory funnel, wash the ethereal solution with a 10% sodium hydroxide solution (10 mL), then with a saturated sodium chloride solution (3 × 10 mL). Dry the organic phase on anhydrous sodium sulfate. Filter and remove the diethyl ether on a rotary evaporator to obtain 590 mg of a yellow oil exhibiting two main spots by TLC ($R_f = 0.35$ and 0.15) and a small amount of residual sclareol ($R_f = 0.80$).

4. Flash chromatograph on a silica gel column (30 g) using diethyl ether then increasing concentrations (5 to 33%) of ethyl acetate in diethyl ether as solvent to elute in the following order: sclareol (52 mg), 3β-hydroxysclareol (2, 317 mg, 62%), and unseparated 6α- and 18α-hydroxysclareol (41 mg, 8%).

5. Crystallize 3β-hydroxysclareol from ethyl acetate–pentane; m.p. 169–171°C. $[\alpha]_D^{22} = -9.7°$ ($c = 1.9$, MeOH).

[a] Commercially available sclareol is contaminated with about 3–5% of its 13β-epimer. The products of 13-episclareol hydroxylation are easily separated from the main products in the chromatographic step and the subsequent crystallization.
[b] The continuous diethyl ether extraction can be replaced advantageously by extraction with dichloromethane on a shaking table.

4.2 Hydroxylation of synthetic compounds

Biohydroxylation is not limited to natural substances. The regio- and stereoselectivity of microbial hydroxylation reactions can be also exploited to functionalize simple or complex synthetic compounds, in order to use them as chiral synthons or synthetic intermediates.[6] For example, enantiomeric enones with various B-ring sizes [39–43] have been examined using various fungal strains, and particularly *Mucor plumbeus* CBS 110-16, in order to produce hydroxylated derivatives[44–46] specifically oxidized in the B-ring (V. Goubaud, C. Kiefl, J.-P. Girault, and R. Azerad, unpublished results). Some results of this study, summarized in Fig. 5.1.2, illustrate in this series the balance between stereoelectronic effects in the substrate (favouring allylic hydroxylation) and the influence of the binding mode of such molecules (in both enantiomeric forms) on the active site of the hydroxylating enzyme(s) (oriented delivery of the

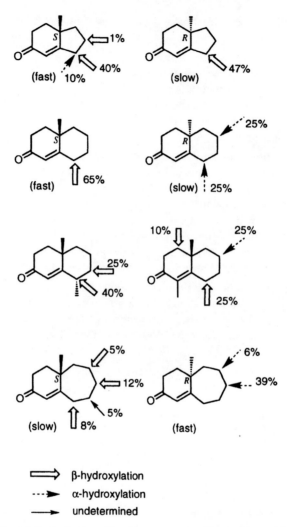

Fig. 5.1.2 Biohydroxylation of some bicyclic enones by *M. plumbeus* CBS 110–16. The data (%) correspond to the average observed yields of the main purified and identified hydroxylated derivatives.

hydroxyl group) (C. Kiefl and R. Azerad, unpublished data). A slight alteration in the substrate structure and geometry may result in a significant change in regio- and stereospecificity, a general feature of the microbial hydroxylation reactions. Nevertheless, this reaction represents a simple and unique way to perform the difficult functionalization of unactivated positions of the B-ring of some isomers, as shown in Protocol 4.

5: Oxidation using a biocatalyst

Protocol 4.
Hydroxylation of (R)-(−)-4a-methyl-4,4a,5,6,7,8-hexahydro-2(3H)-naphthalenone (7) by *Mucor plumbeus* CBS 110-16 (Scheme 5.1.4)

Caution! Carry out all procedures in a well-ventilated hood and wear disposable gloves and safety goggles.

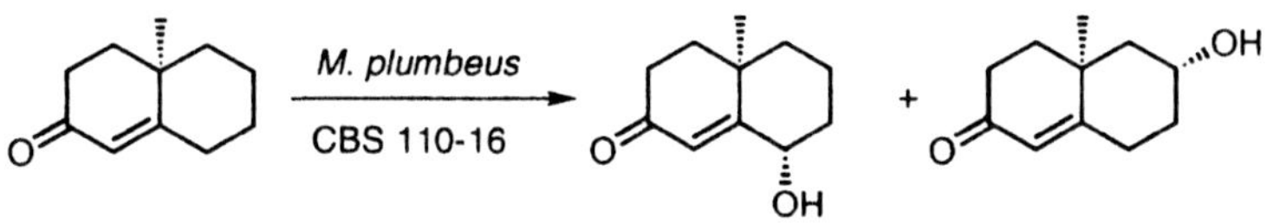

Scheme 5.1.4

The following bioconversion is described with a washed fungal biomass ('resting cells'), incubated in a buffer, and affording particularly clean products after extraction. However, very similar results would be obtained by adding the substrate directly into a grown culture and carrying on the incubation under the same conditions (see Protocol 3).

Equipment

- Bench-top fermenter (7.5 L)
- Autoclave
- Laminar air-flow cabinet
- Sterile pipettes
- Buchner funnel (1 L or larger)
- Pharmaceutical gauze
- Two Erlenmeyer flasks (2 L)
- Thermostated rotary shaker
- Sintered glass filter funnel (8-cm diameter, porosity 3)
- Separating funnel (2 L)
- Flash chromatography column (2.5 × 40 cm)

Materials

- (R)-(−)-4a-Methyl-4,4a,5,6,7,8-hexahydro-2(3H)-naphthalenone (7),[a] 370 mg
- Freshly grown and sporulated agar slope of *M. plumbeus* CBS 110-16
- Potassium phosphate buffer, 0.1 M, pH 7.00, 1.5 L
- Ethanol   **flammable**
- Cotton wool
- Ethyl acetate   **flammable, irritant**
- Celite
- Sodium chloride
- Anhydrous sodium sulfate
- Silica gel 60 (Merck, 230–400 mesh), 25 g

Method

1. Prepare, in a 7.5-L bench-top fermenter, 3 L of complete liquid medium and autoclave for 20 min at 120 °C.
2. Inoculate with a spore suspension of *M. plumbeus* CBS 110-16 to obtain approximately 2×10^6 spores per litre[b] and grow at 27 °C and 350 r.p.m with an air flow of 0.6 L/min. After 65 h, the maximum biomass (236 g fresh

Protocol 4. *Continued*

weight = 30.6 g dry weight) is obtained as small pale-yellow pellets (diameter 1–2 mm).

3. Filter the whole fermenter content[c] on three layers of pharmaceutical gauze in a large Buchner funnel, and resuspend the recovered biomass into water (1 L) to wash it. Filter again and resuspend the washed biomass into two 2-L Erlenmeyer flasks containing 0.1 M potassium phosphate buffer, pH 7.00 (2 × 0.75 L).[d]

4. Add in each flask (*R*)-(−)-4a-methyl-4,4a,5,6,7,8-hexahydro-2(3*H*)-naphthalenone **7** (185 mg dissolved in 2 mL of ethanol). Plug the flasks with cotton wool and incubate in the thermostated rotary shaker adjusted at 27 °C and 250 r.p.m. The progress of bioconversion can be monitored by taking an aliquot (0.5 mL) of the incubation supernatant, extracting with ethyl acetate, and carrying out TLC (cyclohexane–ethyl acetate, 6:4) of the extract on silica gel F_{254} (Merck).

5. After 3 days, filter the incubation mixture on a Celite pad in a sintered glass funnel. Saturate the filtrate by stirring with solid sodium chloride, filter again on Celite and extract the filtrate three times with ethyl acetate (3 × 400 mL). Dry the organic phase on anhydrous sodium sulfate, filter and remove the solvent under vacuum on a rotary evaporator to obtain a yellow oil (380 mg).

6. Flash chromatograph on a silica gel column (25 g) eluted with cyclohexane–ethyl acetate (1:1) to separate residual substrate **7** (R_f = 0.55, 57 mg), the 8-hydroxy derivative **8** (R_f = 0.35, 83 mg, 20% yield), m.p. 72 °C (from diethyl ether–hexane), $[\alpha]_D^{21} = -95°$ (*c* = 1.04, $CHCl_3$), and the 6-hydroxy derivative **9** (R_f = 0.20, 91 mg, 22% yield), m.p. 129–130 °C (from diethyl ether–hexane), $[\alpha]_D^{21} = -210°$ (*c.* = 1.025, $CHCl_3$).

[a] The *R*-enantiomer of the starting enone (**7**, 85% e.e.) may be prepared in three steps from (±)-2-methylcyclohexanone and methyl vinyl ketone, using (*S*)-1-phenylethylamine as a chiral auxiliary.[42,43] A 95–99% enantiomeric excess can be obtained using a single low-temperature crystallization.
[b] Data resulting from spore counting of a diluted suspension under the microscope, using a micrometric (Malassez) cell.
[c] Beyond this step, sterile conditions are no longer necessary.
[d] In this experiment, the biomass is concentrated twice (about 150 g/L fresh weight = 20 g/L dry weight) compared to the final growth conditions.

Microbial hydroxylation of dibenzosuberone (**10**) (Protocol 5) illustrates a stereospecific benzylic hydroxylation reaction leading to the desymmetrization of a symmetrical substrate. With most of the fungal microorganisms used, no aromatic hydroxylation took place. Similarly optically active 10-hydroxylated derivatives of related tricyclic antidepressants (imipramine, desipramine, amitryptiline, nortriptyline, etc.) formed in the animal and human metabolism of the drugs have been shown recently to be highly bioactive compounds.[47] 10-Hydroxydibenzosuberone enantiomers have been used to prepare the cor-

5: Oxidation using a biocatalyst

responding derivatives[48] of one of them (amineptine) and to help in the determination of the absolute configuration of its 10-hydroxylated metabolites (J.-Y. Beaumal, F. Lefoulon, M. Maurs, and R. Azerad, unpublished data).

Protocol 5.
Biohydroxylation of dibenzosuberone (10,11-dihydro-5H-dibenzo[a,d]cycloheptene-5-one) (10) by *Absidia cylindrospora* LCP 57.1569: (−)-(10,11-dihydro-10-hydroxy-5H-dibenzo[a,d]cycloheptene-5-one) (Structure 11, Scheme 5.1.5)

Caution! Carry out all procedures in a well-ventilated hood and wear disposable gloves and safety goggles.

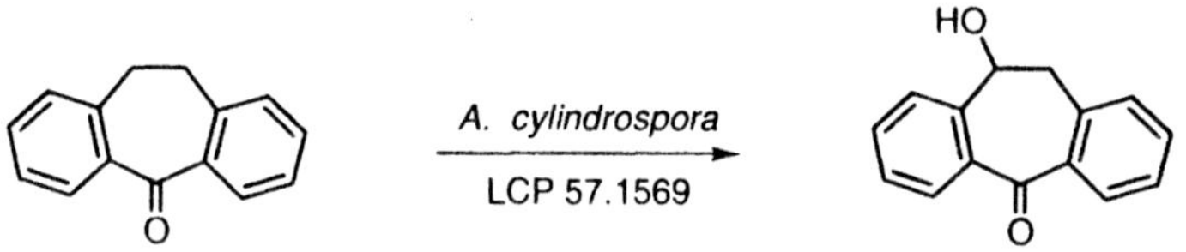

10 → *A. cylindrospora* LCP 57.1569 → (−)-11 (22%; 85% ee) (>98% ee after crystallization)

Scheme 5.1.5

This protocol illustrates a biohydroxylation experiment run in a classical fermenter and corresponding to a relatively slow reaction, with recycling of the biomass with residual substrate adsorbed. Extraction of products from the incubation medium was performed by adsorption on a hydrophobic polymeric adsorbent, to minimize solvent volumes and extraction work-up.

Equipment
- Bench-top fermenter (7.5 L)
- Autoclave
- Buchner funnel (1 L or larger)
- Pharmaceutical gauze
- Erlenmeyer flask (6 L)
- Chromatography column (4 × 50 cm)
- Flash chromatography column (2.5 × 40 cm)
- Sintered glass filter funnel (diameter 8 cm, porosity 3)
- Glass tubes (25 mL)

Materials
- Freshly grown and sporulated agar slope of *A. cylindrospora* LCP 57.1569
- Dibenzosuberone (98%), 3.5 g
- Potassium dihydrogenphosphate, 68 g
- Potassium hydroxide — **corrosive**
- Amberlite XAD-2 (20–50 mesh), 500 g
- Methanol, 2 L — **flammable, toxic**
- Acetone, 2 L — **flammable, irritant**
- Ethyl acetate — **flammable, irritant**
- Cyclohexane — **flammable, irritant**
- Silica gel (Merck 60, 230–400 mesh), 150 g
- 2-Propanol, 99% — **flammable, irritant**

Protocol 5. *Continued*

Method

1. Sterilize at 120°C for 30 min a 7.5-L bench-top fermenter equipped with a helical (propellor-shaped) turbine and containing 5 L of liquid medium.
2. Inoculate with 3 mL of a spore suspension of *Absidia cylindrospora* LCP 57.1569, prepared from a freshly grown agar slope to obtain approximately 2×10^6 spores per litre.[a] Adjust the fermenter at 27°C, with an air flow of 0.8 L/min, and stir at 250 r.p.m. for 68 h.
3. Add pure dibenzosuberone (3 mL, 3.47 g, 16.7 mmol) and increase stirring to 300 r.p.m. Progress of the bioconversion can be monitored by sampling a 1-mL aliquot from the medium, extracting with ethyl acetate and carrying out TLC of the extract (solvent: cyclohexane–ethyl acetate, 6:4).
4. After 6 days incubation, filter the whole fermenter content[b] on three layers of pharmaceutical gauze in a large Buchner funnel, keep the filtrate for extraction of product(s) and resuspend the recovered biomass (containing unchanged dibenzosuberone) in the same fermenter with 5 L of a phosphate buffer [KH_2PO_4 (68 g) in water (4.5 L), adjusted to pH 7.0 with solid KOH, and made up to 5 L with water]. Run a bioconversion cycle (27°C, 300 r.p.m.) for 5 days and again filter the whole fermenter content.
5. Both filtrates are percolated on an XAD-2 column (4×30 cm). Rinse with deionized water (1.5 L), dry with air, then elute with methanol (1 L) then acetone (500 mL).
6. Extract the pressed biomass[c] by magnetic stirring with acetone (250 mL) and filter on a Buchner funnel. Treat the residual biomass twice with acetone. Evaporate acetone from the filtrates and dissolve the residue in warm ethyl acetate. Remove the insoluble part on a sintered glass filter funnel and pool the filtrate and the organic eluates from the XAD-2 column (step 5).
7. Evaporate the solvents and flash chromatograph the residue (4.32 g) on a silica gel column (150 g), using ethyl acetate–cyclohexane (2:8) as solvent. The effluent is collected in fractions of about 20 mL. Pool the fractions containing unchanged dibenzosuberone (2.0 g) as shown by TLC ($R_f = 0.75$). Pool the fractions containing 10-hydroxydibenzosuberone ($R_f = 0.45$) and evaporate the solvents. Yield: 858 mg, 3.83 mmol, 22%.[d]
8. Crystallize the (−)-10-hydroxydibenzosuberone fraction by dissolving it into a small volume of hot 2-propanol (about 8 mL). Water is slowly added (6 mL) with cooling in an ice-box. Scrape the flask with a glass rod until some turbidity appears and store the solution overnight at 4°C for crystallization. Yield: 592 mg, 2.64 mmol, 15.8%. M.p. 114–114.5°C. $[\alpha]_D^{22} = -62.5°$ (c = 1.095, MeOH). HPLC on a Chiralcell OD column indicates >98% ee.[e]

[a] Data resulting from spore counting of a diluted suspension under the microscope, using a micrometric (Malassez) cell.

b Beyond this step, sterile conditions are no longer necessary.
c This step can be omitted if unchanged dibenzosuberone is not to be recovered. However, a small amount of 10-hydroxydibenzosuberone is present in the biomass and will be lost if this step is omitted. Adding a third recycling of the biomass only affords a minute amount of hydroxylation product.
d At this stage, (−)-10-hydroxydibenzosuberone is approximately 90–95% pure and the observed $[\alpha]_D^{22}$ is −45.9° (c = 1.075, MeOH). HPLC on a Chiralcell OD column indicates 85% ee.
e A very similar procedure, using *Cunninghamella echinulata* NRRL 3655, affords a comparable yield of the pure enantiomeric (+)-10-hydroxydibenzosuberone.

4.3 Hydroxylation of drugs

Determination of the pharmacological activity and potential toxic effects of the biotransformation products of a drug in animal and in man is an essential step in its pharmaceutical evaluation. Most of the transformations are accomplished in the liver and generally consist of oxidation/hydroxylation reactions (Phase I), followed (in the liver and kidney) by conjugation reactions (Phase II). In order to supply such metabolites, usually obtained in small yields through costly *in vivo* animal experiments or liver cell cultures, or after time-consuming chemical synthesis, alternative or complementary systems have been required. The use of microbial systems, and particularly fungi, which are thought to possess equivalent detoxification enzymes compared to the liver cell, has been recommended: their advantages are low cost, easy handling and scaling-up capability. In the past 20 years, several examples have validated the so-called concept of 'microbial models of mammalian metabolism' first coined by Smith and Rosazza and successfully developed by other authors.[49–57]

A preparative application of this concept is illustrated by the production of (3*S*)-3-hydroxydihydroquinidine **13**[58,59] which has been identified as a minor human metabolite of dihydroquinidine and was considered as a potential new antiarrythmic agent, tentatively developed in France by Laboratoire Nativelle (Procter & Gamble Pharmaceuticals). The chemical hydroxylation methods available for its synthesis were expensive, due to the lack of stereospecificity of the oxidation reaction and the low yield, and gave an opportunity for investigating the feasibility of the scaling up of a stereospecific microbial hydroxylation. Among a number of strains tested,[60–62] *M. plumbeus* CBS 110-16 was the most efficient.[20] The process was optimized to produce large amounts of (3*S*)-hydroxydihydroquinidine (P. Wirsta, K. Regnard, M. C. Huet, B. Bartet, F. Deschamp, T. Ogerau, and R. Azerad, unpublished data). *O*-Acetyl-dihydroquinidine **12** was selected as a better hydroxylation substrate than dihydroquinidine itself or any other quinine–quinidine derivative, even though the acetyl derivative was slowly hydrolysed during the course of the incubation. A partial answer to this problem was to introduce the substrate by portions into the reactor (see Protocol 6), at a rate comparable to the rate of the hydroxylation reaction.

Protocol 6.
Biohydroxylation of 9-O-acetyl-10,11-dihydroquinidine (12) by M. plumbeus CBS 110-16: (3S)-3-hydroxy-10,11-dihydroquinidine (Structure 13, Scheme 5.1.6)

Caution! Carry out all procedures in a well-ventilated hood and wear disposable gloves and safety goggles.

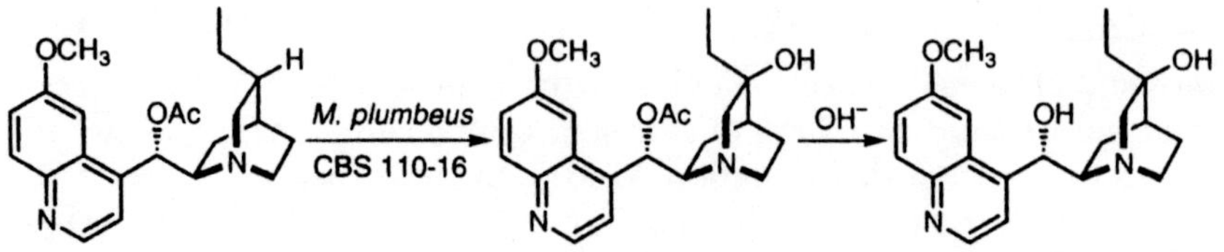

Scheme 5.1.6

This protocol illustrates the use of washed pellets of *M. plumbeus* for hydroxylation in a simple airlift bioreactor (Fig. 5.1.3), to prevent excessive shearing of the fungal biomass and ensure optimal access of oxygen.

Equipment

- Bench-top fermenter (7.5 L)
- Airlift bioreactor (see Fig. 5.1.3) (1.2 L)
- Buchner funnel (1 L or larger)
- Magnetic stirrer with Teflon bar
- Chromatography column (1.2 × 70 cm)
- Flash chromatography column (2 × 50 cm)

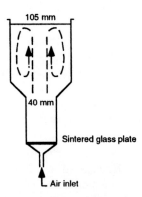

Fig. 5.1.3 Schematic view of a 1.2-L airlift bioreactor. The air bubbles-promoted convection motion of the pellets is shown by dashed arrows.

Materials

- Corn steep liquor (Solulys 48 L, Roquette, France), 100 g
- Glucose, 300 g
- Potassium dihydrogenphosphate (KH_2PO_4), 10 g
- Potassium hydrogenphosphate (K_2HPO_4), 20 g

5: Oxidation using a biocatalyst

- Sodium nitrate, 20 g — oxidizer, toxic
- Potassium chloride, 5 g
- Magnesium sulfate heptahydrate, 5 g
- Iron(II) sulfate heptahydrate, 0.2 g — irritant
- Antifoaming agent (Pluriol®, Eurane, France) — toxic
- Freshly grown and sporulated agar slope of *M. plumbeus* CBS 110-16
- Potassium phosphate buffer, 0.1 M, pH 7.0, 1 L
- 9-O-Acetyl-10,11-dihydroquinidine, 3.265 g[a] — toxic
- Sodium hydroxide — corrosive
- Dichloromethane — toxic
- Methanol — flammable, toxic
- Concentrated aqueous ammonium hydroxide solution (28%) — corrosive
- Amberlite XAD-16, 30 g
- Aluminium oxide (basic) for chromatography, 35 g

Method

1. Sterilize at 120 °C for 30 min a 7.5-L bench-top fermenter containing 5 L of double-strength liquid medium (corn steep liquor 20 g, glucose 60 g, KH_2PO_4 2 g, K_2HPO_4 4 g, sodium nitrate 4 g, potassium chloride 1 g, magnesium sulfate heptahydrate, 1 g, iron(II) sulfate heptahydrate, 0.04 g per litre) and 2–3 drops of an antifoaming agent (Pluriol®).

2. Inoculate with 5 mL of a spore suspension of *M. plumbeus* CBS 110-16, prepared from a freshly grown agar slope, to obtain approximately 10^7 spores per litre.[b] Adjust the fermenter at 25 °C, with an air flow of 2.5 L/min for 24 h, then 5 L/min, and stir at 350 r.p.m.

3. After 48 h, collect the pellets (0.3–0.5 mm in diameter) by filtration and wash them three times by resuspending in deionized water. The yield of biomass obtained is 500–750 g fresh weight (about 19–24 g/L dry weight).[c,d]

4. Resuspend the biomass (250 g fresh weight = 40 g/L dry weight) in 0.1 M potassium phosphate buffer, pH 7.0 (1 L), in the airlift bioreactor maintained at 25 °C, and adjust the aeration at 0.4 L/min. Add immediately acetyldihydroquinidine (815 mg, 2.5 mmol) then, by portions, 350 mg (1.2 mmol) per day for 7 days (roughly corresponding to the consumption of substrate). The total duration of the biohydroxylation phase is 10 days.

5. Collect the biomass by filtration[e] and wash it with slightly acidified water (0.2 L). Clarify the biohydroxylation filtrate (1.2 L) by centrifugation and bring it to pH 11.5 by addition of sodium hydroxide (8 g). Stir magnetically for 24 h at room temperature and collect the precipitate by filtration. Extract the precipitate three times with dichloromethane–methanol (1:1, 200 mL), dry the extract on anhydrous sodium sulfate and evaporate on the rotary evaporator. The crude residue is crystallized from methanol to yield 1.2 g of dihydroquinidine arising from the hydrolysis of acetyldihydroquinidine.

6. The hydrolysed incubation filtrate is poured onto a column of Amberlite XAD-16, at a flow rate of about 1–2 mL/min. Rinse with water and elute with methanol (120 mL).

Protocol 6 *Continued*

7. Evaporate the methanol solution and the mother liquors of dihydroquinidine crystallization (step 5) on a rotary evaporator and flash chromatograph the residue (2.2 g) on an aluminium oxide column (35 g) eluted with dichloromethane–methanol–conc. ammonium hydroxide (99:1:0.2). Dihydroquinidine (0.28 g) is eluted first, followed by pure (3S)-3-hydroxydihydroquinidine (1.45 g). Yield: 42%. $[\alpha]_D^{22} = +238°$ ($c = 1.14$, ethanol).

[a] 9-O-acetyldihydroquinidine was easily prepared by acetylation of dihydroquinidine with acetic anhydride in dichloromethane.
[b] Data resulting from spore counting of a diluted suspension under the microscope, using a micrometric (Malassez) cell.
[c] The pellets can be stored at 4°C as a filtration cake (15–20% dry matter) for at least 1 week, without significant loss of activity.
[d] Beyond this step, sterile conditions are no longer necessary.
[e] A second biohydroxylation cycle can be performed with the recovered biomass. However, the hydroxylation rate is reduced to 30–40% of the initial value.

References

1. Murray, H. C.; Peterson, D. H. **1952**, US Patent 2 602 769.
2. Peterson, D. H.; Murray, H. C. *J. Am. Chem. Soc.* **1952**, *74*, 1871.
3. Peterson, D. H.; Murray, H. C.; Eppstein, S. H.; Reineke, L. M.; Weintraub, A.; Meister, P. D.; Leigh, H. M. *J. Am. Chem. Soc.* **1952**, *74*, 5933.
4. Fonken, G. S.; Johnson, R. A. *Chemical Oxidations with Microorganisms*; Marcel Dekker: New York, **1972**.
5. Kieslich, K. *Biotransformations (Biotechnology. A Comprehensive Treatise in Eight Volumes)*; Rehm, H.-J.; Reed, G., ed.; Verlag Chemie: Weiheim, **1984**; vol. 6a.
6. Holland, H. L. *Organic Synthesis with Oxidative Enzymes*; VCH Publishers, Inc.: New York, **1992**.
7. Charney, W.; Herzog, H. L. *Microbial Transformations of Steroids*; Academic Press: New York, **1967**.
8. Iizuka, H.; Naito, A. *Microbial Conversion of Steroids and Alkaloids*; University of Tokyo Press, Springer-Verlag: Berlin, **1981**.
9. Mahato, S. B.; Mukherjee, A. *Phytochemistry* **1984**, *23*, 2131.
10. Mahato, S. B.; Banerjee, S. *Phytochemistry* **1985**, *24*, 1403.
11. Sedlaczek, L. *CRC Crit. Rev. Biotechnol.* **1988**, *7*, 187.
12. Mahato, S. B.; Banerjee, S.; Podder, S. *Phytochemistry* **1989**, *28*, 7.
13. Mahato, S. B.; Majumdar, I. *Phytochemistry* **1993**, *34*, 883.
14. Holland, H. L. *The Alkaloids*; Manske, R. H. F.; Rodrigo, R. G. A., ed.; Academic Press: New York, **1981**, p. 323.
15. Sebek, O. K. *Mycologia* **1983**, *75*, 383.
16. Sariaslani, F. S.; Rosazza, J. P. N. *Enzyme Microb. Technol.* **1984**, *6*, 242.
17. Krasnobajew, V. *Biotechnology. A Comprehensive Treatise in Eight Volumes*; vol. 6a, Kieslich K., ed.; Verlag Chemie: Weiheim, **1984**, p. 31.
18. Lamare, V.; Furstoss, R. *Tetrahedron* **1990**, *46*, 4109.
19. Rosazza, J. P., ed., *Microbial Transformations of Bioactive Compounds*; CRC Press: Boca Raton, FL., **1982**.

20. Azerad, R. *Chimia* **1993**, *47*, 93.
21. Sariaslani, F. S. *Adv. Appl. Microbiol.* **1991**, *36*, 133.
22. Onions, A. H. S.; Allsop, D.; Eggins, H. O. W. *Smith's Introduction to Industrial Mycology*; E. Arnold; London, **1986**.
23. Kirsop, B. E.; Doyle, A., ed.; *Maintenance of Microorganisms and Cultured Cells*; Academic Press: London, **1991**.
24. Hieda, T.; Mikami, Y.; Obi, Y.; Kisaki, T. *Agric. Biol. Chem.* **1982**, *46*, 2249.
25. Hieda, T.; Mikami, Y.; Obi, Y.; Kisaki, T. *Agric. Biol. Chem.* **1983**, *47*, 243.
26. Arias, J. M.; Garcia-Granados, A.; Jimenez, M. B.; Martinez, A.; Rivas, F.; Onorato, M. E. *J. Chem. Res. (S)* **1988**, 277.
27. Kouzi, S. A.; McChesney, J. D. *Helv. Chim. Acta* **1990**, *73*, 2157.
28. Aranda, G.; Lallemand, J. Y.; Hammoumi, A.; Azerad, R. *Tetrahedron Lett.* **1991**, *32*, 1783.
29. Aranda, G.; ElKortbi, M. S.; Lallemand, J. Y.; Neuman, A.; Hammoumi, A.; Facon, I.; Azerad, R. *Tetrahedron* **1991**, *47*, 8339.
30. Kouzi, S. A.; McChesney, J. D. *J. Nat. Prod.* **1991**, *54*, 483.
31. Abraham, W.-R. *Phytochemistry* **1994**, *36*, 1421.
32. Hanson, J. R.; Hitchcock, P. B.; Nasir, H.; Truneh, A. *Phytochemistry* **1994**, *36*, 903.
33. Herlem, D.; Ouazzani, J.; Khuong-Huu, F. *Tetrahedron Lett.* **1996**, *37*, 1241.
34. Hollinshead, D. M.; Howell, S. C.; Ley, S. V.; Mahon, M.; Ratcliffe, N. M.; Worthington, P. A. *J. Chem. Soc., Perkin Trans. 1* **1983**, 1579.
35. Aranda, G.; Facon, I.; Lallemand, J. Y.; Leclaire, M.; Azerad, R.; Cortes, M.; Lopez, J.; Ramirez, H. *Tetrahedron Lett.* **1992**, *33*, 7845.
36. Ramirez, H. E.; Cortes, M.; Agosin, E. *J. Nat. Prod.* **1993**, *56*, 762.
37. Aranda, G.; Lallemand, J.-Y.; Azerad, R.; Maurs, M.; Cortes, M.; Ramirez, H.; Vernal, G. *Synth. Commun.* **1994**, *24*, 2525.
38. Aranda, G.; Bertranne-Delahaye, M.; Azerad, R.; Maurs, M.; Cortes, M.; Ramirez, H.; Vernal, G.; Prangé, T. *Synth. Commun.* **1997**, *27*, 45.
39. Pfau, M.; Revial, G.; Guingant, A.; D'Angelo, J. *J. Am. Chem. Soc.* **1985**, *107*, 273.
40. D'Angelo, J.; Revial, G.; Volpe, T.; Pfau, M. *Tetrahedron Lett.* **1988**, *29*, 4427.
41. Revial, G. *Tetrahedron Lett.* **1989**, *30*, 4121.
42. Revial, G.; Pfau, M. *Org. Synth.* **1991**, *70*, 35.
43. Goubaud, V.; Azerad, R. *Synth. Commun.* **1996**, *26*, 915.
44. Holland, H. L.; Auret, B. J. *Can. J. Chem.* **1975**, *53*, 2041.
45. Hammoumi, A.; Girault, J.-P.; Azerad, R.; Revial, G.; D'Angelo, J. *Tetrahedron: Asymmetry* **1993**, *4*, 1295.
46. Goubaud, V.; Hammoumi, A.; Girault, J.-P.; Revial, G.; D'Angelo, J.; Azerad, R. *Tetrahedron: Asymmetry* **1995**, *6*, 2811.
47. Nordin, C.; Bertilsson, L. *Clin. Pharmacokinet.* **1995**, *28*, 26.
48. Grislain, L.; Gelé, P.; Bromet, N.; Luijten, W.; Volland, J. P.; Mocaer, E.; Kamoun, A. *Eur. J. Drug Metabol. Pharmacokinet.* **1990**, *15*, 339.
49. Smith, R. V.; Rosazza, J. P. *J. Pharm. Sci.* **1975**, *11*, 1737.
50. Smith, R. V.; Rosazza, J. P. *Biotechnol. Bioeng.* **1975**, *17*, 785.
51. Rosazza, J. P.; Smith, R. V. *Adv. Appl. Microbiol.* **1979**, *25*, 169.
52. Smith, R. V.; Rosazza, J. P. *J. Nat. Prod.* **1983**, *46*, 79.
53. Clark, A. M.; McChesney, J. D.; Hufford, C. D. *Med. Res. Rev.* **1985**, *5*, 231.
54. Davis, P. J. *Dev. Ind. Microbiol. (J. Ind. Microbiol. Suppl. No. 3)* **1988**, *29*, 197.

55. Griffiths, D. A.; Best, D. J.; Jezequel, S. G. *Appl. Microbiol. Biotechnol.* **1991**, *35*, 373.
56. Clark, A. M.; Hufford, C. D. *Med. Res. Rev.* **1991**, *11*, 473.
57. Azerad, R., Microbial models for drug metabolism, in *Adv. Biochem. Eng./Biotechnol.*; Faber, K., ed., Springer Verlag: Berlin, vol. 63, **1999**; pp. 169–218.
58. Fenard, S.; Koenig, J.-J.; Jaillon, P.; Jarreau, F. X. *J. Pharmacol. (Paris)* **1982**, *13*, 129.
59. Carroll, F. I.; Abraham, P.; Gaetano, K.; Mascarella, S. W.; Wohl, R. A.; Lind, J.; Petzoldt, K. *J. Chem. Soc., Perkin Trans. I* **1991**, 3017.
60. Eckenrode, F. M. *J. Nat. Prod.* **1984**, *47*, 882.
61. Ogerau, T.; Jarreau, X.; Azerad, R. **1985**, Fr. Patent No. 85.11221.
62. Siebers-Wolff, S.; Arfmann, H.-A.; Abraham, W.-R.; Kieslich, K. *Biocatalysis* **1993**, *8*, 47.

5.2 Oxidation of alcohols

G. FANTIN and P. PEDRINI

1. Introduction

The asymmetric oxidation of alcohols by purified enzymes or microrganisms, if compared with the reduction of ketones, is less studied mainly because: (i) the oxidation is thermodynamically unfavourable; (ii) the oxidation products (aldehydes) inhibit the enzyme; (iii) the oxidation must be carried out at pH 8–9 and this is unsuitable both for the products and for the recycling of the coenzyme $NAD(P)^+$; (iv) the oxidation of secondary alcohols involves the destruction of a valuable asymmetric centre.

The asymmetric application of this reaction is substantially limited to the kinetic resolution of racemic secondary alcohols and to the asymmetrization of prochiral or *meso*-diols to chiral lactones. These reactions exploit the enantiomeric and the enantiotopic specificity of the biocatalyst.

2. Enzymatic oxidations

The most common enzymes that operate the oxidation of alcohols together with the reduction of the corresponding ketones are alcohol dehydrogenases (ADHs) (Scheme 5.2.1).

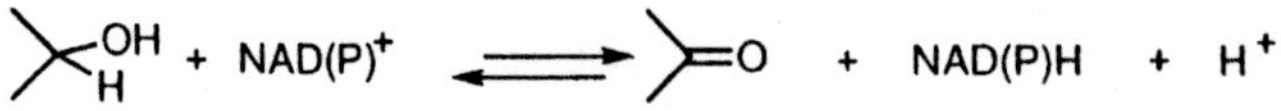

$$\ce{>CH(OH) + NAD(P)^+ <=> >C=O + NAD(P)H + H^+}$$

Scheme 5.2.1

The asymmetric synthesis potential of an ADH is very great but the high cost of nicotinamide cofactor $NAD(P)^+$ dictates the necessity to use a catalytic

5: Oxidation using a biocatalyst

amount of the coenzyme in conjunction with an auxiliary system capable of continuously regenerating it in active form.

In the past 20 years horse liver alcohol dehydrogenase (HLADH), in particular, has been widely used because of its commercial availability and well-defined and predictable stereospecificity.[1] Various mono-,[2] bi-,[3] and polycyclic[4-6] racemic alcohols have been kinetically resolved using HLADH with a flavin mononucleotide (FMN) recycling system. A typical example is shown in Scheme 5.2.2.

Protocol 1.
Kinetic resolution of (±)-4-*exo*-twistanol (Structure 1, Scheme 5.2.2)

Caution! Carry out all procedures in a well-ventilated hood and wear disposable gloves and safety glasses.

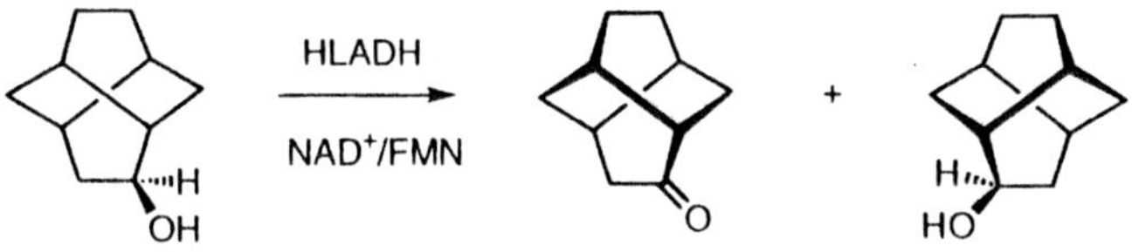

(±)-*exo*-1   (−)-2   (+)-*exo*-1

Scheme 5.2.2

This reaction is a representative kinetic resolution of racemic alcohols with HLADH, NAD$^+$ as coenzyme, and FMN as recycling reagent (recycling system with one enzyme).

Equipment

- Flask (2 L)
- Magnetic stirrer
- Lyophilizer
- Separating funnel (2 L)
- GLC column (10% Carbowax 20 M on Chromosorb W or 15% silicone DC QF-1 on Uniport B)
- FID detector
- Sintered glass filter funnel (porosity 2)
- Column for chromatography (20 cm × 1 cm)
- Rotary evaporator
- Sublimator

Materials

- Glycine–sodium hydroxide (0.05 M) buffer solution (pH 9),[a] 1 L — corrosive
- (±)-4-*exo*-Twistanol,[b] 100.9 mg, 0.67 mmol
- Nicotinamide–adenine dinucleotide (NAD$^+$),[c] 49.8 mg, 0.075 mmol — irritant
- Flavin mononucleotide (FMN) (Sigma), 750 mg, 1.5 mmol
- Horse liver alcohol dehydrogenase (HLADH) (1.37–1.95 U/mg),[d] 7.44 mg
- Diethyl ether, 620 mL — flammable, explosive
- Anhydrous magnesium sulfate — harmful
- Pentane, 190 mL — flammable, irritant
- Neutral alumina for chromatography (Woelm, activity III), 5.5 g — irritant by inhalation

Method

1. To a flask (2 L) equipped with a magnetic stirrer bar add glycine–NaOH

Protocol 1 *Continued*

buffer solution pH 9 (1 L), (±)-4-*exo*-twistanol (100.9 mg, 0.67 mmol), NAD⁺ (49.8 mg, 0.075 mmol), FMN (750 mg, 1.5 mmol), and HLADH (7.44 mg).

2. Allow the reaction to prooceed at 25°C until GLC monitoring indicates 53% oxidation (about 15.5 h).

3. Transfer the mixture to a separating funnel (2 L), extract with diethyl ether (3 × 200 mL), and dry over anhydrous magnesium sulfate (*ca.* 50 g). Filter the mixture using a sintered filter funnel and evaporate the filtrate under reduced pressure on a rotary evaporator.

4. Prepare a column for chromatography using neutral alumina (activity III, Woelm) (5.5 g) and pentane as eluent. Dissolve the dried residue (*ca.* 100 mg) in pentane (10 mL) and elute the column with pentane (120 mL). Evaporate the pentane eluates on a rotatory evaporator to yield a residue which must be purified by sublimation *in vacuo* (70–75°C, 25 mmHg) to give (−)-4-twistanone **2**, m.p. 159–161°C, $[\alpha]_D^{27} = -270°$ (c = 0.97, EtOH), 36 mg, 36%. Elute the column with pentane–diethyl ether 4:1 (70 mL) and after evaporation of the eluates obtain a residue which must be purified by sublimation *in vacuo* (80–85°C, 25 mmHg) to give (+)-4-*exo*-twistanol **1**, m.p. 192–194°C, $[\alpha]_D^{25} = +383.6°$ (c = 0.60, CHCl₃), 28 mg, 28%.

[a] The glycine–NaOH buffer pH 9 was obtained by mixing 124 mL of 0.1 M NaOH and 876 mL of 0.1 M glycine (7.505 g of glycine and 5.85 g of NaCl for 1 L of water).
[b] The racemic 4-*exo*-twistanol (m.p. 190–191°C in a sealed tube) is prepared by LiAlH₄ reduction of (±)-4-twistanone[49] followed by chromatographic separation of isomeric *exo* and *endo* compounds.
[c] NAD⁺/3H₂O was obtained from Kohjin Co. Ltd., Tokyo, Japan.
[d] HLADH was purchased from Boehringer (Mannheim) as a crystalline suspension in phosphate buffer containing 10% ethanol. Immediately before each experiment the enzyme suspension was freeze-dried to a powder whose average activity was found to be 1.37–1.95 U/mg.

A similar procedure has also been used for the kinetic resolution of (R,S)-1-hydroxymethyl-2-methyl-ferrocene[7] to the (1R)-aldehyde (86% ee) and (R,S)-1-hydroxyethylferrocene[7] to leave the (R)-alcohol (>97% ee), while racemic hydroxy-thiolanes, thianes, and thiepanes[8] are oxidized but the recovered alcohols show poor enantiomeric excesses. All these reactions are carried out in water but in some cases this gives rise to practical drawbacks such as the insolubility of most of the substrates, which necessitates working in emulsion or in suspension. In these cases the immobilization on glass beads[9] has been developed for the use of HLADH in organic solvents: (±)-*trans*-3-methylcyclohexanol and (±)-*cis*-2-methylcyclopentanol are oxidized to the corresponding enantiomerically pure ketones in isopropyl ether.

Other NAD(H)-dependent oxidoreductases are used for the asymmetric oxidation of 1,2- and 1,3-diols. Immobilized glycerol dehydrogenase (GDH)[10] from *Cellulomonas sp.* oxidizes enantioselectively 1,2-cyclohexanediol and 1,2-butanediol, secondary alcohol dehydrogenase (SADH)[11] purified from

5: Oxidation using a biocatalyst

Geotrichum sp., 1,3-butanediol and galactose oxidase[12] from *Dactylium dendroides*, and 3-halo-1,2-propanediols. On the other hand, the *R*-enantiomers of the racemic alkyl methyl carbinols are stereospecifically oxidized by fatty alcohol oxidase (FAOD)[13] from *Pichia guilliermondii* while alcohol oxidase (AOX)[14] from *Candida boidinii*, in the oxidation of the racemic 2-methyl-1-butanol, enriches the *R*-enantiomer.

Bovine liver catalase[15] also shows high enantioselectivity in the oxidation of some chiral alcohols (i.e. 2,3-butanediol, citronellol, and menthol) in organic solvents. The reactions are carried out in tetrahydrofuran, dioxane, and acetone all containing 1–3% of water and are several times faster than in water.

Recently, *endo*-bicyclo-heptenols and octenols were resolved by a new unpurified alcohol dehydrogenase from *Bacillus stearothermophilus* (BSADH)[16] both in water and in a biphasic system (10% heptane as organic solvent) (Scheme 5.2.3).

Protocol 2.
Kinetic resolution of (±)-6-*endo*-bicyclo[3.2.0]hept-2-en-6-ol (Structure 3, Scheme 5.2.3)

Caution! Carry out all procedures in a well-ventilated hood and wear disposable gloves and safety glasses.

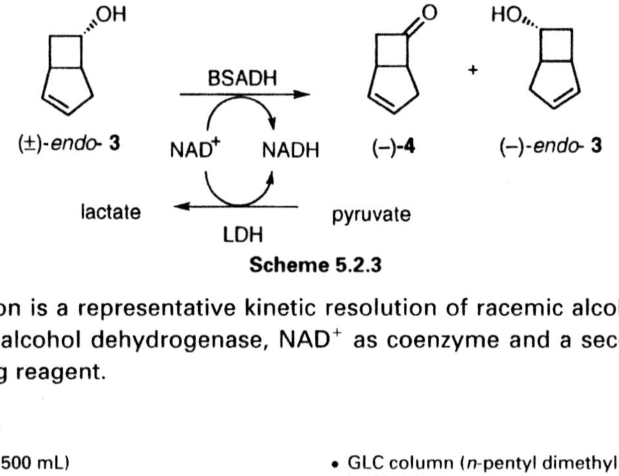

Scheme 5.2.3

This reaction is a representative kinetic resolution of racemic alcohols with an unpurified alcohol dehydrogenase, NAD^+ as coenzyme and a second enzyme as recycling reagent.

Equipment
- Four Flasks (500 mL)
- Flask (250 mL)
- Moist-heat sterilizer
- Thermostated reciprocal shaker
- Bottle centrifuge
- Magnetic stirrer

- GLC column (*n*-pentyl dimethyl β-cyclodextrine in OV 1701)
- FID detector
- Continuous liquid extractor
- Column chromatography
- Rotary evaporator

Materials
- Hydrophobic cotton, 100 g
- Sucrose, 40 g
- Peptone (Oxoid), 20 g

Protocol 2 Continued

- Yeast extract, 10 g
- Sodium hydrogenphosphate heptahydrate, 6.8 g — irritant
- Potassium sulfate, 2.6 g — harmful
- Sodium chloride, 0.15 M, 200 mL — irritant
- Tinfoil
- *Bacillus stearothermophilus* ATCC (American type culture collection) 2027
- Triethanolamine hydrochloride (TEA) buffer,[a] 200 mL — irritant
- Lysozyme (Sigma), 40 mg
- Nicotinamide-adenine dinucleotide (NAD^+), (Sigma), 20 mg — irritant
- Porcine heart lactic dehydrogenase (PHLDH), (Sigma), 2 mg
- (±)-6-*endo*-Bicyclo-[3.2.0]hept-2-en-6-ol,[b] 1.54 g, 14 mmol
- Sodium pyruvate (Sigma), 1.54 g, 14 mmol
- Dimethyl sulfoxide (DMSO), 7 mL — irritant
- Diethyl ether, 400 mL — flammable, explosive
- Sodium sulfate anhydrous, 10 g — irritant
- Silica gel 60 (70-230 Mesh ASTM, Merck) — irritant by inhalation
- Petroleum ether, 0.7 L — flammable, toxic, irritant

Method

1. Prepare four portions (250 mL) of a nutrient broth in four flasks (500 mL) respectively dissolving in water (1 L) sucrose (40 g), peptone (20 g), yeast extract (10 g), $Na_2HPO_4 \cdot 7H_2O$ (6.8 g), and K_2SO_4 (2.6 g). Stopper the flasks with hydrophobic cotton stoppers, wrap in tinfoil and sterilize them in a moist-heat sterilizer at 120 °C for 20 min. Allow to cool to room temperature and then inoculate the broth with a loopful[c] of *Bacillus stearothermophilus* ATCC (American type culture collection) 2027.

2. Put the flasks into a reciprocal shaker at 39 °C for 48 h to grow the cells.

3. Harvest by centrifugation (8000 r.p.m./15 min) 20 g of wet cells (obtained from four portion of 250-mL cultures), wash with 0.15 M NaCl (200 mL), suspend the cells in TEA buffer (100 mL) and treat with lysozyme (40 mg) for 1 h at 22 °C with stirring. Use the supernatant for the enzymatic oxidation without further purification.[d]

4. To a flask (250 mL) equipped with a magnetic stirrer add 50 mM TEA buffer (pH 7.5, 100 mL),[e] NAD^+ (20 mg), the enzyme solution prepared as above (13 units, 3.7 mL), PHLDH (2 mg), the racemic alcohol **3** (1.54 g, 14 mmol) dissolved in DMSO (7 mL), and sodium pyruvate (1.54 g, 14 mmol). Continue stirring at room temperature until GLC monitoring[f] indicates about 50% oxidation (about 18 h).

5. Extract the reaction mixture with diethyl ether (100 mL) in a continuous liquid extractor, dry the organic layer with anhydrous Na_2SO_4 and remove the solvent under reduced pressure. Separate the products by silica gel column chromatography using as eluent petroleum ether/diethyl ether 7/3. Evaporate the eluates to obtain the ketone (−)-**4** $[\alpha]_D^{25} = -63°$ (c = 1.2, $CHCl_3$) (43%, 98% ee) and the unreacted alcohol (−)-**3** $[\alpha]_D^{25} = -68°$ (c = 1.1, $CHCl_3$) (52%, 92% ee).

5: Oxidation using a biocatalyst

[a] TEA buffer is composed of 50 mM triethanolamine–HCl buffer (pH 7.5) containing 0.1 mM EDTA and 1 mM β-mercaptoethanol.
[b] The alcohol **3** was obtained by reduction of the corresponding commercially available ketone (Merck) with NaBH$_4$ followed by chromatographic separation of isomeric *exo* and *endo* compounds.
[c] The neck of the flask and the loopful are carefully flame-sterilized before each operation.
[d] Enzyme assay is carried out at 22°C by monitoring the absorbancy change at 340 nm of diphosphopyridine dinucleotide coenzyme involved in the oxidation reaction using 10 μL of the supernatant in 1 mL of TEA buffer containing 0.2 mM NAD$^+$ and 5 mM alcohol **3**.
[e] In the biphasic system the reaction is carried out using 90 mL of TEA buffer and 10 mL of heptane.
[f] Carrier gas : helium 0.8 atm; temp. 100–200 (1.5°C/min). The enantiomeric excess of the alcohol is determined after derivatization with acetic anhydride and pyridine: withdraw 1 mL of the reaction mixture, extract with 1 mL of diethyl ether, dry over anhydrous MgSO$_4$, filter the sulfate and evaporate the solvent in a nitrogen stream. Add a drop of acetic anhydride and pyridine, respectively, and allow to stand 1 h. Add 1 mL of diethyl ether, wash with 5% HCl (2 × 1 mL) and dry the ethereal solution as above. The ethereal solution can be analysed by GLC.

The enzymatic enantiotopic differentiation is widely exploited by HLADH in the oxidation of acyclic, monocyclic, and bicyclic *meso*-diols.[17-24] The enantiotopic differentiation of the hydroxy functions affords chiral aldehydes that cycles to hemiacetals which are further oxidized to the enantiomerically pure lactones. In all cases the yields and the enantiomeric excesses of the γ- and δ-lactones are excellent. A typical example of this reaction is shown in Scheme 5.2.4.

Protocol 3.
Synthesis of (+)-(1*R*,6*S*)-8-oxabicyclo[4.3.0]nonan-7-one (Structure 6, Scheme 5.2.4)

Caution! Carry out all procedures in a well-ventilated hood and wear disposable gloves and safety glasses.

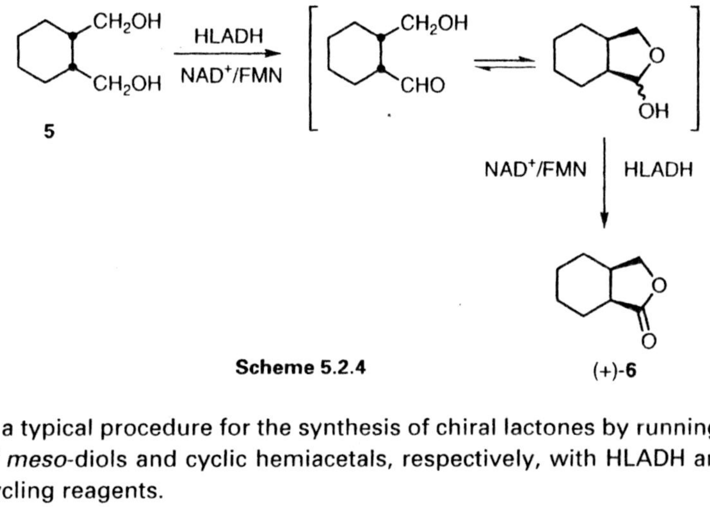

Scheme 5.2.4   (+)-**6**

This is a typical procedure for the synthesis of chiral lactones by running oxidation of *meso*-diols and cyclic hemiacetals, respectively, with HLADH and FMN as recycling reagents.

Protocol 3 *Continued*

Equipment
- Erlenmeyer flask (1 L)
- Magnetic stirrer
- Liquid extractor
- Kugelrohr
- Rotatory evaporator

Materials
- Distilled water, 475 mL
- Glycine, 3.75 g, 0.05 mol
- Aqueous sodium hydroxide, 10%   corrosive
- cis-1,2-Bis(hydroxymethyl)cyclohexane,[a] 2 g, 13.87 mmol
- Nicotinamide–adenine dinucleotide (NAD$^+$) (Sigma), 0.58 g, 0.852 mmol   irritant
- Flavin mononucleotide (FMN) (Sigma), 7.8 g, 16.2 mmol
- Horse liver alcohol dehydrogenase 1.5 u/mg (HLADH),[b] 54 mg, 80 u
- Aqueous sodium hydroxide solution, 50%, 20 mL   corrosive
- Chloroform, 200 mL   harmful
- Hydrochloric acid, 2 M   corrosive, irritant
- Charcoal, 0.5 g   harmful, irritant
- Magnesium sulfate, 20 g   harmful
- Celite   harmful, irritant

Method

1. Add in a 1-L Erlenmeyer flask distilled water (475 mL) and glycine (3.75 g, 0.05 mol) and adjust the pH to 9 by careful addition of aqueous 10% sodium hydroxide.

2. To the buffer solution add cis-1,2-bis(hydroxymethyl)cyclohexane **5** (2 g, 13.87 mmol), NAD$^+$ (0.58 g, 0.852 mmol), and FMN (7.8 g, 16.2 mmol). To the clear orange solution add further HLADH (54 mg).

3. Swirl the solution gently by magnetic stirrer for 1 min, readjust the pH to 9 and keep the reaction mixture at room temperature (not below 20°C) with the mouth of the flask covered by a watchglass. After a few minutes the colour of the solution begins to darken and after several hours becomes an opaque green–brown.

4. Readjust the pH to 9 after 6, 12, 24, 48, and 72 h by the careful addition of aqueous 10% sodium hydroxide.

5. After 4 days bring the pH of the reaction mixture to 13.3 by adding 20 mL of aqueous 50% sodium hydroxide and leave under stirring for 1 h.

6. Extract the mixture continuously for 10 h with chloroform (100 mL), mainly to remove traces of unreacted diol. Discard the chloroform extract.

7. Acidify the aqueous layer to pH 3 with 2 M HCl and re-extract continuously (15 h) with chloroform (100 mL).

8. Add to the green–orange solution charcoal (0.5 g) and magnesium sulfate. Filter the dried and partially decolourized solution on a Celite bed and remove the chloroform under reduced pressure.

5: Oxidation using a biocatalyst

9. Distil the residue (orange–green oil) in a Kugelrohr to obtain 1.5 g (77%) of the lactone (+)-**6** (>97% ee) as a colourless oil: b.p. 85–100°C (0.1–0.05 mmHg), m.p. 26–29°C, $[\alpha]_D^{22} = +51.3°$ ($c = 1.1$, $CHCl_3$).

^a cis-1,2-Bis(hydroxymethyl)cyclohexane **5** (m.p. 42–43°C) is prepared by $LiAlH_4$ reduction of cis-cyclohexane-1,2-dioic acid anhydride (Aldrich).
^b HLADH is the crystalline preparation (>98% protein) sold by Sigma Chemical Company. The activity may be determined prior to use by the Dalziel assay method.[50]

A recent application of this procedure is the synthesis of the pheromone of the spined citrus bug *Biprorulus bibax*.[25] HLADH also shows enantiotopical pro-*S* selectivity in the synthesis of (3*S*)-3-substituted valerolactones (up to 78% ee) by oxidation of 3-substituted pentane-1,5-diols.[26] On the other hand, HLADH is also effective in catalysing chemoselective oxidation of only the primary alcohol functions of several cyclohexane[27] and cyclopentane[28] substrates possessing both primary and secondary hydroxyl groups. The subsequent cyclization of the hydroxy aldehydes and the further oxidation of the hemiacetals afford, also in this case, the corresponding lactones although with poor enantiomeric excesses.

In conclusion, in the oxidation reaction with HLADH the common regeneration system of the nicotinamide cofactors NAD(P) using FMN with dioxygen as the ultimate oxidizing agent is certainly convenient because the reaction is energetically favourable and FMN is innocuous to enzymes. On the other hand, this non-catalytic regeneration system is not efficient enough for large-scale synthesis because of the large quantities of FMN required and the difficult separation of the products from the cofactors. Various improvements of the regeneration system that utilize a second enzyme as FMN reductase[29] and glutamate dehydrogenase (GluDH)[30] are reported. In both cases the enzymes are coimmobilized in polyacrylamide gel (Scheme 5.2.5).

Protocol 4.
Synthesis of (−)-(3*S*)-3-methylvalerolactone with immobilized enzymes (Structure 8, Scheme 5.2.5)

Caution! Carry out all procedures in a well-ventilated hood and wear disposable gloves and safety glasses.

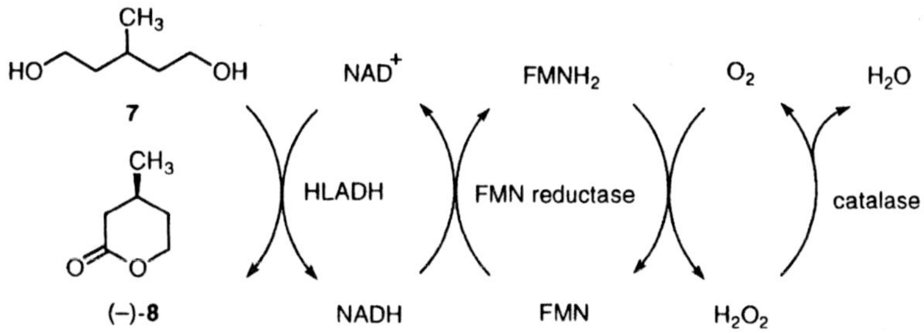

Scheme 5.2.5

Protocol 4 Continued

This is a typical procedure for an HLADH large-scale oxidation of prochiral diols with recycling enzymes coimmobilized in polyacrylamide gel.

Equipment
- Beaker (50 mL)
- Magnetic stirrer
- Mortar and pestle
- Centrifuge
- Flask (2 L)
- Rotatory evaporator
- Liquid–liquid extractor
- Distillation apparatus
- Vacuum pump

Materials for immobilization of FMN reductase
- Poly(acrylamide-co-N-acryloxysuccinimide) (PAN 1000),[a] 3 g
- N-(2-Hydroxyethyl)piperazine-N'-2-(ethanesulfonic acid) (HEPES, Merck) buffer, 0.2 M (pH 7.5) (HEPES buffer),[b] 12 mL
- Nicotinamide–adenine dinucleotide (NAD$^+$), 1.5 mM (Sigma), 12 mg, 0.018 mmol — **irritant**
- Flavin mononucleotide (FMN), 0.5 mM (Sigma), 2.7 mg, 0.006 mmol
- Magnesium chloride, 15 mM, 17 mg, 0.18 mmol — **irritant**
- 1,4-Dithiothreitol (DTT), 0.5 M (Sigma), 150 µL — **irritant, harmful**
- Triethylenetetramine (TET), 0.5 M (Sigma), 2.55 mL — **harmful, corrosive**
- Flavin mononucleotide (FMN) reductase (Boehringer), 10 units
- HEPES buffer, 50 mM (pH 7.5),[b] 150 mL
- Ammonium sulfate, 50 mM, 0.99 g, 7.5 mmol — **harmful, irritant**

Materials for oxidation with immobilized enzymes
- 3-Methylpentane-1,5-diol,[c] 8.9 g, 75 mmol
- Nicotinamide–adenine dinucleotide (NAD$^+$), 0.114 g, 0.15 mmol — **irritant**
- Flavin mononucleotide (FMN), 0.755 g, 1.5 mmol
- Glycine–sodium hydroxide (0.05 M) buffer, pH 9,[d] 1.5 L — **corrosive**
- Horse liver alcohol dehydrogenase (HLADH) immobilized on PAN,[a] 16 units
- Catalase immobilized on PAN,[a] 100 units
- Aqueous sodium hydroxide solutions, 4 M and 10 M — **corrosive**
- Distilled water, 400 mL
- Chloroform, 100 mL — **harmful**
- Hydrochloric acid, 12 M — **corrosive, irritant**

Immobilization of FMN reductase

1. Place PAN 1000 (3 g) in a 50-mL beaker and add 12 mL of 0.2 M HEPES buffer, pH 7.5, containing 1.5 mM NAD$^+$ (12 mg), 0.5 mM FMN (2.7 mg), and 15 mM MgCl$_2$ (17 mg). The polymer should be dissolved within 1 min by magnetic stirring and mechanical grinding with a spatula or glass rod.

2. Add under stirring a solution of 0.5 M DTT (150 µL) and 2.55 mL of 0.5 M TET. After 10 s add a solution (1 mL) containing FMN reductase (10 units): in about 45 s, a resilient gel is formed.

3. Allow the gel to stand at room temperature for 1 h and then transfer to a mortar. Grind the gel with a pestle and suspend the particles in 150 mL of 50 mM HEPES buffer, pH 7.5, containing 50 mM ammonium sulfate (0.99 g) to wash the gel.

4. Agitate the suspension with magnetic stirring for 15 min, transfer to centri-

5: Oxidation using a biocatalyst

fuge tubes and separate the gel by centrifugation at 2000 rpm. Repeat washing twice with HEPES buffer without ammonium sulfate. Measure the enzyme activity, which should be 3.2 units.[e] Apply a similar procedure[a] for immobilizing HLADH (4.3 mg of protein for 1 g of PAN 500) and catalase (10 mg of protein for 1 g of PAN 500).

Oxidation with immobilized enzymes

1. In a flask (2 L) dissolve 3-methylpentane-1,5-diol **7** (8.9 g, 75 mmol), NAD^+ (0.114 g, 0.15 mmol) in 0.05 M glycine–NaOH buffer (1.5 L, pH 9) and add the immobilized enzymes HLADH (16 units), FMN reductase (9 units), and catalase (100 units).
2. Agitate the reaction with a magnetic stirring at room temperature and occasionally readjust the pH of the solution to 9 by adding 4 M NaOH. Monitor the progress of the reaction with enzyme assay.[f]
3. After 14 days (the assay procedure shows little substrate remaining) centrifuge the reaction suspension and decant the liquid from the enzyme-containing gel.
4. Wash the gel with 200 mL of distilled water and again separate the suspension by centrifugation. Repeat the operation with a second portion of water (200 mL) and combine the washes. Evaporate the solution to 400 mL under reduced pressure and raise the pH of the solution to 12 with 10 M NaOH.
5. Extract the solution by continuous liquid–liquid extraction with $CHCl_3$ (100 mL) for 2 days, acidify the aqueous solution with 12 M HCl to pH 3 and continue the extraction with the same $CHCl_3$ for a further 1 day.
6. Evaporate the extract and distil (−)-(3S)-3-methylvalerolactone **8** (5.85 g, 68%) under reduced pressure: b.p. 121–123°C (0.2 torr); $[\alpha]_D^{27} = -24.8°$ (c = 5.6, $CHCl_3$) (90% optical purity).

[a] PAN 1000 (containing 1000 μmol/g of active ester) was prepared by free polymerization of acrylamide and N-acryloxysuccinimide in THF solution using thermal initiation with azo-bis(isobutyronitrile) (AIBN) according to Ref. 51.
[b] 1 M HEPES buffer is prepared by dissolving 4.766 g HEPES in 1 mL NaOH (30%) and 5 mL deionized water, adjusting the pH to 7.5 with aqueous ammonia solution (25%) and diluting to exactly 20 mL with deionized water. Appropriate dilutions are made to obtain HEPES 0.2 M and 50 mM.
[c] 3-Methylpentane-1,5-diol is prepared by reduction of 3-methylglutaric anhydride (Aldrich) with $LiAlH_4$ (Ref. 29).
[d] The glycine–NaOH buffer (pH 9) is obtained by mixing 124 mL of 0.1 M NaOH and 876 mL of 0.1 M glycine (7.505 g of glycine and 5.85 g of NaCl for 1 L of water).
[e] Assay the enzyme activity using 0.11 mM NADH and 0.21 mM FMN in 0.05 M glycine–NaOH buffer at pH 9 by measuring the change in absorbance at 340 nm. A dilute suspension containing gel particles (30–50 μm) is added to the assay solution in the spectrophotometer cuvette. The cuvette is stoppered with a Teflon stopper, shaken for a few seconds, and put into the cell compartment, and the increase or decrease in the absorbance is recorded for 30 s; during this 30-s interval the gel particles do not settle appreciably and because the particles are transparent they do not interfere with the spectrophotometric analysis. After the 30-s interval the cuvette is taken out of the cell compartment and shaken again for a few seconds, and the absorbance is recorded for another 30 s.
[f] Remove a 2-μL sample and add to 1 mL of glycine buffer (pH 9) containing 1.4 M NAD in a cuvette. Add to the cuvette the enzyme HLADH (1 unit) and measure the subsequent increase in absorbance at 340 nm. Calculate from this the concentration of the substrate remaining in the reactor.

Minimization of the product inhibition is also achieved by using a water–organic solvent biphasic system.[31]

3. Microbial oxidations

All the problems concerning the regeneration of the cofactors are clearly overcome by the use of microorganisms: in the cells there are, in fact, enzymes, cofactors, and the appropriate regenerating systems. However, in comparison with the reduction of ketones, there are few reports of microbial asymmetric oxidation of alcohols. Baker's yeast, moreover, which was extensively employed for the reduction of ketones,[32] has only recently been used in the kinetic resolution, via oxidation, of 1-aryl- and 1-heteroarylethanols.[33] While the reduction of the ketones (fermenting conditions, a few days) generally affords the S-enantiomer (according to the Prelog rule),[32] the oxidation of the racemic alcohols (non-fermenting conditions, several days) leaves the unreacted R-enantiomer in excellent ee. In some cases the enantioselectivity can be reversed by varying the energy source supplemented to the yeast.[34] 1-Aryl- and 1-heteroaryl ethanols are also resolved by oxidation with growing cells of *Bacillus stearothermophilus*.[35] This microorganism, however, shows its synthetic potential in the resolution of various racemic secondary alcohols, i.e. endo-bicyclic heptenols,[36,37] endo-bicyclic octenols,[36] and aliphatic unsaturated alcohols.[38] A typical example is shown in Scheme 5.2.6.

Protocol 5.
Kinetic resolution of (±)-5-hexen-2-ol (Structure 9, Scheme 5.2.6)

Caution! Carry out all procedures in a well-ventilated hood and wear disposable gloves and safety glasses.

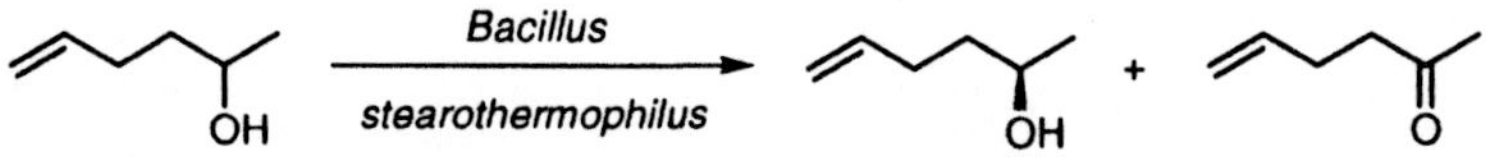

$$(\pm)\text{-}9 \xrightarrow{\text{Bacillus stearothermophilus}} (-)\text{-}(R)\text{-}9 \;+\; 10$$

Scheme 5.2.6

This is a typical example of kinetic resolution by growing cells of *Bacillus stearothermophilus*.

Equipment

- Flask (500 mL)
- Moist-heat sterilizer
- Thermostated reciprocal shaker
- Continuous liquid extractor
- GLC column (n-pentyl dimethyl β-cyclodextrin in OV 1701)
- Column chromatography
- Rotary evaporator

Materials

- Hydrophobic cotton
- Bactotryptone, 2 g
- Yeast extract, 1 g

5: Oxidation using a biocatalyst

- Sodium oxalate, 0.3 g   **harmful**
- Distilled water, 100 mL
- *Bacillus stearothermophilus* ATCC (American type culture collection) 2027
- Tinfoil
- (±)-5-Hexen-2-ol,[a] 0.3 g, 3 mmol
- Dimethyl sulfoxide (DMSO), 1 mL   **irritant**
- Diethyl ether, 300 mL   **flammable, explosive**
- Sodium sulfate anhydrous, 20 g   **irritant**
- Silica gel 60 (70–230 mesh ASTM Merck)   **harmful, irritant**
- Petroleum ether, 800 mL   **flammable, toxic, irritant**

Method

1. In a 500-mL flask prepare the nutrient broth by dissolving bactotryptone (2 g), yeast extract (1 g) and sodium oxalate (0.3 g, 3 mmol) in 100 mL of distilled water. Stopper the flask with hydrophobic cotton, wrap it in tinfoil and sterilize the nutrient broth in a moist-heat sterilizer at 120 °C for 20 min. Allow the broth to cool to room temperature and then inoculate (in this operation sterility is necessary)[b] the broth with a loopful of *Bacillus stearothermophilus* ATCC 2027.

2. Put the flask, stoppered with hydrophobic cotton, into a reciprocal shaker at 39 °C for 48 h to grow the cells.

3. After 48 h, to the suspension of the grown cells add the racemic 5-hexen-2-ol **9** (0.3 g) dissolved in DMSO (1 mL) and put the flask into a reciprocal shaker at 39 °C. Withdraw periodically (every hour) aliquots (1 mL) of the suspension, extract with diethyl ether (1 mL), dry over anhydrous Na_2SO_4 and analyse the crude reaction products by GLC on a chiral column[c] (this operation has to be performed to get over 50% conversion).

4. After 4 h (51% conversion by GLC) extract the reaction mixture with diethyl ether (100 mL) with a continuous liquid–liquid extractor, dry the organic layer over anhydrous Na_2SO_4 and concentrate the solvent under reduced pressure.

5. Purify the products by silica gel column chromatography using petroleum ether/diethyl ether 8:2 as eluent. Evaporate the eluates to obtain 0.135 g (46%) of 5-hexen-2-one and 0.132 g (44%) of the (*R*)-5-hexen-2-ol **9**[d] (100% ee) as a colourless oil: $[\alpha]_d^{25} = -12.1°$ (c =4.6, $CHCl_3$).

[a] (±)-5-Hexen-2-ol is obtained by reduction with sodium borohydride of the corresponding hexenone (Aldrich).
[b] The neck of the flask and the loopful are carefully flame-sterilized before each operation.
[c] Carrier gas: helium 0.9 atm; temp. 80 °C; retention time in min for the ketone **10** 4.76, for (*S*)-**9** 8.7, for (*R*)-**9** 8.95.
[d] Ref. 52.

B. stearotherm. is also surprisingly active in the resolution of racemic alcohols in an organic solvent. The cells, harvested from water by centrifugation, show viability for 24 h in heptane and are able to resolve kinetically bicyclic secondary alcohols with the same efficiency as in water and with immobilized cells.[39]

Improved kinetic resolution of racemic 1-arylethanols is also achieved with *Geotrichum candidum* entrapped with a water-adsorbent polymer.[40] In all cases the *S*-enantiomer is selectively oxidized and the recovered unreacted *R*-enantiomer is enantiomerically pure.

Kinetic resolution is also attained in the oxidation of (±)-1-phenylsulfenyl-2-propanol (*R*-enantiomer is recovered intact)[41] and of *syn*- and *anti*-1-phenyl-propane-1,2-diols[42] with *Corynebacterium equi*. In the last case a high regioselectivity towards the benzylic hydroxy group is coupled with a high enantioselectivity: only the *R*-enantiomer is oxidized. On the other hand, cultured cells of *Nicotiana tabacum* oxidize enantioselectively the hydroxy group of cyclic terpenes as neoisopinocampheol[43] and 2-hydroxy-*p*-menthanes.[44]

The resolution of the racemic citronellol is achieved by utilizing cells of *Rhodococcus equi* and the accumulation of the *R*-enantiomer at high concentration is observed with a water–decane liquid-phase system.[45]

Corynosporium cassiicola gives the optically pure (+)-*trans*-cyclohexane-1,2-diol by oxidation of both the corresponding racemate and the *meso-cis*-diol.[46] This process probably involves a double stereoinversion process catalysed by two dehydrogenases.

Gluconobacter roseus shows an enantiotopic selectivity for the pro-(*R*) hydroxymethyl group of 2-methylpropane-1,3-diol.[47] Other α,ω-diols[48] with prochiral centres are oxidized by *Gluconobacter* with distinction of the pro-*R* and pro-*S* sites of the molecules: the substituents on the prochiral centre seriously affected the rate of oxidation and the optical yield.

References

1. Jones, J. B.; Beck, J. F. In *Applications of Biochemical Systems in Organic Chemistry, Part I*; Jones, J. B. ; Sih, J. ; Perlman, D., ed.; Wiley: New York, **1976**; pp. 107–401.
2. Jones, J. B.; Takemura, T. *Can. J. Chem.* **1984**, *62*, 77.
3. Irwin, A. J.; Jones, J. B. *J. Am. Chem. Soc.* **1976**, *98*, 8476.
4. Nakazaki, M.; Chikamatsu, H.; Naemura, K.; Suzuki, T.; Iwasaki, M.; Sasaki, Y.; Fujii, T. *J. Org. Chem.* **1981**, *46*, 2726.
5. Nakazaki, M.; Chikamatsu, H.; Sasaki, Y. *J. Org. Chem.* **1983**, *48*, 2506.
6. Nakazaki, M.; Chikamatsu, H.; Fujii, T.; Sasaki, Y.; Ao, S. *J. Org. Chem.* **1983**, *48*, 4337.
7. Yamazaki, Y.; Uebayasi, M.; Hosono, K. *Eur. J. Biochem.* **1989**, *184*, 671.
8. Jones, J. B.; Schwartz, H. M. *Can J. Chem.* **1981**, *59*, 1574.
9. Grunwald, J.; Wirz, B.; Scollar, M. P.; Klibanov, A. M. *J. Am. Chem. Soc.* **1986**, *108*, 6732.
10. Lee, L. G.; Whitesides, G. M. *J. Org. Chem.* **1986**, *51*, 25.
11. Mori, T.; Sakimoto, M.; Kagi, T.; Sakai, T. *Biosci. Biotechnol. Biochem.* **1996**, *60*, 1191.
12. Klibanov, A. M.; Alberti, B. N.; Marletta, M. A. *Biochem. Biophys. Res. Commun.* **1982**, *108*, 804.
13. Muller, H.-G.; Guntheberg, H.; Drechsler, H.; Mauersberger, S.; Kortus, K.; Oehme, G. *Tetrahedron Lett.* **1991**, *32*, 2009.

14. Clark, D. S.; Geresh, S.; Di Cosimo, R. *Biorg. Med. Chem. Lett.* **1994**, *4*, 1745.
15. Magner, E.; Klibanov A. M. *Biotechnol. Bioeng.* **1995**, *46*, 175.
16. Giovannini, P. P.; Hanau, S.; Rippa, M.; Bortolini, O.; Fogagnolo, M.; Medici, A. *Tetrahedron* **1996**, *52*, 1669.
17. Goodbrand, H. B.; Jones, J. B. *J. Chem. Soc., Chem. Commun.* **1977**, 469.
18. Jakovac, I. J.; Ng, G.; Lok, K. P.; Jones, J. B. *J. Chem. Soc., Chem. Commun.* **1980**, 515.
19. Jakovac, I. J.; Goodbrand, H. B.; Lok, K. P.; Jones, J. B. *J. Am. Chem. Soc.* **1982**, *104*, 4659.
20. Jones, J. B.; Finch, M. A. W.; Jakovac, I. J. *Can. J. Chem.* **1982**, *60*, 2007.
21. Ng, G. S. Y.; Yuan, L.-C., Jakovac, I. J.; Jones, J. B. *Tetrahedron* **1984**, *40*, 1235.
22. Jones, J. B.; Francis, C. J. *Can. J. Chem.* **1984**, *62*, 2578.
23. Lok, K. P.; Jakovac, I. J.; Jones, J. B. *J. Am. Chem. Soc.* **1985**, 107, 2521.
24. Bridges, A. J.; Raman, P. S.; Ng, G. S. Y.; Jones, J. B. *J. Am. Chem. Soc.* **1984**, *106*, 1461.
25. Mori, K.; Amaike, M.; Oliver, J. E. *Liebigs Ann. Chem.* **1982**, 1185.
26. (a) Jones, J. B.; Lok, K. P. *Can. J. Chem.* **1979**, *57*, 1025. (b) Irwin, A. J.; Jones, J. B. *J. Am. Chem. Soc.* **1977**, *99*, 556.
27. Jones, J. B.; Goodbrand, H. B. *Can. J. Chem.* **1977**, *55*, 2685.
28. Irwin , A. J.; Jones, J. B. *J. Am. Chem. Soc.* **1977**, *99*, 1625.
29. Drueckhammer, D. G.; Riddle, V. W.; Wong, C.-H. *J. Org. Chem.* **1985**, *50*, 5387.
30. Wong, C.-H.; Matos, J. R. *J. Org. Chem.* **1985**, *50*, 1992.
31. Matos, J. R.; Wong, C.-H. *J. Org. Chem.* **1986**, *51*, 2388.
32. Csuk, R.; Glanzer, B. I. *Chem. Rev.* **1991**, *91*, 49.
33. Fantin, G.; Fogagnolo, M.; Medici, A.; Pedrini, P.; Poli, S.; Sinigaglia, M. *Tetrahedron Lett.* **1993**, *34*, 883.
34. Fantin, G.; Fogagnolo, M.; Guerzoni, M. E.; Medici, A.; Pedrini, P.; Poli, S. *J. Org. Chem.* **1994**, *59*, 924.
35. Fantin, G.; Fogagnolo, M.; Medici, A.; Pedrini, P.; Poli, S.; Gardini, F. *Tetrahedron: Asymmetry* **1993**, *4*, 1607.
36. Fantin, G.; Fogagnolo, M.; Medici, A.; Pedrini, P.; Rosini, G. *Tetrahedron: Asymmetry* **1994**, *5*, 1635.
37. Fantin, G.; Fogagnolo, M.; Marotta, E.; Medici, A.; Pedrini, P.; Righi, P. *Chem. Lett.* **1996**, 511.
38. Fantin, G.; Fogagnolo, M.; Giovannini, P. P.; Medici, A.; Pedrini, P. *Tetrahedron: Asymmetry* **1995**, *6*, 3047.
39. Fantin, G.; Fogagnolo, M.; Giovannini, P. P.; Medici, A.; Pedrini, P.; Poli S. *Tetrahedron Lett.* **1995**, *36*, 441.
40. Nakamura, K.; Inoue, Y., Ohno, A. *Tetrahedron Lett.* **1994**, *35*, 4375.
41. Ohta, H.; Kato, Y.; Tsuchihashi, G. *J. Org. Chem.* **1987**, *52*, 2735.
42. Ohta, H.; Yamada, H.; Tsuchihashi, G. *Chem. Lett.* **1987**, 2325.
43. Suga, T.; Hamada, H.; Hirata, T. *Chem. Lett.* **1987**, 471.
44. Suga, T.; Hamada, H.; Hirata, T.; Izumi, S. *Chem. Lett.* **1987**, 903.
45. Oda, S.; Kato, A.; Matsudomi, N.; Ohta, H. *Biosci. Biotechnol. Biochem.* **1996**, *60*, 83.
46. Carnell, A. J.; Iacazio, G.; Roberts, S. M.; Willetts, A. J. *Tetrahedron Lett.* **1994**, *35*, 331.
47. Ohta, H.; Tetsukawa, H. *Chem. Lett.* **1979**, 1379.

48. Ohta, H.; Tettukawa, H.; Noto, N. *J. Org. Chem.* **1982**, *47*, 2400.
49. Gauthier, J.; Deslongchamps, P. *Can. J. Chem.* **1967**, *45*, 297.
50. Dalziel, K. *Acta Chem. Scand.* **1957**, *11*, 397.
51. Pollak, A.; Blumenfeld, H.; Wex, M.; Baughn, R. L.; Whitesides, G. M. *J. Am. Chem. Soc.* **1980**, *102*, 6324.
52. Keinan, E.; Seth, K. K.; Lamed, R.; Ghirlando, R.; Singh, S. P. *Biocatalysis* **1990**, *100*, 545.

5.3 Asymmetric Baeyer–Villiger oxidation using biocatalysis

V. ALPHAND and R. FURSTOSS

1. Introduction

The Baeyer–Villiger (BV) oxidation of cyclic or non-cyclic ketones is one of the major reactions of organic chemistry. This oxidation, first described in 1899, transforms a ketone into its corresponding ester. The cyclic ketones thus lead to cyclic esters, called lactones. Although the stereochemical features of this type of reaction have been studied extensively over the years, ways to achieve BV oxidation in an asymmetric manner using conventional chemistry—i.e. metal-catalysed reactions—have only been described very recently with some success (see Chapter 3, Section 3.1). On the other hand, microbiological or enzymatic BV oxidations are a very efficient way (and, until recently, the only way) to obtain enantiopure lactones from racemic or prochiral cyclic ketones. Most of these results have been summarized in two recent and complementary reviews.[1,2]

In fact, the involvement of BV oxidation in the metabolic or catabolic pathway of many microorganisms has been known for a long time. As early as 1953, several strains were described to cleave the C-17 side-chain of steroids.[1] The first indication of asymmetric activity was reported by Shaw in 1966, the second by Schwab in 1983,[1] but it was only in 1987 that this property was used to synthesize an optically active lactone on a laboratory scale. Thus, Azerad and co-workers[3] incidentally observed that, in the presence of the fungus *Curvularia lunata*, the racemic ketol **1** led to a mixture of the (unreacted) optically pure (*R*)-ketol **1** and the reaction product, i.e. the (*S*)-lactone **2**, formed from spontaneous rearrangement of the corresponding seven-membered intermediate (Scheme 5.3.1).

However, the real breakthrough that triggered the development of numerous studies aimed to exploit the potentialities of these enzymatic reactions for asymmetric synthesis was the description of the BV oxidation of bicyclo[3.2.0]hept-2-ene-6-one **3** by whole cells of the bacteria *Acinetobacter* TD63.[4] Surprisingly, this ketone led to the formation of the two regioisomeric

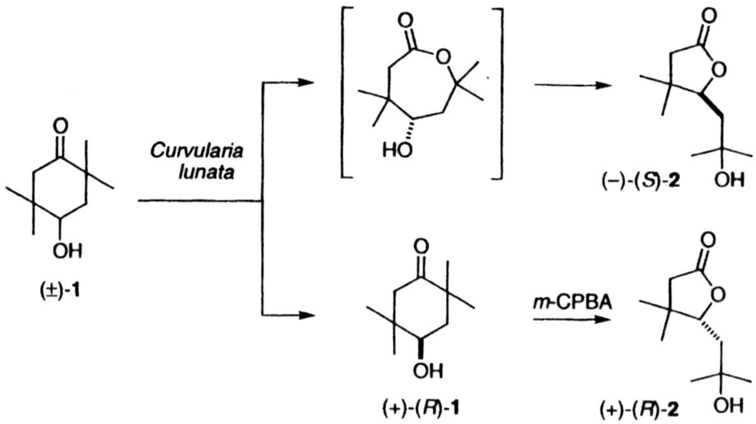

5: Oxidation using a biocatalyst

Scheme 5.3.1

lactones **4** and **5** (in 1:1 proportion and almost quantitative yield), both these lactones being almost enantiomerically pure (Scheme 5.3.2). This is the result of a different (and nearly absolute) regioselectivity during the oxidation of each one of the two enantiomers of the substrate. Lactone **4** arises from the insertion of the oxygen atom between the carbonyl group and the *more* substituted carbon atom—which is generally the 'normal' figure observed for chemical BV oxidation—whereas lactone **5**, the so-called 'abnormal' lactone, is the outcome of the oxygen insertion into the *less* substituted carbon atom bond. This is particularly interesting because it is not possible to obtain this 'abnormal' selectivity using conventional chemistry, which makes these enzymatic reactions even more attractive for organic chemistry.

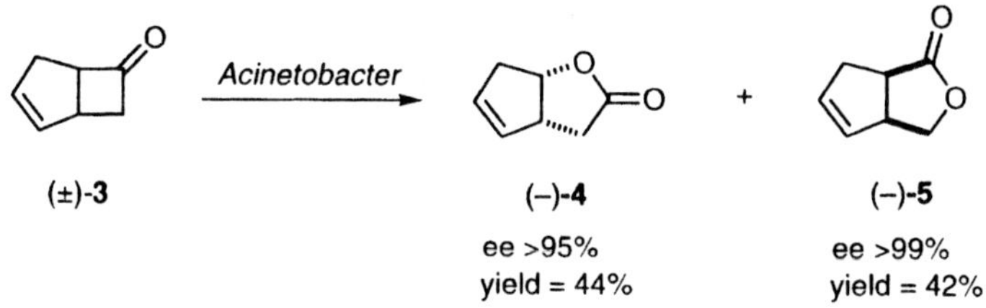

Scheme 5.3.2

2. Mechanism and classification

The first microorganism extensively studied and used for its capacity to achieve asymmetric BV oxidation was the cyclohexanol-grown *Acinetobacter calcoaceticus*.[5,6] The enzyme involved is a cyclohexanone monooxygenase (CHMO), a NADPH-dependent flavoprotein.* The proposed mechanism implies the

---
* This means that the enzymatic system is composed of the enzyme, a flavin coenzyme (a FAD group) and a cofactor (NADPH) as a reducing agent.

formation of an intermediate hydroperoxyflavin **7** (the biological equivalent of the peracids generally used for chemical BV oxidation) which reacts as a nucleophile with the carbonyl group of the substrate* (Scheme 5.3.3).

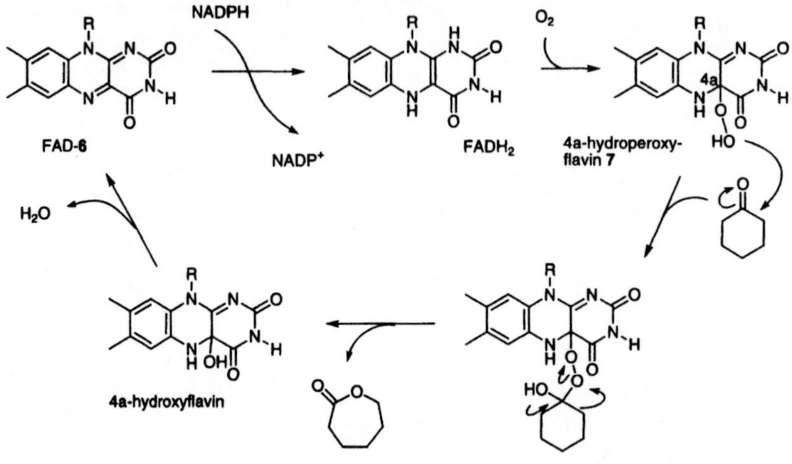

Scheme 5.3.3

About ten other enzymes able to perform BV oxidation (nowadays called 'Baeyer–Villigerases') have been characterized recently[2]—mostly from bacterial sources—and seven of them are now commercialized by Fluka Company. They have been classified as belonging to two different classes: Class I or Class II, depending on their enzymatic characteristics. Thus, Class I Baeyer–Villigerases are FAD binding and NADPH-dependent, whereas Class II enzymes are FMN binding and NADH-dependent.[2]

3. Purified enzymes versus whole cells

Because these oxidations are coenzyme dependent, biotransformations with purified enzymes necessitate the use of a coupled enzyme system in order to recycle the expensive cofactor: NADPH/glucose dehydrogenase or NADH/formate dehydrogenase is classically used.[2] It is also possible to start from the corresponding alcohol instead of the ketone and to operate *in situ* a closed-loop recycling procedure[2,9] (see Scheme 5.3.4). However, this technique is limited by the substrate reactivity of the recycling enzyme. A continuously operating process using a membrane reactor-based technology was also described recently but has not yet been adapted for routine laboratory synthesis.[8] Thus, the use of purified Baeyer–Villigerases appears to be limited at present to microscale analytical studies. Progress in this direction should,

---

* In addition to the nucleophilic attack of a ketone, this 4a-hydroperoxyflavin intermediate can also react as an electrophile and asymmetrically oxidize nucleophilic substrates such as sulfides.[7,8]

5: Oxidation using a biocatalyst

however, appear in the future, since one of these enzymes has recently been cloned and overexpressed.[10,11]

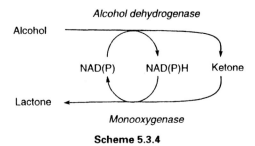

Scheme 5.3.4

On the contrary, biotransformations with whole-cell cultures are much easier to carry out than chemists imagine.[12] In general, with the restriction that only a single Baeyer–Villigerase is present in the cell and/or that the other enzymes do not catalyse some undesired side-reactions, the use of a whole-cell culture is more appropriate to synthetic purposes. For this reason, we will focus here on examples using this type of technique.

One of the important drawbacks of biotransformations is the fact that the outcome of such a reaction is very difficult to predict. This also applies to BV oxidation of course. However, the accumulation of numerous results reported in the literature has allowed authors to build up different 'active-site models' which might enable such predictions to be made.[2] Although these models are quite rough and not very accurate, they nevertheless can be used as a preliminary approach.

Further work is currently under way in certain laboratories to establish the X-ray structure of some of these enzymes, producing results which will certainly help (but not totally solve!) the prediction problem. At present, the best way to cope with this question is to perform the biotransformation on an analytical scale!

4. Some examples of biocatalysed BV oxidations

In principle, as for all biotransformations, different types of substrates can be oxidized: (a) enantiopure, (b) racemic, and (c) prochiral compounds.

4.1 BV oxidation of enantiopure ketones

Biooxidation of an enantiopure ketone is interesting when it leads to 'abnormal' lactones impossible to obtain by chemical BV oxidation. Thus, the biotransformation of commercially available (+)-dihydrocarvone **8** by whole cells of *A. calcoaceticus* NCIMB 9871 allows the 'abnormal' lactone **9** (in the form of its hydroxy acid **10**) to be prepared as a single reaction product[13] (see Scheme 5.3.5, Protocol 1).

Protocol 1.
Biotransformation of (+)-dihydrocarvone with whole cells of *A. calcoaceticus* NCIMB 9871 (Structures 8, 9, and 10, Scheme 5.3.5)

Caution! Carry out all procedures in a well-ventilated hood and wear disposable gloves and safety goggles.

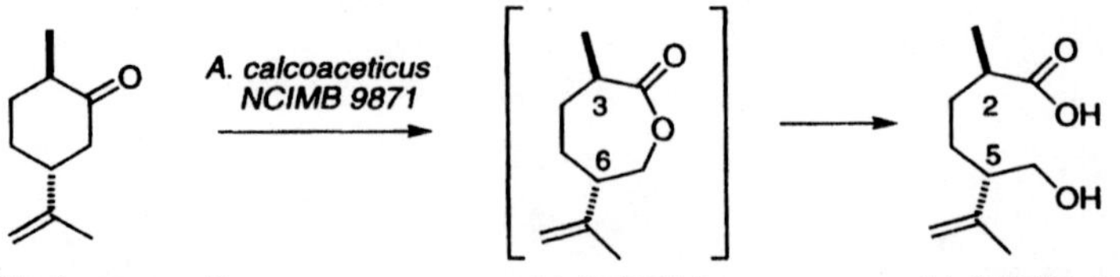

(+)-Dihydrocarvone **8** → [(−)-(3*R*,6*S*)-**9**] → (−)-(2*R*,5*S*)-**10**

Scheme 5.3.5

This is an example of a biotransformation with a growing bacterial culture.

Equipment

- Inoculating loop
- Autoclave
- Bunsen burner
- Sterile syringes and needles (>5 mL)
- Thermostated (orbital or reciprocating) shaker
- Two autoclavable culture tubes with cap
- Erlenmeyer or shake flasks (2 × 3 L) with cotton plug
- Erlenmeyer flasks (3 × 25 mL) with cotton plug
- Gas chromatography apparatus equipped with an FID detector
- Water-cooled condenser
- Round-bottomed flask (250 mL)
- Heating mantle
- Liquid–liquid continuous extractor for heavy solvents (1 L or 2 × 500 mL)
- Rotary evaporator
- Bulb-to-bulb distillation apparatus

Materials

• *Acinetobacter calcoaceticus* NCIMB 9871 strain from the National Collection of Industrial and Marine Bacteria (UK)	**Class 2 pathogen**
• Distilled or deionized water, 1.5 L	
• Yeast extract (Difco), 0.2 g	
• Sodium hydrogenphosphate (Na_2HPO_4), 4 g	
• Potassium dihydrogenphosphate (KH_2PO_4), 2 g	
• $(NH_4)_2SO_4$, 3 g	
• Magnesium sulfate, 0.5 g	
• Calcium chloride, 0.1 g	
• Iron(II) sulfate, 0.01 g	**irritant**
• *cis/trans*-1,2-Cyclohexanediol, 2.5 g (or cyclohexanol)	
• Sodium hydroxide, 5 N	**corrosive**
• Ethanol, 5 mL	**flammable**
• (+)-Dihydrocarvone **8**, 800 mg (5.3 mmol)	**irritant**
• An Optima 1701-type capillary column (30 m × 0.25 mm, Macherey–Nagel)	
• Hydrochloric acid, 2 M	**corrosive, toxic**
• Dichloromethane, 0.5 or 1 L	**harmful by inhalation**

Culture

1. Prepare solution A containing yeast extract, Na_2HPO_4, KH_2PO_4, ammonium sulfate and 1 L distilled water. Transfer 10 mL of this solution into a culture

5: Oxidation using a biocatalyst

tube and 2 × 500 mL into two 3-L flasks sealed with a cotton plug. Prepare solution B composed of magnesium sulfate and iron(II) sulfate in 5 mL distilled water. Prepare solution C containing calcium chloride in 5 mL distilled water and solution D containing *cis/trans*-1,2-cyclohexanediol (carbon source) in 25 mL distilled water (see Fig. 5.3.1). Sterilize all these solutions (autoclave 110 bar, 15 min).

2. After cooling, divide the B and C solutions between the 3-L flasks under sterile conditions.[a] Add 10 mL of solution D to each 3-L flask (see Fig. 5.3.1) and 0.2 mL to the culture tube.
3. Under sterile conditions, inoculate the culture tube from an agar slope of *A. calcoaceticus* NCIMB 9871,[b] allow it to stand 6 h at 30 °C and divide this inoculum between the two 3-L flasks (see Fig. 5.3.2).
4. Place the flasks on an orbital or a reciprocating shaker at 30 °C and 150 r.p.m. for *ca.* 15 h.

Biotransformation

1. Add 2.5 mL of solution D to each flask.[c] After 1 h stirring, adjust the pH to 7, then add to each flask 2.5 mL of an ethanolic solution containing 1 g of substrate **8** solubilized in 5 mL ethanol (see Fig. 5.3.2). Keep on stirring in the culture conditions.
2. Take periodically a 1-mL aliquot. Extract it with 1 mL of ethyl acetate and analyse the organic phase by GC until complete disappearance of the substrate **8** (7–24 h).[d,e]
3. Combine the two biotransformation media and quench the reaction medium with HCl solution adjusting the pH to *ca.* 3. Extract the aqueous medium with dichloromethane continuously for 24 h (see Fig. 5.3.2).
4. Remove the solvent on a rotary evaporator under reduced pressure.
5. If necessary, purify the residue on a silica gel chromatography column (eluent pentane–ethyl acetate 1:1 v/v)[f] and/or by bulb-to-bulb distillation to yield 780 mg of (3R,6S)-(−)-3-methyl-6-isopropenyl-2-oxooxepane in its hydroxy acid form **10** as a yellow oil (80% yield). The ^{13}C NMR spectrum in CDCl$_3$ is 144.4 (C), 114.0 (CH$_2$), 63.9 (CH$_2$), 49.8 (CH), 39.2 (CH), 31.0 (CH$_2$), 26.5 (CH$_2$), 18.6 (CH$_3$), 16.6 (CH$_3$), and the optical rotation $[\alpha]_D^{29} = -6.6°$ (c = 1.8, CHCl$_3$).

[a] Near a Bunsen burner, with sterile syringes.
[b] Preparation of agar slopes: suspend 2.3 g Bacto Nutrient (Difco) in 10 mL distilled or deionized water and heat to boiling to dissolve completely. Place *ca.* 10 mL of this solution into culture tubes sealed with cotton. Sterilize, then cool them in an inclined position.
[c] Work under sterile conditions is not necessary any more.
[d] At this point of the experiment, the reaction products may be scarcely visible by GC analysis because they may be inside the cells.
[e] The time and the success of this experiment depend on the quality of the aeration. Too low oxygenation may lead to the partial formation of corresponding dihydrocarveols.
[f] GC conditions: oven 180 °C, 1.2 bar He, $t_{R\ ketone}$ = 4.6 min, $t_{R\ lactone}$ = 11 min, $t_{R\ hydroxy\ acid}$ = 16 min; TLC conditions: SiO$_2$, eluent ethyl acetate/pentane/EtOH 15:20:1, stain reagent I$_2$, $R_{f\ ketone}$ = 0.9, $R_{f\ alcohol}$ = 0.85, $R_{f\ lactone}$ = 0.8, $R_{f\ hydroxy\ acid}$ = 0.4.

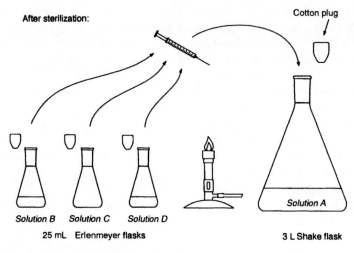

Fig. 5.3.1 Preparation of the culture medium.

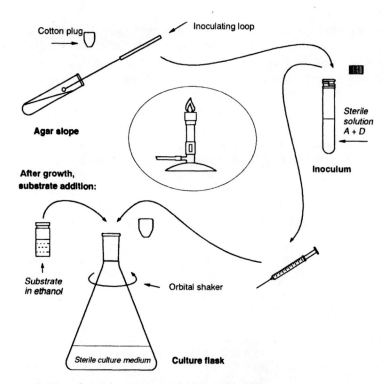

Fig. 5.3.2 Culture of *A. calcoaceticus* and biotransformation.

5: Oxidation using a biocatalyst

It is interesting to note that the product thus obtained, which results from the transformation of a natural substrate by a natural catalyst (the enzyme), can still be regarded—at least in view of the European legislation concerning food additives (aromas, flavours, etc.)—as being 'natural'. This is a very important issue for the food and flavour industry since there obviously is a high preference from customers for ingredients labelled 'natural' instead of 'synthetic'.

4.2 BV oxidation of racemic ketones

According to the substrate and the microorganism, different types of reaction can be obtained when starting from a racemic compound. One can observe the formation of (a) a mixture of two different regioisomeric lactones (see Scheme 5.3.2), (b) a mixture of a single optically active lactone (resulting from the preferential—if not total—oxidation of one enantiomer) and of the residual ketone which can, in certain cases, be reduced into the corresponding alcohol (see examples in Scheme 5.3.6[14] and Protocol 3).

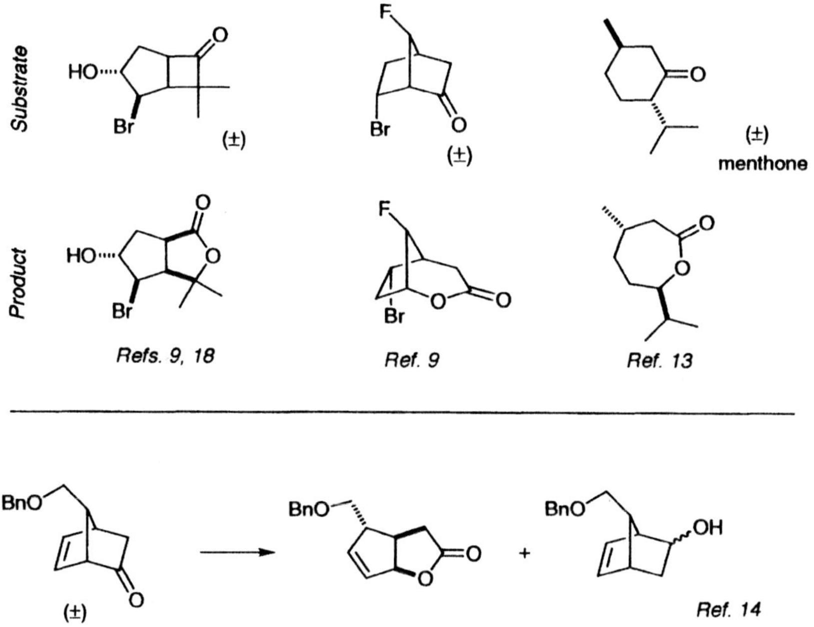

Scheme 5.3.6

The former case (a) is illustrated, for example, by the biooxidation of **3** by the bacteria *Acinetobacter* TD63,[4,15] *A. calcoaceticus*,[15] or different purified enzymes.[9] As pointed out previously, this transformation, which we described in 1989, has become over the years *the* standard reaction to test Baeyer–Villigerase activity in many microorganisms[12,16,17] (see Protocol 2). More generally, several bicyclic [n.2.1] compounds (i.e. bearing additional substituents such as alkyl groups α of the carbonyl group[18,19] and/or an oxygen atom or a double bond in the five- or six-membered ring[15,20]), as well as the racemic dihydrocarvone,[13] were tested with different microorganisms or enzymes and reacted in a similar manner.

Protocol 2.
Analytical biotransformation of bicyclo[3.2.0]hept-2-ene-6-one with A. calcoaceticus[a] (Structures 3, 4, and 5, Scheme 5.3.2)

Caution! Carry out all procedures in a well-ventilated hood and wear disposable gloves and safety goggles.

Equipment

- Inoculating loop
- Autoclave
- Bunsen burner
- Sterile syringes and needles (>5 mL)
- Thermostated (orbital or reciprocating) shaker
- Two autoclavable culture tubes with cap
- Erlenmeyer or shake flasks (2 × 3 L) with cotton plug

- Erlenmeyer flasks (3 × 25 mL) with cotton plug
- Gas chromatography apparatus equipped with an FID detector
- Water-cooled condenser
- Round-bottomed flask (250 mL)
- Heating mantle
- Liquid–liquid continuous extractor for heavy solvents (1 L or 2 × 500 mL)

Materials

- *Acinetobacter calcoaceticus* NCIMB 9871 strain from the National Collection of Industrial and Marine Bacteria (UK)   **Class 2 pathogen**
- Distilled or deionized water, 1.5 L
- Yeast extract (Difco), 0.2 g
- Sodium hydrogenphosphate (Na_2HPO_4), 4 g
- Potassium dihydrogenphosphate (KH_2PO_4), 2 g
- $(NH_4)_2SO_4$, 3 g
- Magnesium sulfate, 0.5 g
- Calcium chloride, 0.1 g
- Iron(II) sulfate, 0.01 g   **irritant**
- *cis/trans*-1,2-Cyclohexanediol, 2.5 g (or cyclohexanol)
- Sodium hydroxide, 5 N   **corrosive**
- Ethanol, 5 mL   **flammable**
- (±)-Bicyclo[3.2.0]hept-2-en-6-one (Merck), 1 g
- Ethyl acetate, 10 mL   **flammable**
- A chiral Lipodex D capillary column (0.25 mm × 50 m, Macherey–Nagel).

Culture

As described in Protocol 1.[b]

Biotransformation

1. As 'Biotransformation, step 1' in Protocol 1.
2. Take periodically a 1 mL aliquot. Extract it with 1 mL of ethyl acetate[c] and analyse the organic phase by GC (oven: 170°C, injector and detector 250°C,

5: Oxidation using a biocatalyst

carrier gas He 1.4 bar; Retention times: 13.5 min for the 'normal' (−)-lactone **4** and 12 min for the 'abnormal' (−)-lactone **5**).[d]

[a] This experiment may be carried out with other microorganisms as long as the appropriated culture medium is chosen.
[b] It is possible to carry out this experiment on a smaller scale (for example, 100 mL of culture medium in a 500-mL shake flask).
[c] The concentrations of different compounds can be calculated if an internal standard is present in ethyl acetate.
[d] With other microorganisms, lactones with the opposite absolute configuration (t_R = 13 min for the 'normal' (+)-lactone **4** and t_R = 15 min for the 'abnormal' (+)-lactone **5**) could be obtained.

The case (b), i.e. the formation of one single enantiopure lactone from a racemic compound, is illustrated by the oxidation of bicyclo[3.2.0]hept-2-en-6-one **3** by the fungus *Cunninghamella echinulata*.[21] Interestingly, this only leads to the 'abnormal' lactone **5** in >95% ee and 30% yield. This can be quite advantageous compared to the oxidation by *A. calcoaceticus* since the purification procedure is made much easier* (Scheme 5.3.7, Protocol 3).

Protocol 3.
Biotransformation of bicyclo[3.2.0]hept-2-ene-6-one with *C. echinulata* (Structures 3 and 5, Scheme 5.3.7)

Caution! Carry out all procedures in a well-ventilated hood and wear disposable gloves and safety goggles.

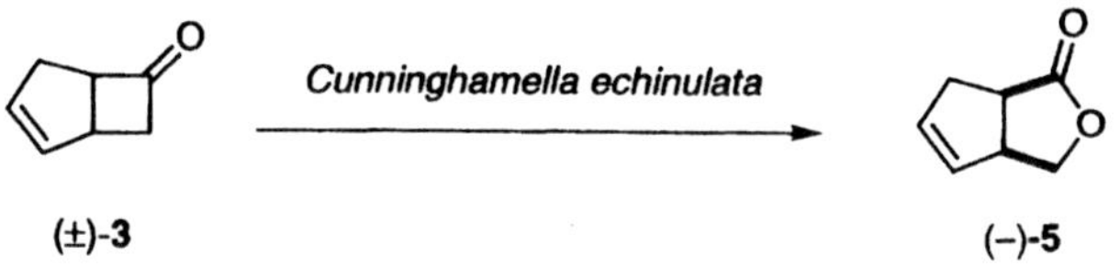

(±)-3   *Cunninghamella echinulata* →   (−)-5

Scheme 5.3.7

This is an example of a biotransformation with fungal resting cells.

Equipment

- The same equipment as Protocol 1 except for the 25-mL Erlenmeyer flasks
- Glass beads (1–2 mm, 5 g)
- Paper filters
- Filtering flask
- Porcelain Buchner filter funnel
- Glass chromatography column

Materials

- *Cunninghamella echinulata* NRRL 3655 strain from the Northern Regional Research Laboratories (USA)
- Tween 80, aqueous solution 0.5%, 20 mL
- Corn steep liquor (Roquette), 20 g
- Glucose, 5 g
- Potassium hydrogenphosphate (K_2HPO_4), 2 g
- Potassium dihydrogenphosphate (KH_2PO_4), 1 g

---

*These two regioisomeric lactones are difficult to separate by column (or flash) chromatography.

Protocol 3. Continued

- (±)-Bicyclo[3.2.0]hept-2-ene-6-one (Merck), 0.5 g, 4.6 mmol
- A chiral Lipodex D capillary column (0.25 mm × 50 m, Macherey–Nagel).
- Dichloromethane, 0.5 L   **harmful by inhalation**
- Ethanol, 5 mL   **flammable**
- Pentane   **flammable**
- Ethyl acetate   **flammable**
- Hydrochloric acid, 2 M   **corrosive**
- Silica gel for flash chromatography   **harmful by inhalation**

Culture

1. Prepare a 20-mL aqueous solution of 0.5% Tween 80 in a sealed culture tube. Prepare a solution containing corn steep liquor and glucose in 1 L of tap water. Divide it between the two 3-L flasks, seal them with a cotton plug. Sterilize them (autoclave 110 bar, 15 min).
2. Under sterile conditions, add glass beads and the Tween solution into the tube containing a 3-week-old agar slope of *C. echinulata*, shake it gently by hand to detach the spores from the solid agar, then, with a sterile syringe, divide this inoculum between the two 3-L flasks.
3. Place the flasks on an orbital or a reciprocating shaker at 27 °C and 100 r.p.m. for *ca.* 60 h.

Biotransformation

1. Filter the culture medium through two superposed paper filters on a Buchner funnel and wash the mycelium with tap water. Scramble this solid fungal cake into a 3-L shake flask with 0.5 mL of phosphate buffer (Na_2HPO_4, KH_2PO_4, pH 7). Add 0.5 g of bicycloheptenone **3** solubilized in 2.5 mL ethanol.[a] Keep on stirring (24–30 h) in the culture conditions.
2. Take periodically a 1 mL aliquot. Extract it with 1 mL of ethyl acetate and analyse the organic phase by GC (see Protocol 2) until constant concentrations of the final products.[b]
3. Follow steps 3 and 4 of Protocol 1.
4. Prepare a silica gel chromatography column and load the column with pentane–ethyl acetate mixture (4:1 v/v) until elution of the desired lactone (eluates analysed by GC). Evaporate the eluates containing the lactone and purify the residue by bulb-to-bulb distillation to yield 270 mg of (−)-(1*R*,5*S*)-**5** as a colourless oil [30% yield, >95% ee, $[\alpha]_D^{20} = -68°$ ($c = 2$ $CHCl_3$)].[c]

[a] By increasing the number of culture flasks, higher quantities of product can be prepared.
[b] (+)-(1*S*,4*R*)-**3** remains unchanged in the medium generally, but corresponding alcohols may be formed during the reaction.
[c] The purification step must be done with care because of the extreme volatility of the products.

It should be noted that, in the classical manner for such a biocatalysed resolution, the enantiomeric excess (ee) of the lactone formed and of the residual

5: Oxidation using a biocatalyst

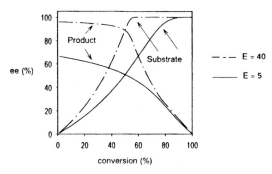

Fig. 5.3.3 Curves showing the enantiomeric excesses of a substrate and a product versus the conversion ratio (% disappeared substrate) for different enantiomeric ratios, E.

ketone changes as the reaction proceeds. This evolution is described by Sih's equations[22] and is characterized by the 'E-value', called the enantiomeric ratio (see Fig. 5.3.3).

One of the main limitations of the use of a whole-cell culture procedure is the possible presence in the microorganism of other enzymes that could metabolize the product. Thus, cyclohexanol- or cyclohexanediol-grown *A. calcoaceticus* contains a lactone hydrolase able to hydrolyse some α-substituted δ-valerolactones[23,24] or ε-caprolactones.[25] This problem has been solved by (a) the use of a microorganism lacking this enzyme, such as cyclohexanediol-grown *Acinetobacter* TD63[23–25] or *Pseudomonas putida* NCIMB 10007,[9] (b) addition into the reaction mixture of an inhibitor of this enzyme,[23,24] and (c) the use as biocatalyst of a microorganism 'engineered' to overexpress the cyclohexanone monooxygenase.[10,11]

4.3 BV oxidation of prochiral ketones

As far as synthetic goals are concerned, the best outcome of any asymmetric synthesis should, of course, be obtained using prochiral substrates. Indeed, this would allow the production of a theoretical 100% yield of the product. Thus, the ketones depicted in Scheme 5.3.8[26,27,28,29] were transformed in good yields and ees using whole cells or purified enzymes. For example, starting from various 3-substituted cyclobutanones (which can be prepared in two steps from commercial material) and using *C. echinulata*, it is possible to prepare some enantiopure 3-substituted γ-butyrolactones that are highly valuable synthons for further synthesis of biologically active products such as Baclofen®[30] and β-proline.[31]

5. Conclusion

In conclusion, it appears that the biocatalysed Baeyer–Villiger oxidation of ketones is at present an excellent method—if not the best—for performing this type of reaction in an asymmetric manner. It can be carried out using with purified enzymes or whole cells:

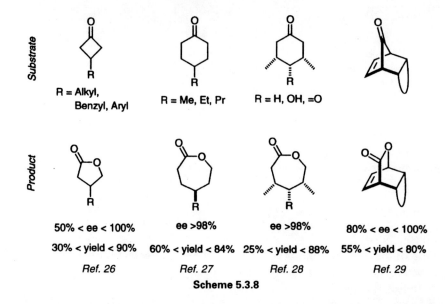

Scheme 5.3.8

either whole cells or purified enzymes, but the whole-cell culture method is undoubtedly the best at the present time for preparative-scale synthesis. Owing to the relative simplicity of growing bacteria or fungi, these reactions can also be used and applied by organic chemists, provided they invest some time learning elementary and simple microbiology techniques.

References

1. Alphand, V.; Furstoss, R. Baeyer–Villiger monooxygenases. In *Enzyme Catalysis in Organic Synthesis*; Drauz, K.; Waldmann, H., eds.; VCH: Weinheim, **1995**, pp. 755–762.
2. Willetts, A. *Trends Biotechnol.* **1997**, *15*, 55–62.
3. Ouazzani-Chahdi, J.; Buisson, D.; Azerad, R. *Tetrahedron Lett.* **1987**, *28*, 1109–1112.
4. Alphand, V.; Archelas, A.; Furstoss, R. *Tetrahedron Lett.* **1989**, *30*, 3663–3664.
5. Donoghue, N. A.; Trudgill, P. W. *Eur. J. Biochem.* **1975**, *60*, 1–7.
6. Walsh, C. T.; Chen, Y.-C. J. *Angew. Chem., Int. Ed. Engl.* **1988**, *27*, 333 and references therein.
7. Alphand, V.; Gaggero, N.; Colonna, S.; Pasta, P.; Furstoss, R. *Tetrahedron* **1997**, *53*, 9695–9706 and references therein.
8. Pasta, P.; Carrea, G.; Gaggero, N.; Grogan, G.; Willetts, A. *Biotechnol. Lett.* **1996**, *18*, 1123–1128.
9. Gagnon, R.; Grogan, G.; Levitt, M. S.; Roberts, S. M.; Wan, P. W. H.; Willetts, A. J. *J. Chem. Soc., Perkin Trans. 1* **1994**, 2537–2543.
10. Stewart, J. D.; Reed, K. W.; Kayser, M. M. *J. Chem. Soc., Perkin Trans. 1* **1996**, 755–757.

11. Stewart, J. D.; Reed, K. W.; Zhu, J.; Chen, G.; Kayser, M. M. *J. Org. Chem.* **1996**, *61*, 7652–7653.
12. Roberts, S. M.; Turner, N. J.; Willetts, A. J.; Turner, M. K. *Introduction to Biocatalysis using Enzymes and Microorganisms*; Cambridge University Press: Cambridge, **1995**.
13. Alphand, V.; Furstoss, R. *Tetrahedron: Asymmetry* **1992**, *3*, 379–381.
14. Konigsberger, K.; Alphand, V.; Furstoss, R.; Griengl, H. *Tetrahedron Lett.* **1991**, *32*, 499–500.
15. Alphand, V.; Furstoss, R. *J. Org. Chem.* **1992**, *57*, 1306–1309.
16. Shipston N. F., Lenn M. J., Knowles C. J. *J. Microbiol. Methods* **1992**, *15*, 41–52.
17. Kelly, D. R.; Knowles, C. J.; Mahdi, J. G.; Wright, M. A.; Taylor, I. N.; Roberts, S. M.; Wan, P. W. H.; Grogan, G.; Pedragosa-Moreau, S.; Willetts, A. J. *J. Chem. Soc., Chem. Commun.* **1996**, 2333–2334.
18. Lenn M. J., Knowles C. J. *Enzyme Microb. Technol.* **1994**, *16*, 964–969.
19. Carnell, A. J.; Roberts, S. M.; Sik, V.; Willetts, A. J. *J. Chem. Soc., Perkin Trans. 1* **1991**, 2385–2389.
20. Petit, F.; Furstoss, R. *Tetrahedron: Asymmetry* **1993**, *4*, 1341–1352.
21. Lebreton, J.; Alphand, V.; Furstoss, R. *Tetrahedron* **1997**, *53*, 145–160.
22. Chen, C. S.; Fujimoto, Y.; Sih, C. J. *J. Am. Chem. Soc.* **1981**, *103*, 3580.
23. Alphand, V.; Archelas, A.; Furstoss, R. *Biocatalysis* **1990**, *3*, 73–83.
24. Alphand, V.; Archelas, A.; Furstoss, R. *J. Org. Chem.* **1990**, *55*, 347–350.
25. Alphand, V.; Furstoss, R.; Pedragosa-Moreau, S.; Roberts, S. M.; Willetts, A. J. *J. Chem. Soc., Perkin Trans. 1* **1996**, 1867–1871.
26. Gagnon, R.; Grogan, G.; Groussain, E.; Pedragosa-Moreau, S.; Richardson, P. F.; Roberts, S. M.; Willetts, A. J.; Alphand, V.; Lebreton, J.; Furstoss, R. *J. Chem Soc., Perkin Trans. 1* **1995**, 2527–2528.
27. Taschner, M. J.; Black, D. J.; Chen, Q.-Z. *Tetrahedron: Asymmetry* **1993**, *4*, 1387–1390.
28. Taschner, M. J.; Black, D. J. *J. Am. Chem. Soc* **1988**, *110*, 6892–6893.
29. Taschner, M. J.; Peddada, L. *J. Chem. Soc., Chem Commun.* **1992**, 1384–1385.
30. Mazzini, C.; Lebreton, J.; Alphand, V.; Furstoss, R. *Tetrahedron Lett.* **1997**, *38*, 1195–1198.
31. Mazzini, C.; Lebreton, J.; Alphand, V.; Furstoss, R. *J. Org. Chem.* **1997**, *62*, 5215–5218.

5.4 Oxidation of sulfides

S. COLONNA, N. GAGGERO, G. CARREA, and P. PASTA

1. Introduction

Sulfoxides are versatile and convenient chiral synthons in asymmetric synthesis, in particular for enantioselective carbon–carbon bond formation.[1] Their value is further illustrated by the involvement of the sulfinyl functionality in different biological activities and in products of great pharmaceutical interest.[2] In recent years, good to excellent enantioselectivities have been achieved for sulfoxidation catalysed by isolated enzymes such as FAD-dependent monooxygenase from pig-liver[3] or from *Pseudomonas sp.*,[4] toluene and naphthalene dioxygenases[5] and a dioxygenase from *Pseudomonas putida*.[6] Biotransformations with whole cells have mainly employed fungi such as *Aspergillus niger*,[7] *Mortierella isabellina*,[8] *Helminthosporium sp.*,[9] the bacterium *Corinebacterium equi*[10] and, very recently, baker's yeast.[11] The physiological role of enzymes which effect sulfide oxidation is to metabolize xenobiotics which may be harmful to the host organism; sulfoxides are more hydrophilic than sulfides and are excreted more readily. Monooxygenases are either cytochrome P-450 or flavin-dependent enzymes and mediate the transfer of oxygen to the substrate, although the detailed mechanism by which this occurs has yet to be firmly established.

In this chapter we will describe two enzymatic approaches in the oxidation of organic sulfides to optically active sulfoxides based on the use of chloroperoxidase and cyclohexanone monooxygenase.

2. Chloroperoxidase-catalysed oxidation of methyl *p*-tolyl sulfide to (*R*)-methyl *p*-tolyl sulfoxide

Chloroperoxidase (CPO) is a heme glycoprotein, isolated from the marine fungus *Caldariomyces fumago*, which exhibits a wide reactivity,[12] including the peroxidative formation of carbon–halogen bonds, the oxidation of primary aryl amines to hydroxylamines and the epoxidation of olefins. Following a study of Kobayashi *et al.*[13] we investigated the use of this enzyme in the asymmetric sulfoxidation of numerous alkyl aryl sulfides that are converted to the corresponding sulfoxides with high enantioselectivity.[14] The versatility of CPO in promoting enantioselective sulfoxidations has also been exploited recently with a series of bicyclic sulfides[15] and, even more significantly, with numerous dialkyl sulfides.[16]

The general oxidation procedure is exemplified by the preparation of (*R*)-methyl *p*-tolyl sulfoxide (Scheme 5.4.1), a commercially used source of α-sulfinyl carbanion that is then reacted with electrophiles.

Protocol 1.
Synthesis of (*R*)-methyl *p*-tolyl sulfoxide (Scheme 5.4.1)

Caution! Carry out all procedures in a well-ventilated hood and wear disposable gloves and safety goggles.

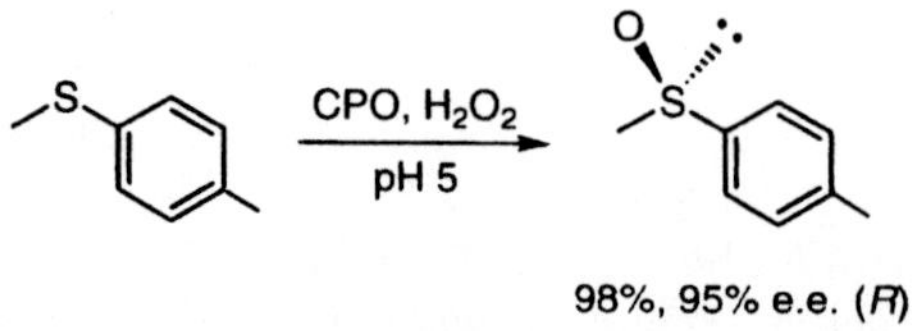

98%, 95% e.e. (*R*)

Scheme 5.4.1

5: Oxidation using a biocatalyst

Equipment
- Magnetic stirrer
- Gilson pipette (100 µL)
- Disposable plastic syringe (1 mL)

Materials
- Methyl p-tolyl sulfide (Aldrich), 69 mg, 0.5 mmol — **stench, liquid**
- Chloroperoxidase from *Caldariomyces fumago*, Sigma suspension (RZ 0.6), 80 µL, 300 units
- Hydrogen peroxide (30 wt% solution in water), 0.1 mL, 1 mmol — **oxidizer, corrosive**
- Sodium citrate buffer, 0.1 M, pH 5, 45 mL
- Saturated aqueous sodium sulfite
- Diethyl ether — **flammable**
- Dichloromethane — **irritant**

Method

1. Place methyl p-tolyl sulfide and the buffer solution (45 mL) in a 100-mL round-bottomed flask equipped with a magnetic stirrer bar. Stir the mixture, then add the chloroperoxidase suspension using the Gilson pipette.[a]

2. Dilute the hydrogen peroxide solution (0.1 mL) with buffer solution (4.5 mL). Add this solution in 13 aliquots of 0.35 mL using the disposable syringe, at 5 min intervals.[b]

3. Once all the hydrogen peroxide has been added, stir for a further 5 min, then add aqueous sodium sulfite (10 mL) to quench any excess oxidant. Extract the mixture with diethyl ether (2 × 50 mL) and dichloromethane (2 × 50 mL), dry over sodium sulfate and concentrate *in vacuo* to give a white solid.

4. Purify the material by flash chromatography on silica gel, eluting with ethyl acetate, to give (*R*)-methyl p-tolyl sulfoxide as an amorphous white solid (75 mg, 98% yield), m.p. 75°C, Lit.[17] 73–75°C.; $[\alpha]_D^{25} = +185°$ (c = 1.19, CHCl$_3$), Lit.[18] +192° (c = 1.2, CHCl$_3$); ^{1}H NMR (300 MHz, CDCl$_3$): 7.54 (2H, d, J = 8.2 Hz), 7.33 (2H, d, J = 8.2 Hz), 2.70 (3H, s), 2.41 (3H, s).

[a] The sulfide does not dissolve completely in the aqueous buffer, but becomes dispersed as oily droplets when the mixture is stirred.
[b] To monitor the reaction by TLC, remove 0.2 mL of the reaction mixture and shake with 0.5 mL of ether. Analyse the top layer, developing the plate in ethyl acetate. The starting material has an R_f of 0.73 and the product an R_f of 0.19 (vizualization by UV at 254 nm).

The chiral purity of (*R*)-methyl p-tolyl sulfoxide was determined by chiral HPLC using a Chiralcel OB column (25 cm × 0.46 cm i.d.; Daicel) and a Jasco HPLC instrument (model PU-980 pump and model UV-975 detector). The mobile phase was *n*-hexane–2-propanol 85:15 and the flow rate 1 mL/min, with reading at 254 nm. The data were computed by a HP-3395 integrator (Hewlett Packard). Figure 5.4.1 shows that the product obtained had a high ee value (95%). The absolute configuration was assigned by chiral HPLC comparison with an authentic sample obtained from Aldrich.

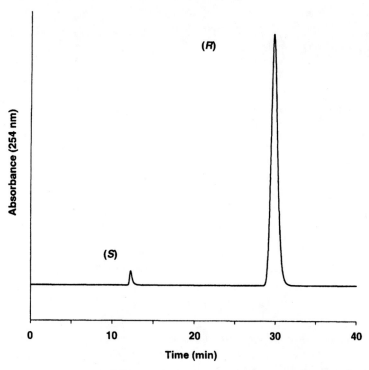

Fig. 5.4.1 Chiral HPLC of (R)-methyl p-tolyl sulfoxide obtained by CPO catalysis.

The enzymatic procedure is easy to carry out and provides a mild and convenient alternative to non-enzymatic asymmetric oxidation methods. The key to success when carrying out this reaction is to maintain a low concentration of hydrogen peroxide by adding this reagent slowly. In this way the uncatalysed oxidation, producing racemic sulfoxide, is largely suppressed.

3. Cyclohexanone monooxygenase-catalysed oxidation of 1,3-dithiane to (R)-1,3-dithiane monosulfoxide

Cyclohexanone monooxygenase (CHMO) from *Acinetobacter calcoaceticus* NCIMB 9871 is a flavoenzyme active as a monomer which contains one firmly but non-covalently bound FAD unit per monomer.[19] As reported in Section 5.3 it has wide potential application in organic synthesis based on the Baeyer–Villiger reaction.[19] Walsh and Chen[19] described the synthesis of (S)-ethyl p-tolyl sulfoxide using CHMO (64% ee).

We have extended the study of stereochemistry of oxidation at sulfur using numerous alkyl aryl sulfides,[20] the optical purity of products ranging from 98% ee and (R)-configuration to 93% ee and (S)-configuration. The versatility of cyclohexanone monooxygenase from *Acinetobacter* is further exemplified by its ability to promote enantioselective oxidation of functionalized sulfides,[21] alkyl benzyl sulfides,[22] and dialkyl sulfides.[16] We have proposed an active-site model of the enzyme to explain the stereoselectivity of the sulfoxidation and to predict the absolute configuration of the prevailing sulfoxide.[23]

The general oxidation procedure is exemplified by the preparation of 1,3-dithiane monosulfoxide (Scheme 5.4.2) which is obtained with 100% optical purity.[24] This is a major finding since 1,3-dithioacetal monosulfoxides serve as chiral acyl anion equivalents in a range of synthetically useful transformations such as enolate alkylation and amination, Mannich reactions, organometallic additions, and so on.[25]

Protocol 2.
Synthesis of (*R*)-1,3-dithiane monosulfoxide (Scheme 5.4.2)

Caution! Carry out all procedures in a well-ventilated hood and wear disposable gloves and safety goggles.

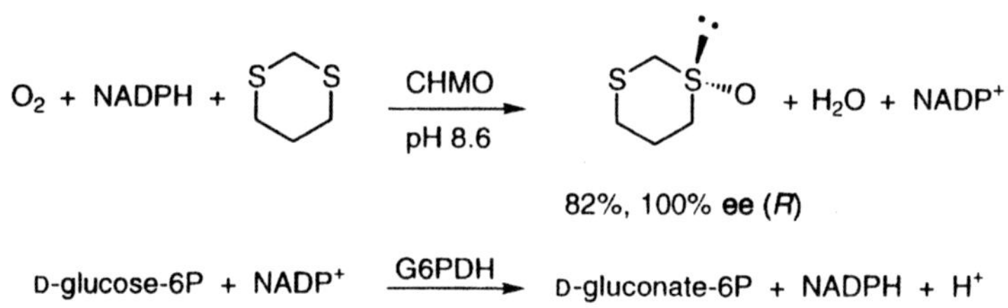

$$O_2 + NADPH + \text{1,3-dithiane} \xrightarrow[\text{pH 8.6}]{\text{CHMO}} \text{sulfoxide} + H_2O + NADP^+$$

82%, 100% ee (*R*)

$$\text{D-glucose-6P} + NADP^+ \xrightarrow{\text{G6PDH}} \text{D-gluconate-6P} + NADPH + H^+$$

Scheme 5.4.2

Equipment
- Magnetic stirrer
- Gilson pipette (100 µL)
- Disposable plastic syringe (1 mL)

Materials
- 1,3-Dithiane (Aldrich), 60 mg, 0.5 mmol — **flammable solid**
- Cyclohexanone monooxygenase from *Acinetobacter calcoaceticus*, purified as described by Latham and Walsh,[26] 10 mg, 60 units
- Glucose-6-phosphate dehydrogenase from *Leuconostoc mesenteroides*, Sigma lyophilized powder, 2 mg, 400 units
- Glucose-6-phosphate disodium salt (Sigma), 450 mg, 1.5 mmol
- NADP disodium salt, 15 mg, 0.02 mmol — **irritant**
- Tris(hydroxymethyl)aminomethane (Tris)–HCl buffer, 0.05 M, pH 8.6, 15 mL — **irritant**
- Diethyl ether — **flammable**

Method

1. Place cyclohexanone monooxygenase, glucose-6-phosphate dehydrogenase,[a] glucose-6-phosphate,[a] NADP, and buffer solution (15 mL) in a 50-mL round-bottomed flask equipped with a magnetic stirrer bar. Stir the mixture, then add 1,3-dithiane.[b]

Protocol 2 *Continued*

2. Stir for 16 h, then freeze-dry and extract the residue with 2-propanol (3 × 30 mL), dry over sodium sulfate and concentrate *in vacuo*.[c]

3. Purify the material by flash chromatography on silica gel, eluting with diethyl ether–methanol 9:1 to give (*R*)-1,3-dithiane monosulfoxide as an amorphous solid (55 mg, 82% yield); $[\alpha]_D^{25} = +214°$ ($c = 1.0$, CH_2Cl_2), Lit.[27] +210° ($c = 0.97$, CH_2Cl_2); ^{1}H NMR (60 MHz, $CDCl_3$) 3.84 (2H, q), 3.2 (1H, m), 2.8 (1H, m), 2.3 (4H, m).

---

[a] Glucose-6-phosphate and glucose-6-phosphate dehydrogenase serve to regenerate the coenzyme.[20]
[b] 1,3-Dithiane is only partially soluble in the aqueous buffer and becomes dispersed when the mixture is stirred.
[c] The conversion of 1,3-dithiane to monosulfoxide and monosulfone was determined by GLC with a CP-cyclodextrin-β-2,3,6 M19 capillary column (50 m × 0.25 mm i.d., Chromopack). Conditions: oven temperature from 100°C (initial time 1 min) to 190°C with a heating rate of 5°C/min, H_2 as carrier gas. The retention times for 1,3-dithiane, 1,3-dithiane monosulfoxide and 1,3-dithiane monosulfone were 12.0, 26.1, and 66.4 min, respectively.

The chiral purity of (*R*)-1,3-dithiane monosulfoxide was determined by chiral HPLC using a Chiralcel OD column (25 cm × 0.46 cm i.d.; Daicel) and a Jasco HPLC instrument (model PU-980 pump and model UV-975 detector).

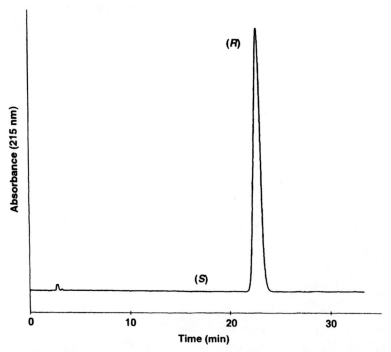

Fig. 5.4.2 Chiral HPLC of (*R*)-1,3-dithiane monosulfoxide obtained by CHMO catalysis.

5: Oxidation using a biocatalyst

The mobile phase was n-hexane–2-propanol 80:20 and the flow rate 1 mL/min, with reading at 215 nm. The data were computed by an HP-3395 integrator (Hewlett Packard). Figure 5.4.2 shows that the optical purity of the (R)-product was practically complete as the presence of the enantiomer with (S)-configuration was not detected. The absolute configuration was assigned by chiral HPLC comparison with a specimen of (R)-configuration prepared according to Kagan's procedure (see Chapter 4, Section 4.1).[28]

The enzymatic oxidation of 1,3-dithiane by CHMO which gives the (R)-monosulfoxide in 82% chemical yield compares favourably with chemical process in terms of enantioselectivity. Indeed, Kagan obtained (R)-1,3-dithiane monosulfoxide in 20% ee.[28] The oxidation of 1,3-dithiane to the corresponding monosulfoxide in the presence of growing cultures of *Aspergillus foetidus*, *Helmintosporium* species, and *Mortierella isabellina* occurred in 17–24, 13–15, and 0% ee, respectively.[29] Interestingly, the prevailing enantiomers obtained with the two fungi have different absolute configuration.[29]

With racemic 1,3-dithiane monosulfoxide the time course of the reaction with cyclohexanone monooxygenase was monitored by GLC (conversion) and HPLC (ee). The (S)-enantiomer was oxidized to the corresponding monosulfone faster than the (R)-enantiomer, the E value being 20 (the enantiomer ratio, E, was calculated according to Chen *et al.*[30]). As a consequence, the asymmetric synthesis [$v_R/v_S = 12$, i.e. the ratio of the rates of formation of (R)- and (S)-sulfoxides, determined at low degrees of conversion by chiral HPLC] accompanied by kinetic resolution (see Scheme 5.4.3) led to enantiomerically pure (R)-1,3-dithiane monosulfoxide.

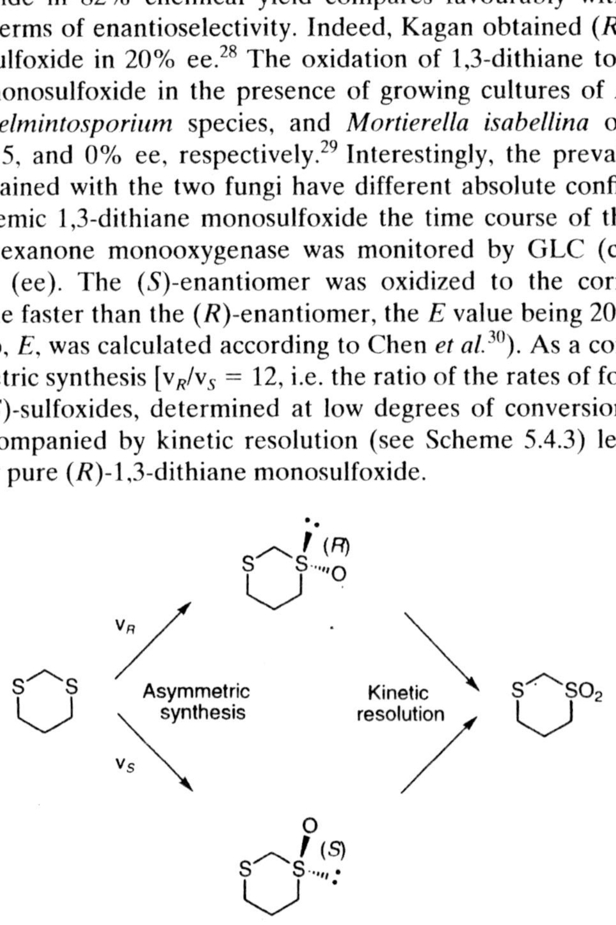

Scheme 5.4.3 Pathway of the CHMO-catalysed oxidation of 1,3-dithiane to 1,3-dithiane monosulfoxide and 1,3-dithiane monosulfone.

In conclusion, CHMO appears to be the catalyst of choice in the chemical and biochemical repertory for the enantioselective monosulfoxidation of 1,3-dithiane.

Acknowledgements

This work was partially supported by C.N.R. M.U.R.S.T. (Ministero dell'Università e della Ricerca Scientifica e Tecnologica), Programmi di Ricerca Scientifica di Interesse Nazionale 'Processi efficienti per l'ossidazione controllata di composti organici.'

References

1. Walker, A. J. *Tetrahedron: Asymmetry* **1992**, *3*, 961–998 and references therein.
2. Carreño, M. C. *Chem. Rev.* **1995**, *95*, 1717-1760.
3. Light, D. H.; Waxman, D. J.; Walsh, C. T. *Biochemistry* **1982**, *21*, 2490–2498.
4. Katopodis, A. G.; Smith, H. A.; May, S. W. *J. Am. Chem. Soc.* **1988**, *110*, 897–899.
5. Klee, K.; Brand, J. M.; Gibson, D. T. *Biochem. Biophys. Res. Commun.* **1995**, *212*, 9–15.
6. Allen, C. C. R.; Boyd, D. R.; Dalton, H.; Sharma, N. D.; Haughey, S. A.; McMordie, R. A. S.; McMurray, B. T.; Sheldrake, G. N.; Sproule, K. *J. Chem. Soc, Chem. Commun.* **1995**, 119–120.
7. Holland, H. L. *Chem. Rev.* **1988**, *88*, 473–485.
8. Holland, H. L.; Pöpperl, H.; Ninniss, R. W.; Chenchaiah, P. C. *Can. J. Chem.* **1985**, *63*, 1118–1120.
9. Abushanab, E.; Reed, D.; Suzuki, F.; Sih, C. J. *Tetrahedron Lett.* **1978**, *37*, 3415–3418.
10. Ohta, H.; Kato, Y.; Tsuchihashi, G. *Chem. Lett.* **1986**, 581–584.
11. Tang, J.; Brackenridge, I.; Roberts, S. M.; Beecher, J.; Willetts, A. J. *Tetrahedron* **1995**, *51*, 13217–13238.
12. Thomas, J. A.; Morris, D. R.; Hager, L. P. *J. Biol. Chem.* **1970**, *245*, 3129–3134.
13. Kobayashi, S.; Nakano, M.; Kimura, T.; Schaap, A. P. *Biochemistry* **1987**, *26*, 5019–5022.
14. Colonna, S.; Gaggero, N.; Manfredi, A.; Casella, L.; Gullotti, M.; Carrea, G.; Pasta, P. *Biochemistry* **1990**, *29*, 10465–10468; Colonna, S.; Gaggero, N.; Casella, L.; Carrea, G.; Pasta, P. *Tetrahedron: Asymmetry* **1992**, *3*, 95–106.
15. Allenmark, S. G.; Andersson, M. A. *Tetrahedron: Asymmetry* **1996**, *7*, 1089–1094.
16. Colonna, S.; Gaggero, N.; Carrea, G.; Pasta, P. *J. Chem. Soc., Chem. Commun.* **1997**, 439–440.
17. Aldrich *Catalogue Handbook of Fine Chemicals*, **1996–1997**.
18. Solladié, G.; Hutt, J.; Girardin, A. *Synthesis* **1987**, 173.
19. Walsh, C. T.; Chen, Y. C. J. *Angew. Chem., Int. Ed. Engl.* **1988**, *27*, 333–343.
20. Carrea, G.; Redigolo, B.; Riva, S.; Colonna, S.; Gaggero, N.; Battistel, E.; Bianchi, D. *Tetrahedron: Asymmetry* **1992**, *3*, 1063–1068.
21. Secundo, F.; Carrea, G.; Dallavalle, S.; Franzosi, G. *Tetrahedron: Asymmetry* **1993**, *4*, 1981–1982.
22. Pasta, P.; Carrea, G.; Holland, H.L.; Colonna, S.; Dallavalle, S. *Tetrahedron: Asymmetry* **1995**, *6*, 933–936.
23. Ottolina, G.; Pasta, P.; Carrea, G.; Colonna, S.; Dallavalle, S.; Holland, H. L. *Tetrahedron: Asymmetry* **1995**, *6*, 1375–1386.
24. Colonna, S.; Gaggero, N.; Pasta, P.; Ottolina, G. *J. Chem. Soc., Chem. Commun.* **1996**, 2303–2306 and references therein.
25. Page, P. C. B.; Allin, S. M.; Colligton, E. W.; Caw, R. E. *Tetrahedron Lett.* **1994**, *35*, 2607–2608 and references therein.
26. Latham, J. A.; Walsh, C. T. *J. Am. Chem. Soc.* **1987**, *109*, 3421–3427.

5: Oxidation using a biocatalyst

27. Page, P. C. B.; Gareh, M. T.; Porter, R. A. *Tetrahedron: Asymmetry* **1993**, *4*, 2139–12142.
28. Samuel, O.; Ronan, B.; Kagan, H. B. *J. Orgnomet. Chem.* **1989**, *370*, 43–50.
29. Auret, B. J.; Boyd, D. R.; Breen, F.; Greene, R. M. E.; Robinson, P. M. *J. Chem. Soc., Perkin Trans. 1* **1981**, 930–933.
30. Chen, C. S.; Fujimoto, Y.; Girdankas, G.; Sih, C. J. *J. Am. Chem. Soc.* **1982**, *104*, 7294–7299.

List of Suppliers

Aldrich Chemical Co. Inc.
USA: 1001 West Saint Paul Avenue, Milwaukee, WI 53233. Tel. (414) 273-3850; 800 558-9160; Fax: (414) 273-4979.

Aldrich-Chemie GmbH & Co KG
Germany: Mail: Post Box 1120, 89552 Steinheim. Ship: Riedstrasse 2; 89555 Steinheim. Tel. +49 7329 97 1230; Fax: +49 7329 97 1160.

Machinery-Nagel Society
Germany: PO Box 101352, D-52348 Düren.

National Collection of Industrial, Food and Marine Bacteria NCIMB Ltd.
UK: Torry Research Station, Aberdeen AB24 3RY, Scotland.

Northern Regional Research Laboratories
USA: Agricultural Research Service, US Department of Agriculture, 1815N. University St. Peoria, IL61604.

Sigma-Aldrich Japan K.K.
Japan: JL Nihonbashi Building, 1-10-15 Nihonbashi, Horidome-cho, Chuo-Ku, Tokyo 103. Tel. 81 3 5640 8850; 0120 07 0406; Fax: 81 3 5640 8855.

Index

Reactions

AA reaction 105
AD reaction 81, 84, 85, 87, *89*
aerobic epoxidation 33
allylic hydroxylation 189
allylic oxidation 10
aminohydroxylation 104, 105, 107, 109
asymmetrisation of prochiral or *meso*-diols to chiral lactones 200
autoxidation of enolates 129, 131
aziridination 115, 117, *120*, 122
Baeyer-Villiger (BV) oxidation 73, 75, 147, 214, 225
benzylic hydroxylation 6, 192
biohydroxylations at saturated carbon 183, *185*, 187, 189, 195
biooxidation of sulfides 227
catalytic epoxidation 19, 56, *57*, 59
C-H bond insertion 116
chiral phase-transfer catalysed epoxidation 70
C-H oxidation 5
desymmetrization of symmetrical substrate 192
desymmetrization of *meso* diallyl alcohols 64
dihydroxylation 5, 81, 82, 87, *89*, 105
dihydroxylation of enol ethers 143
dihydroxylation of silyl enol ethers 143
double asymmetric differentiation *48*
enantiotopic differentiation of the hydroxy functions
(desymmetrization of *meso* diols) 205
epoxidation 19, 20, 25, 33, 37, 45, 50, 51, 56, 70, 77
epoxidation of allylic alcohols 50, 51, 61
epoxidation of electron deficient olefins 70, 71
epoxidation of isolated olefins 20, 25, 33
epoxidation of α,β-unsaturated ketones 70, *78*
fungal hydroxylation 181
heteroatom dealkylation 187
hydroxylation 6, 128, 181, 183, 187, *189*, 195
hydroxylation of enolates 128
α-hydroxylation of chiral ketone enolate 128
α-hydroxylation of prochiral ketone enolate 138-140

hydroxylation of enol derivatives 141
hydroxylation of hydrocarbons 6, 181
hydroxylation of silyl enol ether 141
immobilisation 202, *208*
in situ derivatization of epoxy alcohols 59, *60*
Kharasch reaction 10
kinetic resolution 61-69, 175, 201-205, 210, 221-225, 233
kinetic resolution of β-hydroxy amines 175-179
kinetic resolution of phosphines 172
kinetic resolution of racemic α-furuyl toluenesulfonamides 66
kinetic resolution of racemic furyl and thienyl carbinols 66
kinetic resolution of racemic ketones 221-225
kinetic resolution of racemic secondary alcohols 61-69, 201-205
microbial hydroxylation 181
microbial oxidation of alcohols 210
oxidation of alcohols 200
oxidation of heteroatoms 153, 171
oxidation of imines 172, *173*
oxidation of nitrogen-containing compounds 172, 175
oxidation of lithium enolate 129, 132, 140
oxidation of saturated C-H bonds 5, 181
oxidation of selenides 153, 166
oxidation of sulfides 153, 163, 164, 227
N-oxide formation 171, 175, *177*
phase-transfer catalysed epoxidation 70-77
polyamino acid-catalysed chiral epoxidation 71-77
resolution of racemic secondary alcohols 61-69, 200-205, 210
selenoxidation 153, 166
sulfimidation 153
sulfoxidation 154-169, 227-233
sulfoxidations catalysed by chiral vanadium complex 165
sulfoxidations catalysed by titanium reagent 157-164

Index

Reagents

Absidia cylindrospora LCP 57.1569 *194*
Acinetobacter calcoaceticus (*A. calcoaceticus*) 215, *218, 222*, 225
Acinetobacter TD63 214, 221
AD-mix-α *90, 96*, 143
AD-mix-β *90, 96*, 143
alcohol dehydrogenase (ADH) 200, *203*
alkyl hydroperoxide 50, 54
Aspergillus niger 228
Bacillus stearothermophilus ATCC (American type culture collection) 2027 204, *211*
Baeyer-Villigerase 216
baker's yeast 228
benzenesulfonyl azide 116
benzyl carbamate 107, *109*
1,1-bi-2-naphthol(BINOL) 78, 162
(1*S*,2*S*)-bis(2,6-dichlorobenzylidenamino) cyclohexane 124
1,4-bis(dihydroquinidine)anthraquinone [(DHQD)$_2$AQN] 81, *89*, 105
1,4-bis(dihydroquinidine)phthalazine [(DHQD)$_2$PHAL] 81, 82, *89, 93, 95, 97*, 105, *109*
1,4-bis(dihydroquinine)anthraquinone [(DHQ)$_2$AQN] 81, *89*, 105
1,4-bis(dihydroquinine)phthalazine [(DHQ)$_2$PHAL] 81, *89*, 95, 97, *107, 112*
bis(imine) complexes 122
bis(imine) ligands 122, *124*
1,2:4,5-bis-*O*-isopropylidene-β-D-*erythro*-2,3-hexodiulo-2,6-pyranose *47*
bis(oxazoline) 12, 117, *120*
bis(oxazolinyl)pyridine 15
bis-(*S*)-prolinato-copper(II) complex 13
bovine liver catalase 203
N-bromoacetamide 1*12*
1,3-butanediol and galactose oxidase 203
t-butyl hydroperoxide (TBHP) 19, 51, *52, 58, 60, 63, 65, 68*, 70, 81, 157, 158, 163, 176, *178*
t-butyl hypochlorite 108, *109*
t-butyl peroxybenzoate *12, 14*
Caldariomyces fumago 228, *229*
(camphorylsulfonyl)oxaziridine *138*, 140
(+)-(camphorylsulfonyl)oxaziridine 138
(−)-(camphorylsulfonyl)oxaziridine 138
N-carboxyanhydride (NCA) 71, *73*
Caro's reagent 147
Cellulomonas sp. 202
Chloramine-T trihydrate 105, *107*
chloroperoxidase (CPO) 228, *229*
m-chloroperoxybenzoic acid (*m*-CPBA) *32*, 37, *41, 141*
N-chloro-*N*-sodio carbamates 107
N-chloro-*N*-sodio-*p*-toluenesulfonamide (chloramine-T) 105, *107*

cinchona alkaloids (quinine and quinidine) 70
combination of an aliphatic aldehyde and dioxygen 33, 148
copper(II) acetate 1*3*
copper(I) trifluoromethanesulfonate 12, *120*
copper(II) trifluoromethanesulfonate 117
copper(II)-tris(oxazoline) complexes 15
Corinebacterium equi 212, 228
Corynosporium cassiicola 212
Cu(MeCN)$_4$ClO$_4$ 117, 118
cumene hydroperoxide (CHP) *60, 79, 158, 159, 163*
Cunninghamella echinulata (*C. echinulata*) 195, *223*, 225
Curvularia lunata 214
cyclodextrin 71
cyclohexanediol-grown *Acinetobacter* TD63 225
cyclohexanone monooxygenase (CHMO) 215, 230, *231*, 233
cytochrome P450 20, 228
cytochrome P450-monooxygenases 181
dialkyl tartrate 51, 54
(*R,R*)-dialkyl tartrate 51
(−)-*O,O*-dibenzoyl-D-pertartaric acid 171
(+)-[(8,8-dichlorocamphoryl)sulfonyl] oxaziridine 140
1,4-dichlorophthalazine *92, 93, 95*
dicyclododecyl tartrate (DCDT) *64*
dicyclohexyl tartrate (DCHT) *64*
diethyl tartrate (DET) *64, 157, 158, 159, 161, 163*
dihydroquinidine (DHQD) 81, *93, 95, 99*
dihydroquinine (DHQ) 81, 84, *95*, 100
diisopropyl tartrate (DIPT) 57, *60, 63, 65*, 68, 176, *178*
[(8,8-dimethoxycamphoryl)sulfonyl] oxaziridine 140
dimethyl dioxirane (DMDO) 43, *44*, 143
dimethyl tartrate (DMT) *64*
dioxirane 37, 43, 46
dioxygenase 228
2,5-diphenyl-4,6-dichloropyrimidine 97, *99*, *100*
2,5-diphenyl-4,6-bis(9-*O*-dihydroquinidinyl)pyrimidine [(DHQD)$_2$PYR] 81, *89, 99*
2,5-diphenyl-4,6-bis(9-*O*-dihydroquininyl) pyrimidine [(DHQ)$_2$PYR] 81, *89, 100*
ethyl azido formate 115
FAD dependent monooxygenase 228
fatty alcohol oxidase (FAOD) 203
filamentous fungus, *Rhizopus arrhizus* 181
flavin-dependent enzymes 228
flavine mononucleotide (FMN) 201, *201, 205, 207, 208*

Index

fructose-based ketone 46
Geotrichum candidum 212
Gluconobacter 212
Gluconobacter roseus 212
glucose-6-phosphate dehydrogenase 2*31*
glycerol dehydrogenase (GDH) 202
N-haloamides 105
N-halocarbamates 105
N-halosulfonamides 105
Helminthosporium sp. 228
horse liver alcohol dehydrogenase (HLADH) 201, *201*, *205*, *207*
horse liver alcohol dehydrogenase (HLADH) immobilised on PAN *208*
hydrogen peroxide (H_2O_2) 20, 70, 74, 155, *157*, 164, *166*
(*R*)-3-hydroxymethyl-BINOL 78
imine 154, *156*
immobilised enzymes *207*
iodobenzene diacetate 133, *134*, 135
iodosylbenzene 7, 9, 20, 21, *24*, *28*, 33, 135, 164
o-iodosylbenzoic acid 135
iron-porphyrin complex 6, 7, 20, 116
lanthanoid-BINOL complexes *78*
lanthanoid complex 78
lanthanum triisopropoxide 78
manganese-porphyrin complex 22, 116
metalloporphyrin 6, 20, 25, 116
metallosalen complex 25, 26
methanesulfonamide 85, *89*, *91*
(+)-(2*R*,3*S*)-3-[(*S*)-2-methylbutyl)]-2,3-epoxy-1,2-benzisothiazole-1,1-dioxide 49
(*S*)-2,2-(1-methylethylidene)bis[4,5-dihydro-4-phenyloxazole] 1*21*
N-methylmorpholine *N*-oxide 81, 84, *144*
methyl(trifluoromethyl)dioxirane 43
Mn(III)(OAc)$_3$•2H$_2$O *34*
Mn(TPP)Cl 116
molecular sieves 54, 56, *57*, *61*, *63*, *65*, 78, 119, *121*, *125*, 162, *163*, *168*
(+)-monoperoxycamphoric acid 45, *172*
Mortierella isabellina 228, 233
Mucor plumbeus ATCC 4740 *188*
Mucor plumbeus CBS 110–16 190, *191*, 195, *196*
NAD(P)$^+$ 200
Nicotinamide adenin dinucleotide (NAD$^+$) 201, *203*, *206*, *208*
Nicotiana tabacum 212
nitridomanganese(V) porphyrin complex 116
p-nitroperoxybenzoic acid *39*
osmium tetroxide (OsO4) 81, 84, 85, 143, *144*
oxaziridines 45, *48*, 135, *136*, *138*, 154, *155*, 166, *167*
(3-oxocamphorsulfonyl)oxadirizine *172*
oxodiperoxymolybdenum(pyridine)-1,3-dimethyl-3,4,5,6-tetrahydro-2(1H)-pyrimidinone (MoO$_5$•Py•DMPU) (MoOPD) 133

oxodiperoxymolybdenum(pyridine)hexamethylphosphoric triamide (MoO$_5$•Py•HMPA) (MoOPH) 131, *132*
Oxone 43, *46*, *174*
oxo(salen)manganese(V) complex 26
oxygen 33, *35*, *129*, 148, *149*
peroxomonosulfuric acid 147
peroxy acid (peracid) 38, *39*, *41*
peroxylauric acid 40
peroxyphthalic acid 40
phase-transfer agents 70
2-(phenylsulfonyl)-3-phenyloxaziridine 1*36*
poly-*S*-alanine 71
poly-*S*-leucine *74*
potassium hexacyanoferrate(III) [$K_3Fe(CN)_6$] 84, *89*, 97
potassium hypochlorite 70
potassium osmate dihydrate [$K_2OsO_2(OH)_4$] 84, *89*, *97*, *107*, *109*, *112*
potassium superoxide 131
(*S*)- or (*R*)-proline 13
Pseudomonas putida 228
Pseudomonas putida NCIMB 10007 225
Pseudomonas sp. 228
Rhizopus arrhizus 181
Rh$_2$(OAc)$_4$ 115, 116
Rhodococcus equi 212
(salen)manganese complex 8, 9, 26, *28*, *30*, *32*, 77, 164
secondary alcohol dehydrogenase (SADH) 202
sodium hypochlorite 26, *29*
N-sulfonyloxaziridine 37, 45, *48*, 135, *136*, *138*, 140
the Orsay water-modified reagent 157, 158, *159*
the Padova reagent 158, 159, *161*, 162
threitol-strapped manganese(III)-porphyrin 21
2:1 (Ti/tartrate) catalyst 176
titanium reagents using *tert*-butyl hydroperoxide 150
titanium-tartrate complex 56
titanium tetraisopropoxide-diethyl tartrate (DET) 51, *52*, 56, *60*
titanium tetraisopropoxide-diethyl tartrate-water see, the Orsay water-modified reagent
titanium tetraisopropoxide-diisopropyl tartrate (DIPT) 57, *60*, *62*, *64*, *67*, 176, 177
titanium tetraisopropoxide [Ti(OPr-*i*)$_4$] 51, *52*, 56, *57*, *60*, *62*, *63*, *65*, *68*, *159*, *161*, *178*
toluene dioxygenase 228
[*N*-(*p*-toluenesulfonyl)imino]phenyliodinane (PhI=NTs) 116, *121*, *124*
vanadium-schift base complex 164, *165*
vanadyl bis(acetylacetonate) [VO(acac)$_2$] 50, 164, *166*

Index

Subject

abnormal lactone 215, 217, *223*
(5Z,7E)-3β-acetoxy-9,10-secocholesta-
   5,7,10(19)-triene *44*
9-O-acetyl-10,11-dihydroquinidine *196*
3β-acetyl vitamin D₃ 43
alcohols 6-10, 51, 61, 200
   aliphatic unsaturated alcohols *210*
   allylic alcohols 51, 61
   1-arylethanols 212
   endo-bicyclic heptenols 210
   endo-bicyclic octenols 210
   (±)-6-*endo*-bicyclo[3.2.0]hept-2-en-6-ol 203
   (−)-6-*endo*-bicyclo[3.2.0]hept-2-en-6-ol *203*
   cinnamyl alcohol 57, *58*
   citronellol 203, 212
   meso diallyl alcohols 64
   (E)-2-hexen-1-ol 51, *53*
   (R)-5-hexen-2-ol *211*
   (±)-5-hexen-2-ol *210*
   menthol 203
   (±)-*trans*-3-methylcyclohexanol 202
   (±)-*cis*-2-methylcyclopentanol 202
   (E)-3-phenylpropen-1-ol 57
   (±)-1-phenylsulfenyl-2-propanol 212
   primary allylic alcohols 51
   racemic secondary allylic alccohols 61
   (1E,3R)-1-trimethylsilyl-1-octen-3-ol 62, 63, 64
alkaloids 181
allyl aryl selenides *168*
allylic sulfonamide 116
allylsilanes 38
β-amino alcohols 104, 175
(2R,3S)-2-amino-3-hydroxy-3-phenylpropionic
   acid hydrochloride *112*
aziridines 115-126
(S)-benzoin 138
bicyclo[3.2.0]hept-2-ene-6-one 214, *222, 223*
trans-2-butene 82
2-*t*-butyl-3-(*p*-bromophenyl)oxaziridine 172
t-butyl cinnamate 119
4-*t*-butylcyclopentene 40
γ-butyrolactones 225
camphorsulfonic acid *174*
(−)-(camphorsulfonyl)imine *175*
(+)-(2R,8aS)-10-
   (camphorsulfonyl)oxaziridine *173*
carbohydrate glycals 116
chalcones 71, 72, 75
cinnamate esters 119, *120*
conjugated olefins 26, 31
6-cyano-2,2-dimethylchromene 123, *125*
cyclobutanone derivatives 150
cyclohexene 13, *14*
(R)-2-cyclohexenyl benzoate *14*
cyclopentene *11*

(S)-2-cyclopentenyl benzoate *11*
deoxybenzoin *138*
dibenzosuberone (10,11-dihydro-5H-
   dibenzo[a,d]cycloheptene-5-one) *193*
1,4-dichlorophthalazine 92, *93, 95*
(+)-dihydrocarvone *218*
10,11-dihydro-5H-dibenzo[a,d]cycloheptene-
   5-one *193*
(−)-10,11-dihydro-10-hydroxy-5H-
   dibenzo[a,d]cycloheptene-5-one *193*
1,2-dihydronaphthalene *30, 35*, 124
2,2-dimethylchromene 28
N,N-dimethyl(2-hydroxydecyl)amine 177
1,1-dimethylindan 10
(R)-3,3-dimethylindan-1-ol *9*
3-[5-[3-(2,5-dimethyl-1H-pyrrol-1-yl)-2,5-
   dimethoxyphenyl]-5-methoxy-4-methyl-
   2-hydroxy-1-oxopentyl]-4-(1-
   methylethyl)-2-oxazolidinone *136*
3-[5-[3-(2,5-dimethyl-1H-pyrrol-1-yl)-2,5-
   dimethoxyphenyl]-5-methoxy-4-methyl-
   1-oxopentyl]-4-(1-methylethyl)-2-
   oxazolidinone *136*
diols 81, 87-92, 205-207
   acyclic *meso*-diols 205
   bicyclic *meso*-diols 205
   cis-1,2-bis(hydroxymethyl)cyclohexane *206, 207*
   1,2-butanediol 202
   2,3-butanediol 203
   (+)-*trans*-cyclohexane-1,2-diol 212
   1,2-cyclohexanediol 202
   meso diols 200, *205*
   1,2-diols 82, 87
   α,ω-diols 212
   3-halo-1,2-propanediols 203
   3-methylpentane-1,5-diol *209*
   monocyclic *meso*-diols 205
   syn-1-phenylpropane-1,2-diols 212
   anti-1-phenylpropane-1,2-diols 212
   2,5-diphenyl-4,6-dichloropyrimidine 97, *99*
cis-disubstituted olefins 27
1,2-disubstituted olefins (substrates for
   dihydroxylation reaction) 85, *90*
drimane families 187
electron deficient olefins 70, 77
enolates and enol derivatives 128, 141
enolsilanes 116, 143
epoxides (oxiranes) 21, 37, 50, 70
(5Z)-(7R,8R)-3β-acetoxy-7,8-epoxy-9,10-
   secocholesta-5,10(19)-diene *44*
(2S,3S)-epoxycinnamyl alcohol 57
(3S,4S)-3,4-epoxy-2,2-dimethylchromane 28
(2S,3S)-2,3-epoxy-1-hexanol *52*
epoxy ketones 75
(2R,3S)-1,2-epoxy-3-methylpentane 37

242

Index

trans-(2*S*,3*R*)-epoxy-4-methyl-1-
  phenylpentan-1-one *79*
1,2-epoxy-4-penten-3-ol *64*
(1*S*,2*R*)-1,2-epoxy-1-phenylpropane *24*
(*R*)-2,3-epoxypropyl *p*-nitrobenzoate *60*
(1*S*,2*R*)-1,2-epoxy-1,2,3,4-
  tetrahydronaphthalene *29*, *35*
erythro epoxy alcohol *61*
(1*S*,2*R*,4*S*)-limonene oxide *48*
(2*R*-*trans*)-(2-naphthalenyl)-3-
  phenyloxiranylmethanone *73*
(2*S*-*trans*)-3-phenyloxiranemethanol *57*
(*S*)-styrene oxide *32*
(1*S*,2*S*,3*S*)-1-trimethylsilyl-1,2-epoxyoctan-3-ol
  62, *64*
[1*R*-(1α,2α,3β,5α)]-3,6,6-
  trimethylspirobicyclo[3.1.1]heptane-2,2′-
  oxirane *39*
water-soluble epoxy alcohols *56*
(2*S*,4*S*,5*E*,9*R*,10*R*)-9,10-epoxy-2-hydroxy-4-
  isopropyl-7-methylene-5-cyclodecen-1-
  one *132*
(4*S*,5*E*,9*R*,10*R*)-9,10-epoxy-4-isopropyl-7-
  methylene-5-cyclodecen-1-one *132*
ester *214*
cis-ethyl cinnamate *77*
functionalized aziridines *115*
α-furfuryl toluenesulfonamide *66*
2-furyl carbinols *67*
(*R*)-1-(2-furyl)hexan-1-ol *67*
glycidates *75*
glycidic acid esters *73*
5-hexen-2-one *211*
α-hydroxy acids *135*
β-hydroxy amines *175*, *177*
(4*R*,5*R*,7*S*)-5-(α-hydroxybenzyl)-1-methyl-4-
  phenylpyrrolidin-2-one *129*
α-hydroxy carbonyl product *128*, *129*, *131*, *133*
(−)-10-hydroxydibenzosuberone *194*
(3*S*)-3-hydroxy-10,11-dihydroquinidine *196*
6-hydroxy-3-ethoxy-7-methyl-2-
  oxabicyclo[3,3,0]octan-6-carboxaldehyde
  144, *145*
(3*S*,4*R*,5*R*,7*S*)-3-hydroxy-5-(α-
  hydroxybenzyl)-1-methyl-4-
  phenylpyrrolidin-2-one *129*
α-hydroxy ketone *131*, *134*, *135*, *141*, *143*
2-hydroxy-*p*-menthanes *212*
cis-2-hydroxy-6-methoxy-4,7-dimethyltetral-1-
  one *134*
(*R*,*S*)-1-hydroxymethyl-2-methylferrocene
  202
(*R*)-2-hydroxy-2-methyl-1-tetralone *140*
11 α-hydroxyprogesterone *181*
3 β-hydroxysclareol *188*
β-hydroxy tertiary amines *171*, *176*
α-hydroxy tetralone *134*
imines *171*, *172*, *173*

indene *123*, *124*
isolated olefins *19-20*, *37*
isopropyl cinnamate *112*
labdane *187*, *188*
γ-lactones *205*
δ-lactones *205*
lactones (products of Baeyger-Villiger
  reaction) *147*, *149*, *150*, *215*
(*S*)-limonene *48*
6-methoxy-4,7-dimethyltetral-1-one *134*
methyl acrylate *110*
methyl (2*S*)-3-[*N*-(benzyloxycarbonyl)amino]-
  2-hydroxypropanoate *109*
trans-methyl cinnamate *121*
4-methylcyclopentene *40*
3-methylcyclopentene *40*
cis-3-methylcyclopenetene-1,2-oxide *40*
methyl (3aα,6β,7β,7aα)-2,3,3a,6,7,7a-
  hexahydro-3a-hydroxy-6-methyl-3-oxo-7-
  [[2-(trimethylsilyl)ethoxy]methoxy]-4-
  benzofurancarboxylate *141*
(*R*)-(−)-4a-methyl-4,4a,5,6,7,8-hexahydro-
  2(3*H*)-naphthalenone *191*
(3*R*,6*S*)-(−)-3-methyl-6-isopropenyl-2-
  oxooxepane *219*
(*S*)-3-methyl-1-pentene *37*
trans-4-methyl-1-phenyl-2-penten-1-one *79*
trans-β-methylstyrene *46*, *118*, *119*
cis-β-methylstyrene *31*, *119*, *122*
(1*R*,2*R*)-β-methylstyrene oxide *46*
methyl (6β,7β,7aα)-2,3,3a,6,7,7a-tetrahydro-3-
  [(triethylsilyl)oxy]-6-methyl-7-[[2-
  (trimethylsilyl)ethoxy]methoxy]-4-
  benzofurancarboxylate *142*
2-methyl-1-tetralone *140*
methyl 2,4,6-triisopropylphenyl selenide *168*
methyl 2,4,6-triisopropylphenyl selenoxide
  168
(−)-(3*S*)-3-methylvalerolactone *207*, *209*
naphthalchalcone *74*
1,4-naphthoquinones *70*
(+)-(1*R*,6*S*)-8-oxabicyclo[4.3.0]nonan-7-one
  205
oxaziridines *48*, *138*, *140*, *171*, *172*, *173*
N-oxide of *N*,*N*-dimethyl(β-
  hydroxydecyl)amine *178*
penta-1,4-dien-3-ol (divinyl carbinol) *65*
2-pentylcyclopentanone *151*
2-phenylcyclohexanone *148*, *149*
3-phenyl-2-oxepanone *148*
(*Z*)-1-phenylpropene *24*
phosphites *172*
α-pinene *42*
progesterone *181*
racemic 2-phenylcyclohexanone *148*
sclareol *188*
selenides *153*, *166*
selenoxides *153*, *167*

Index

simple olefins without precoordinating functional groups 37
steroids 181
trans-stilbene 107
styrene 21, 31, *32*, 45, 87, 116, 118, 122, 124
substituted chalcones 72
substituted 1,4-naphthoquinones 70
sulfides 153-166, 227
  alkyl aryl sulfides 155-166, 228
  alkyl benzyl sulfides 231
  bicyclic sulfides 228
  dialkyl sulfides 228, 231
  di-*t*-butyl disulfide (*t*-BuS)$_2$ 165
  ethyl 1,3-dithiane-2-carboxylate 1*61*
  1,3-dithiane 230, *231*, 233
  methyl phenyl sulfide *163*
  methyl *p*-tolyl sulfide 155, *159*, 163, *165*, *229*
  2-phenyl-1,3-dithiane 165, *166*
sulfoxides 153-166, 227
  (*R*)-1,3-dithiane monosulfoxide 230, *231*, 233
  1,3-dithioacetal monosulfoxides 231
  trans-2-ethoxycarbonyl-1,3-dithiane dioxide *161*
  (*S*)-ethyl *p*-tolyl sulfoxide 230
  (*R*)-methyl phenyl sulfoxide *163*
  (*S*)-methyl *p*-tolyl sulfoxide *155*, *160*
  (*R*)-methyl *p*-tolyl sulfoxide *228*, 229
  (S_S)-*trans*-2-phenyl-1,3-dithiane monoxide *156*
  2-phenyl-1,3-dithiane monoxide 165
sulfinyl functionality 227
terpenes 181, 187
1,2,3,4-tetrahydronaphthalene 7
(*R*)-1,2,3,4-tetrahydro-1-naphthol 7
tetralone 133
tetrasubstituted olefins (substrates for dihydroxylation reaction) 84, 85, *89*, *90*
2-thienyl carbinols 66
(1*S*,2*S*)-*N*-(*p*-toluenesulfonyl)-2-amino-1,2-diphenylethanol *106*
(2*R*,3*S*)-*N*-*p*-toluenesulfonyl-2-carbomethoxy-3-phenylaziridine *120*, *122*
(3*R*,4*R*)-(+)-(*N*-*p*-toluenesulfonyl)-6-cyano-2,2-dimethylchromane(1,2)imine *124*
3,6,6-trimethyl-2-methylenebicyclo[3.1.1]heptane *39*
6-trimethylsilyloxymethylene-3-ethoxy-7-methyl-2-oxabicyclo[3,3,0]octane *144*
trisubstituted conjugated olefins 26
trisubstituted olefins (substrates for dihydroxylation reaction) 85, *90*, *91*
(±)-4-*exo*-twistanol *201*
(−)-4-twistanone *202*
α,β-unsaturated carbonyl compounds 70
α,β-unsaturated ester 115
α,β-unsaturated ketones 76, 78, *78*